INTERNATIONAL UNION OF PURE AND APPLIED CHEMISTRY

ANALYTICAL CHEMISTRY DIVISION
COMMISSION ON SOLUBILITY DATA

SOLUBILITY DATA SERIES

Volume 35

4-AMINOBENZENESULFONAMIDES

Part II

5-Membered Heterocyclic Substituents

A complete list of volumes published in the Solubility Data Series will be found on p. 343.

SOLUBILITY DATA SERIES

Editor-in-Chief
A.S. KERTES

Volume 35

4-AMINOBENZENESULFONAMIDES

Part II

5-Membered Heterocyclic Substituents

Volume Editors

ANTHONY N. PARUTA
University of Rhode Island
Kingston, RI, USA

RYSZARD PIEKOS
Medical Academy
Gdansk, Poland

PERGAMON PRESS

OXFORD · NEW YORK · BEIJING · FRANKFURT
SÃO PAULO · SYDNEY · TOKYO · TORONTO

U.K.	Pergamon Press plc, Headington Hill Hall, Oxford OX3 0BW, England
U.S.A.	Pergamon Press, Inc., Maxwell House, Fairview Park, Elmsford, New York 10523, U.S.A.
PEOPLE'S REPUBLIC OF CHINA	Pergamon Press, Room 4037, Qianmen Hotel, Beijing, People's Republic of China
FEDERAL REPUBLIC OF GERMANY	Pergamon Press GmbH, Hammerweg 6, D-6242 Kronberg, Federal Republic of Germany
BRAZIL	Pergamon Editora Ltda, Rua Eça de Queiros, 346, CEP 04011, Paraiso, São Paulo, Brazil
AUSTRALIA	Pergamon Press Australia Pty Ltd., P.O. Box 544, Potts Point, N.S.W. 2011, Australia
JAPAN	Pergamon Press, 5th Floor, Matsuoka Central Building, 1-7-1 Nishishinjuku, Shinjuku-ku, Tokyo 160, Japan
CANADA	Pergamon Press Canada Ltd., Suite No. 271, 253 College Street, Toronto, Ontario, Canada M5T 1R5

First edition 1988

The Library of Congress has catalogued this serial title as follows:

Solubility data series.—Vol. 1—Oxford; New York;
Pergamon, c 1979-
v.; 28 cm.
Separately catalogued and classified in LC before no. 18.
ISSN 0191-5622 = Solubility data series.
1. Solubility—Tables—Collected works.
QD543.S6629 541.3'42'05-dc19 85-641351
AACR 2 MARC-S

British Library Cataloguing in Publication Data

4-Aminobenzenesulfonamides
Pt. 2: 5-membered heterocyclic
substituents
1. Sulfonamides. Solubility
I. Paruta, Anthony N. II. Piekos, Ryszard
III. Series
547.7'5045'42

ISBN 0-08-034708-8

Printed in Great Britain by A. Wheaton & Co. Ltd., Exeter

CONTENTS

FOREWORD

If the knowledge is
undigested or simply wrong,
more is not better

How to communicate and disseminate numerical data effectively in chemical science and technology has been a problem of serious and growing concern to IUPAC, the International Union of Pure and Applied Chemistry, for the last two decades. The steadily expanding volume of numerical information, the formulation of new interdisciplinary areas in which chemistry is a partner, and the links between these and existing traditional subdisciplines in chemistry, along with an increasing number of users, have been considered as urgent aspects of the information problem in general, and of the numerical data problem in particular.

Among the several numerical data projects initiated and operated by various IUPAC commissions, the *Solubility Data Project* is probably one of the most ambitious ones. It is concerned with preparing a comprehensive critical compilation of data on solubilities in all physical systems, of gases, liquids and solids. Both the basic and applied branches of almost all scientific disciplines require a knowledge of solubilities as a function of solvent, temperature and pressure. Solubility data are basic to the fundamental understanding of processes relevant to agronomy, biology, chemistry, geology and oceanography, medicine and pharmacology, and metallurgy and materials science. Knowledge of solubility is very frequently of great importance to such diverse practical applications as drug dosage and drug solubility in biological fluids, anesthesiology, corrosion by dissolution of metals, properties of glasses, ceramics, concretes and coatings, phase relations in the formation of minerals and alloys, the deposits of minerals and radioactive fission products from ocean waters, the composition of ground waters, and the requirements of oxygen and other gases in life support systems.

The widespread relevance of solubility data to many branches and disciplines of science, medicine, technology and engineering, and the difficulty of recovering solubility data from the literature, lead to the proliferation of published data in an ever increasing number of scientific and technical primary sources. The sheer volume of data has overcome the capacity of the classical secondary and tertiary services to respond effectively.

While the proportion of secondary services of the review article type is generally increasing due to the rapid growth of all forms of primary literature, the review articles become more limited in scope, more specialized. The disturbing phenomenon is that in some disciplines, certainly in chemistry, authors are reluctant to treat even those limited-in-scope reviews exhaustively. There is a trend to preselect the literature, sometimes under the pretext of reducing it to manageable size. The crucial problem with such preselection - as far as numerical data are concerned - is that there is no indication as to whether the material was excluded by design or by a less than thorough literature search. We are equally concerned that most current secondary sources, critical in character as they may be, give scant attention to numerical data.

On the other hand, tertiary sources - handbooks, reference books and other tabulated and graphical compilations - as they exist today are comprehensive but, as a rule, uncritical. They usually attempt to cover whole disciplines, and thus obviously are superficial in treatment. Since they command a wide market, we believe that their service to the advancement of science is at least questionable. Additionally, the change which is taking place in the generation of new and diversified numerical data, and the rate at which this is done, is not reflected in an increased third-level service. The emergence of new tertiary literature sources does not parallel the shift that has occurred in the primary literature.

With the status of current secondary and tertiary services being as briefly stated above, the innovative approach of the *Solubility Data Project* is that its compilation and critical evaluation work involve consolidation and reprocessing services when both activities are based on intellectual and scholarly reworking of information from primary sources. It comprises compact compilation, rationalization and simplification, and the fitting of isolated numerical data into a critically evaluated general framework.

The *Solubility Data Project* has developed a mechanism which involves a number of innovations in exploiting the literature fully, and which contains new elements of a more imaginative approach for transfer of reliable information from primary to secondary/tertiary sources. *The fundamental trend of the Solubility Data Project is toward integration of secondary and tertiary services with the objective of producing in-depth critical analysis and evaluation which are characteristic to secondary services, in a scope as broad as conventional tertiary services.*

Fundamental to the philosophy of the project is the recognition that the basic element of strength is the active participation of career scientists in it. Consolidating primary data, producing a truly critically-evaluated set of numerical data, and synthesizing data in a meaningful relationship are demands considered worthy of the efforts of top scientists. Career scientists, who themselves contribute to science by their involvement in active scientific research, are the backbone of the project. The scholarly work is commissioned to recognized authorities, involving a process of careful selection in the best tradition of IUPAC. This selection in turn is the key to the quality of the output. These top experts are expected to view their specific topics dispassionately, paying equal attention to their own contributions and to those of their peers. They digest literature data into a coherent story by weeding out what is wrong from what is believed to be right. To fulfill this task, the evaluator must cover *all* relevant open literature. No reference is excluded by design and every effort is made to detect every bit of relevant primary source. Poor quality or wrong data are mentioned and explicitly disqualified as such. In fact, it is only when the reliable data are presented alongside the unreliable data that proper justice can be done. The user is bound to have incomparably more confidence in a succinct evaluative commentary and a comprehensive review with a complete bibliography to both good and poor data.

It is the standard practice that the treatment of any given solute-solvent system consists of two essential parts: I. Critical Evaluation and Recommended Values, and II. Compiled Data Sheets.

The Critical Evaluation part gives the following information:

(i) a verbal text of evaluation which discusses the numerical solubility information appearing in the primary sources located in the literature. The evaluation text concerns primarily the quality of data after consideration of the purity of the materials and their characterization, the experimental method employed and the uncertainties in control of physical parameters, the reproducibility of the data, the agreement of the worker's results on accepted test systems with standard values, and finally, the fitting of data, with suitable statistical tests, to mathematical functions;

(ii) a set of recommended numerical data. Whenever possible, the set of recommended data includes weighted average and standard deviations, and a set of smoothing equations derived from the experimental data endorsed by the evaluator;

(iii) a graphical plot of recommended data.

The Compilation part consists of data sheets of the best experimental data in the primary literature. Generally speaking, such independent data sheets are given only to the best and endorsed data covering the known range of experimental parameters. Data sheets based on primary sources where the data are of a lower precision are given only when no better data are available. Experimental data with a precision poorer than considered acceptable are reproduced in the form of data sheets when they are the only known data for a particular system. Such data are considered to be still suitable for some applications, and their presence in the compilation should alert researchers to areas that need more work.

The typical data sheet carries the following information:

(i) components - definition of the system - their names, formulas and Chemical Abstracts registry numbers;
(ii) reference to the primary source where the numerical information is reported. In cases when the primary source is a less common periodical or a report document, published though of limited availability, abstract references are also given;
(iii) experimental variables;
(iv) identification of the compiler;
(v) experimental values as they appear in the primary source. Whenever available, the data may be given both in tabular and graphical form. If auxiliary information is available, the experimental data are converted also to SI units by the compiler.

Under the general heading of Auxiliary Information, the essential experimental details are summarized:

(vi) experimental method used for the generation of data;
(vii) type of apparatus and procedure employed;
(viii) source and purity of materials;
(ix) estimated error;
(x) references relevant to the generation of experimental data as cited in the primary source.

This new approach to numerical data presentation, formulated at the initiation of the project and perfected as experience has accumulated, has been strongly influenced by the diversity of background of those whom we are supposed to serve. We thus deemed it right to preface the evaluation/compilation sheets in each volume with a detailed discussion of the principles of the accurate determination of relevant solubility data and related thermodynamic information.

Finally, the role of education is more than corollary to the efforts we are seeking. The scientific standards advocated here are necessary to strengthen science and technology, and should be regarded as a major effort in the training and formation of the next generation of scientists and engineers. Specifically, we believe that there is going to be an impact of our project on scientific-communication practices. The quality of consolidation adopted by this program offers down-to-earth guidelines, concrete examples which are bound to make primary publication services more responsive than ever before to the needs of users. The self-regulatory message to scientists of the early 1970s to refrain from unnecessary publication has not achieved much. A good fraction of the literature is still cluttered with poor-quality articles. The Weinberg report (in 'Reader in Science Information', ed. J. Sherrod and A. Hodina, Microcard Editions Books, Indian Head, Inc., 1973, p. 292) states that 'admonition to authors to restrain themselves from premature, unnecessary publication can have little effect unless the climate of the entire technical and scholarly community encourages restraint...' We think that projects of this kind translate the climate into operational terms by exerting pressure on authors to avoid submitting low-grade material. The type of our output, we hope, will encourage attention to quality as authors will increasingly realize that their work will not be suited for permanent retrievability unless it meets the standards adopted in this project. It should help to dispel confusion in the minds of many authors of what represents a permanently useful bit of information of an archival value, and what does not.

If we succeed in that aim, even partially, we have then done our share in protecting the scientific community from unwanted and irrelevant, wrong numerical information.

A. S. Kertes

PREFACE

With few exceptions, these volumes of the solubility data series deal with solubilities of the derivatives of 4-aminobenzenesulfonamide, usually referred to as "sulfanilamide" (sulfanilic acid amide), a name coined in 1937 (1). The history of sulfanilamide begins in 1906, when Schroeter (2) synthesized the molecule containing a 4-acetylaminosulfanilamide portion.

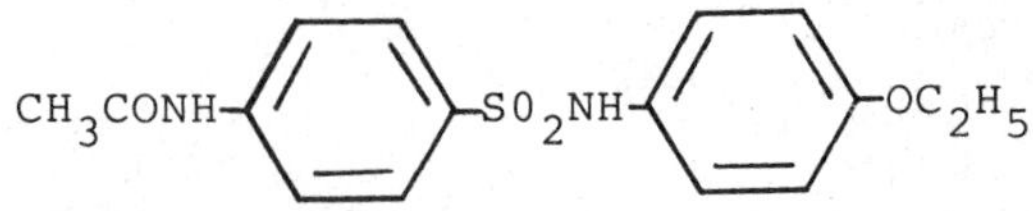

In 1908, Gelmo (3) described sulfanilamide and 13 of its derivatives and gave solubility values for these compounds. In 1935, Domagk (4) detected antibacterial activity of a synthetic azo dye, prontosil, with the structure.

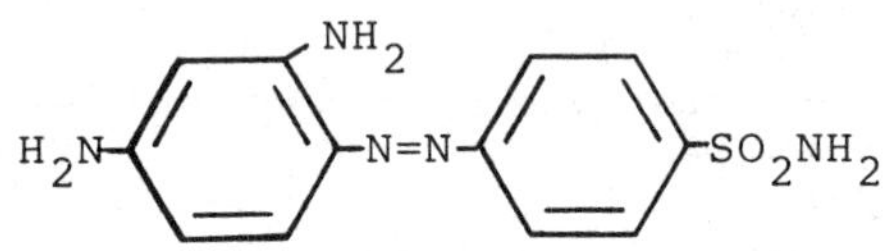

This compound had been tested for antibacterial activity (5), the "sulfanilamide" portion being responsible for its activity. This was confirmed (6) by isolation of sulfanilamide in the urine of patients. Fildes (7) and Wood (8), in 1940, demonstrated that the derivatives of sulfanilamide were antimetabolites of p-aminobenzoic acid (PABA) which is a step in the folic acid synthesis of bacteria. Thus, the structural similarity of PABA and sulfonamides caused interference by competitive antagonism and resulted in a bacteriostatic effect. The discoveries of antibacterial activity led to an exciting flood of research, and thousands of sulfanilamide derivatives have been synthesized. As early as 1948, the number of sulfonamide derivatives (9) was estimated to be several thousand. In the two decades after that, the number of synthesized sulfonamides have gone past 10,000(10)

Clinical trials of these sulfonamides and derivatives have been associated with low solubilities and some renal crystalluria. The low solubility, and its sensitivity to pH, could cause crystalline precipitation in the renal tubules in the filtration of blood into acidic urine. Some of the problems of limited solubility were overcome by complexation or salt formation, and solid state manipulations which in turn have stimulated investigations into solubility of the drugs in water, buffers and some binary solvent system. Analytical methodologies span a wide spectrum of techniques and the relevant references are in pharmaceutical, medical and chemical literature.

In all volumes the chemical structures, registry number and the molecular weight of the compounds considered are collected in the front of each volume. The compounds as they occur on the data sheets are given successively in each volume. In the first volume of this series there are 35 compounds. The second and third volumes have 58 compounds and 108 compounds, respectively.

NOMENCLATURE:

The nomenclature of sulfanilamide derivatives has conventionally been based on the following numbering system: substituents at the nitrogen atom of the amide group ($-SO_2NH_2$) are called $\underline{N}^1$-substituents, whereas substitutents at the 4-amino nitrogen ($4-H_2N-$) are called $\underline{N}^4$-substituents. Substitution in either or both of the two positions lead to compounds referred to as "sulfonamides" (sometimes "sulfanilamides" or even "sulfamides"). Here are illustrative examples of this nomenclature.

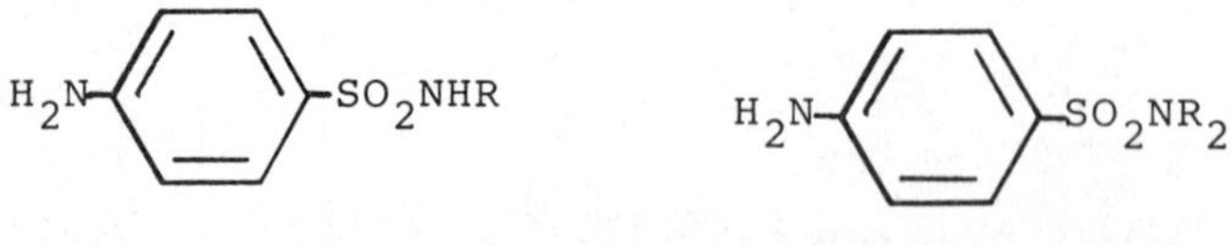

$\underline{N}^1$-substituted sulfonamides

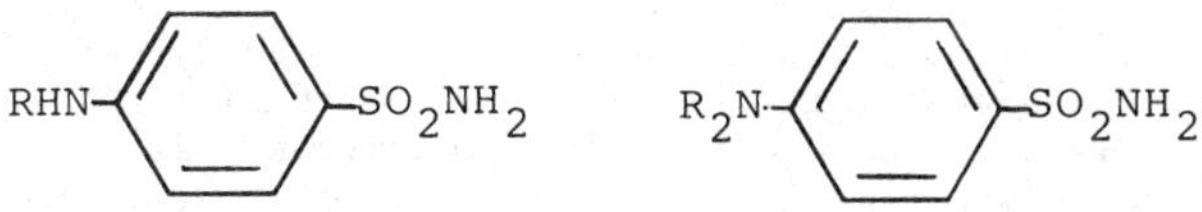

N^4-substituted sulfonamides

N^1,N^4-substituted sulfonamides

The 4-amino group can be diazotized to give derivatives of the formula

As the sulfonamide molecule carries a basic 4-NH_2 group and an acid -SO_2NH_2 one, it is capable of formation the respective salts or complexes, e.g.

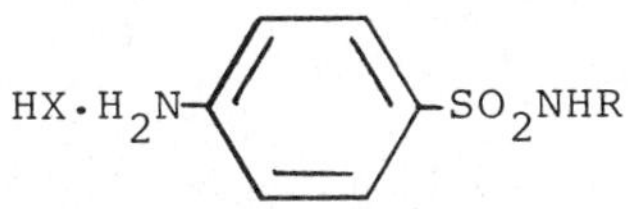

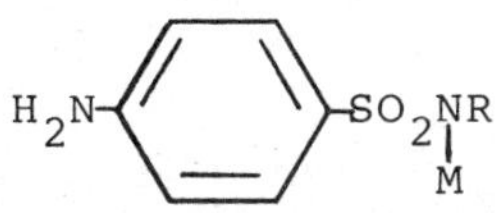

where HX stands for an acid and M is a univalent metal atom.

In common use by health practitioners are nonproprietary names of sulfonamides which are brief and reflect the chemical nature of their molecules. Examples are: sulfacetamide, sulfapyridine, sulfathiazole, sulfadiazine, sulfaguanidine, etc. There are numerous trivial names; for example, sulfanilamide has as many as 140 synonyms, and sulfathiazole has 113. Negwer (11) has compiled an excellent guide to this nomenclature. In chemical literature, systematic names in line either with IUPAC (12) or Chemical Abstract rules (13) are used. The latter has been adopted in these volumes and the systematic name is, where appropriate, followed by the nonproprietary or trivial name.

ORGANIZATION OF THE VOLUMES:

The numerical data on the solubility of 2-aminobenzenesulfonamide, 3-aminobenzenesulfonamide, and 4-aminobenzenesulfonamide and its N^1 and N^4- derivatives, salts and complexes, compiled up to 1985 inclusive, have been divided into three volumes on the basis of chemical structure of the compounds.

The first volume includes the solubility of 2-aminobenzenesulfonamide, 3-aminobenzenesulfonamide, 4-aminobenzenesulfonamide and the derivatives of the last-named compound substituted at either of the nitrogen atoms, or both, with non-cyclic substituents (see System Index at the end of the first volume). The aroyl substituents, -C(:O)aryl, have also been included here. The second volume includes sulfanilamide derivatives substituted with 5-membered heterocyclic rings at either of the nitrogen atoms, and their derivatives. The third volume covers the solubilities of the derivatives substituted with 6-membered rings, mixtures of sulfonamides, and miscellanea. The compilations do not include compounds devoid of the -NH_2, -NHR or -NR_2 group in the benzene ring.

The solvent systems include all solvents with the exception of body fluids. The order of solvents for a particular solute are as follows: water; water-mineral acid; water-mineral base; water-mineral salt; water-miscellaneous mineral components; water-mineral and organic compounds; water-organic components; organic solvents; carboxylic acid and their salts; aliphatic acids; aromatic acids; other acids; alcohols, phenols (mono-, di-, polyhydric); amides; amines; aliphatic amines (primary, secondary, tertiary); aromatic amines (primary, secondary, tertiary); other amines; aminoalcohols; carboxylic acid esters; ethers (excluding tensides); hydrocarbons; aliphatic hydrocarbons; aromatic hydrocarbons; miscellaneous hydrocarbons; halogenated hydrocarbons (flouro-, chloro-, bromo-, iodo-); aliphatic halogenated hydrocarbons; aromatic halogenated hydrocarbons; ketones; tensides (surface-active agents); miscellaneous organic solvents.

SIGNIFICANT FIGURES AND GRAPHICAL DATA:

In most cases, solubility values given in the primary source by various workers are overstated with respect to significant figures. Since the author(s) original values are given on the data sheets, it is difficult to consider significant figures and analytical limitations in a completely consistant fashion. Therefore, the reader should be aware that in most cases the number of significant figures used for calculations was not that given by the original author(s). This was done to maintain coherence and consistency as data were given to varying significant figures. In many cases graphic data of sufficient size and clarity are reproduced. The data can be regarded of sufficient accuracy to serve as a starting point for more precise determinations. In many instances, the effect of additive concentration, pH, temperature, etc. can be depicted.

POLYMORPHISM:

Many sulfonamides exhibit several cyrstalline forms or polymorphs. There are several studies referenced in these volumes that specifically deal with the solubility difference between polymorphic modifications of the same compound. The solubility differences between polymorphs have been found to vary over a large range of values.

AMPHOLYTES:

Solubility of ampholytic sulfonamides as a function of pH varies enormously, sometimes by several orders of magnitude. Unless the pH is known experimentally, the solubility value may be suspect especially at low (1-3) and high pH (10-12) values. In these cases, the solubility is a rapidly changing value, frequently with small incremental changes of pH. The abrupt change of solubility with pH is usually associated with the formation of water soluble anionic and cationic species. Buffers, especially at higher concentrations may alter solubility by salting effect and the pH is also affected by ionic strength.

EQUILIBRATION TIME:

In general, it appears that many of these determined solubilities may not have been under equilibrium conditions. Unfortunately, in too many instances the equilibration time appears too low. Typically, solutes possess low aqueous solubilities and require long dissolution time to reach saturation. Saturation time should be experimentally determined in each case and for each compound. In many cases up to 24 hours may be required.

The editors consider the vast majority of the solubility values given in these volumes as tentative. It should be stressed, however, that they represent a useful starting point for more accurate determinations of a vast array of substituted 4-aminobenzene-sulfonamides, with many structurally and chemically related compound of various types. They amply illustrate the many factors and parameters affecting solubility and the direction and magnitude of these effects.

This compilation and evaluation is not only the result of the joint efforts of the compiler and evaluator, but also of all those who read the manuscripts, expressed their criticism, who procured copies of hard-to-get journals, who translated texts from Japanese as well as of those who in any other way assisted in the compilation and evaluation. We would like to express our gratitude in particular to the following colleagues: Prof. S. Kertes, Dr. M. Salomon, Prof. S. Yalkowsky, Prof. H. Akaiwa, Prof. C. Kalidas, Prof. W. Riess, Prof. A. Guerrero-Laverat, Prof. P. Rohdewald, Prof. J. Pütter, Dr.K. L. Loening, Dr. A. Brodin, Dr. D. Zimma, Mr. K. Hazelton, Dr. R. Fernandez-Prini, and Mr. E. MacMullan.

REFERENCES TO THE PREFACE:

1. Council on Pharmacy and Chemistry, J. Am. Med. Assoc. 1937, *108*, 1888.
2. Schroeter, G. Ber. Dtsch. Chem. Ges. 1906, *39*, 1559.
3. Gelmo, P. J. Prakt, Chem. *[2]*, 1908, *77*, 369.
4. Domagk, G. Dtsch. Med. Wschr. 1935, *61*, 250 and 829.
5. Tréfouël, T. J.; Nitti, F.; Bovet, D. Compt. Rend. Soc. Biol. 1935, *120*, 756.
6. Fuller, A. T. Lancet, 1937, 194.
7. Fildes, P. Lancet, 1940, 955.
8. Woods, D. D. J. Expt. Path. 1940, *21*, 74.
9. Langecker, H. Arch. Exptl. Pat. Pharmakol. 1948, *205*, 291.
10. Rolski, S. Chemia Srodkow Leczniczych (Chemistry of Medicinal Agents), 3rd ed., PZWL, Warsaw, 1968.
11. Negwer, M. Organisch-chemische Arzneimittel und ihre Synonyma (Organic-chemical Drugs and their Synonyms), Akademie-Verlag, Berlin, 1978.
12. Nomenclature of Organic Chemistry, Definitive Rules for Section C. Characteristic Groups Containing Carbon, Hydrogen, Oxygen, Nitrogen, Halogen, Sulfur, Selenium, and/or Tellurium IUPAC Commission on the Nomenclature of Organic Chemistry, London, Butterworths, 1971, rule 641.8.
13. J. Chem. Documentation 1974, *14*, 3.

INTRODUCTION TO THE SERIES ON SOLUBILITY OF SOLIDS IN LIQUIDS: SUBSERIES ON PHARMACEUTICALS

Nature of the Project

The Solubility Data Project (SDP) has as its aim a comprehensive search of the literature for solubilities of gases, liquids, and solids in liquids or solids. Data of suitable precision are compiled on data sheets in a uniform format. The data for each system are evaluated, and where data from different sources agree sufficiently, recommended values are proposed. The evaluation sheets, recommended values, and compiled data sheets are published on consecutive pages.

For *phamaceuticals*, the definitions, thermodynamics and methods of analysis are the same as those for the study of solubility of solids in liquids in general. For this subseries, special sections deal with matters of interest for pharmaceuticals, including discussions of polymorphism, factors influencing the rate of dissolution of drugs, and methods used to inhibit or enhance the rate of dissolution.

Definitions

A *mixture* (1, 2) describes a gaseous, liquid, or solid phase containing more than one substance, when the substances are all treated in the same way.

A *solution* (1, 2) describes a liquid or solid phase containing more than one substance, when for convenience one of the substances, which is called the *solvent*, and may itself be a mixture, is treated differently than the other substances, which are called *solutes*. If the sum of the mole fractions of the solutes is small compared to unity, the solution is called a *dilute solution*.

The *solubility* of a substance *B* is the relative proportion of *B* (or a substance related chemically to *B*) in a mixture which is saturated with respect to solid *B* at a specified temperature and pressure. *Saturated* implies the existence of equilibrium with respect to the processes of dissolution and precipitation; the equilibrium may be stable or meta-stable. The solubility of a substance in metastable equilibrium is usually greater than that of the corresponding substance in stable equilibrium. (Strictly speaking, it is the activity of the substance in metastable equilibrium that is greater.) Care must be taken to distinguish true metastability from supersaturation, where equilibrium does not exist.

Either point of view, mixture or solution, may be taken in describing solubility. The two points of view find their expression in the quantities used as measures of solubility and in the reference states used for definition of activities, activity coefficients and osmotic coefficients.

The qualifying phrase "substance related chemically to *B*" requires comment. The composition of the saturated mixture (or solution) can be described in terms of any suitable set of thermodynamic components. Thus, the solubility of a salt hydrate in water is usually given as the relative proportion of anhydrous salt in solution, rather than the relative proportions of hydrated salt and water.

For *pharmaceuticals*, the solubility of a drug substance in a given medium is of special importance in designing a suitable dosage form for a drug or in determination of a regimen for its administration. The solubility and rate of dissolution will determine the rate of appearance of the drug in various body fluids and at various sites of action. Therefore, the bioavailability of a drug is often determined by its solubility and rate of dissolution.

The solubility is a constant for a given substance in a given medium at constant temperature and pressure. Frequently it is possible to alter the solubility and rate of dissolution dramatically through changes in structure, degree of crystallinity or morphology, or by the addition of a solubilizing agent (cosolvent) to the dissolution medium. The appearance of a drug in adequate concentration at its site of action is a requirement for testing clinical efficiency; thus, enhancement of solubility may be required to render a substance clinically useful.

For reviews of recent literature on solubility and solubilization of drug substances, see (3, 4).

Quantities Used as Measures of Solubility

1. *Mole fraction* of substance *B*, x_B:

$$x_B = n_B / \sum_{s=1}^{c} n_s \qquad [1]$$

where n_s is the amount of substance of *s*, and *c* is the number of distinct substances present (often the number of thermodynamic components in the system). *Mole per cent* of *B* is 100 x_B.

2. *Mass fraction* of substance *B*, w_B:

$$w_B = m_B' / \sum_{s=1}^{c} m_s' \qquad [2]$$

where m_s' is the mass of substance *s*. *Mass per cent* is 100 w_B. The equivalent terms *weight fraction* and *weight per cent* are not used.

3. *Solute mole (mass) fraction* of solute *B* (5, 6):

$$x_{s,B} = n_B / \sum_{s=1}^{c'} n_s = x_B / \sum_{s=1}^{c'} x_s \qquad [3]$$

$$w_{s,B} = m_B' / \sum_{s=1}^{c'} m_s' = w_B / \sum_{s=1}^{c'} w_s \qquad [3a]$$

where the summation is over the solutes only. For the solvent *A*, $x_{S,A} = x_A/(1 - x_A)$, $w_{S,A} = w_A/(1 - w_A)$. These quantities are called *Jänecke mole (mass) fractions* in many papers.

4. *Molality* of solute *B* (1, 2) in a solvent *A*:

$$m_B = n_B / n_A M_A \qquad \text{SI base units: mol kg}^{-1} \qquad [4]$$

where M_A is the molar mass of the solvent.

5. *Concentration* of solute *B* (1, 2) in a solution of volume V:

$$c_B = [B] = n_B / V \qquad \text{SI base units: mol m}^{-3} \qquad [5]$$

The symbol c_B is preferred to [*B*], but both are used. The terms *molarity* and *molar* are not used.

Mole and mass fractions are appropriate to either the mixture or the solution point of view. The other quantities are appropriate to the solution point of view only. Conversions among these quantities can be carried out using the equations given in Table I-1 following this Introduction. Other useful quantities will be defined in the prefaces to individual volumes or on specific data sheets.

In addition to the quantities defined above, the following are useful in conversions between concentrations and other quantities.

6. *Density*: $\rho = m/V$ SI base units: kg m^{-3} [6]

7. *Relative density*: *d*; the ratio of the density of a mixture to the density of a reference substance under conditions which must be specified for both (1). The symbol $d_{t'}^{t}$ will be used for the density of a mixture at t°C, 1 bar divided by the density of water at t'°C, 1 bar. (In some cases, 1 atm = 101.325 kPa is used instead of 1 bar = 100 kPa.)

8. *A note on nomenclature*. The above definitions use the nomenclature of the IUPAC *Green Book* (1), in which a solute is called *B* and a solvent *A* In compilations and evaluations, the first-named component (component 1) is the solute, and the second (component 2 for a two-component system) is the solvent. The reader should bear these distinctions in nomenclature in mind when comparing nomenclature and theoretical equations given in this Introduction with equations and nomenclature used on the evaluation and compilation sheets.

Thermodynamics of Solubility

The principal aims of the Solubility Data Project are the tabulation and evaluation of: (a) solubilities as defined above; (b) the nature of the saturating phase. Thermodynamic analysis of solubility phenomena has two aims: (a) to provide a rational basis for the construction of functions to represent solubility data; (b) to enable thermodynamic quantities to be extracted from solubility data. Both these are difficult to achieve in many cases because of a lack of experimental or theoretical information concerning activity coefficients. Where thermodynamic quantities can be found, they are not evaluated critically, since this task would involve critical evaluation of a large body of data that is not directly relevant to solubility. The following is an outline of the principal thermodynamic relations encountered in discussions of solubility. For more extensive discussions and references, see books on thermodynamics, e.g., (7-14).

Activity Coefficients (1)

(a) *Mixtures*. The activity coefficient f_B of a substance B is given by

$$RT \ln (f_B x_B) = \mu_B - \mu_B^* \qquad [7]$$

where μ_B^* is the chemical potential of pure B at the same temperature and pressure. For any substance B in the mixture,

$$\lim_{x_B \to 1} f_B = 1 \qquad [8]$$

(b) *Solutions*.

(i) *Solute B*. The molal activity coefficient γ_B is given by

$$RT \ln(\gamma_B m_B) = \mu_B - (\mu_B - RT \ln m_B)^{\infty} \qquad [9]$$

where the superscript $^{\infty}$ indicates an infinitely dilute solution. For any solute B,

$$\gamma_B^{\infty} = 1 \qquad [10]$$

Activity coefficients y_B connected with concentrations c_B, and $f_{x,B}$ (called the *rational activity coefficient*) connected with mole fractions x_B, are defined in analogous ways. The relations among them (1, 9) are, where ρ^* is the density of the pure solvent:

$$f_B = (1 + M_A \sum_s m_s)\gamma_B = [\rho + \sum_s (M_A - M_s)c_s]y_B/\rho^* \qquad [11]$$

$$\gamma_B = (1 - \sum_s x_s)f_{x,B} = (\rho - \sum_s M_s c_s)y_B/\rho^* \qquad [12]$$

$$y_B = \rho^* f_{x,B}[1 + \sum_s (M_s/M_A - 1)x_B]/\rho = \rho^*(1 + \sum_s M_s m_s)\gamma_B/\rho \qquad [13]$$

For an electrolyte solute $B \equiv C_{\nu+}A_{\nu-}$, the activity on the molality scale is replaced by (11):

$$\gamma_B m_B = \gamma_{\pm}^{\nu} m_B^{\nu} Q^{\nu} \qquad [14]$$

where $\nu = \nu_+ + \nu_-$, $Q = (\nu_+^{\nu+}\nu_-^{\nu-})^{1/\nu}$, and $\gamma_{\pm}$ is the mean ionic activity coefficient on the molality scale. A similar relation holds for the concentration activity, $y_B c_B$. For the mole fractional activity,

$$f_{x,B} x_B = Q^{\nu} f_{\pm}^{\nu} x_{\pm}^{\nu} \qquad [15]$$

where $x_{\pm} = (x_+ x_-)^{1/\nu}$. The quantities x_+ and x_- are the ionic mole fractions (11), which are:

$$x_+ = \nu_+ x_B/[1 + \sum_s (\nu_s - 1)x_s]; \quad x_- = \nu_- x_B[1 + \sum_s (\nu_s - 1)x_s] \qquad [16]$$

where ν_s is the sum of the stoichiometric coefficients for the ions in a salt with mole fraction x_s. Note that the mole fraction of solvent is now

$$x_A' = (1 - \sum_s \nu_s x_s)/[1 + \sum_s (\nu_s - 1)x_s] \qquad [17]$$

so that

$$x_A' + \sum_s \nu_s x_s = 1 \qquad [18]$$

The relations among the various mean ionic activity coefficients are:

$$f_{\pm} = (1 + M_A \sum_s \nu_s m_s)\gamma_{\pm} = [\rho + \sum_s (\nu_s M_A - M_s)c_s]y_{\pm}/\rho^* \qquad [19]$$

$$\gamma_{\pm} = \frac{(1 - \sum_s x_s)f_{\pm}}{1 + \sum_s (\nu_s - 1)x_s} = (\rho - \sum_s M_s c_s)y_{\pm}/\rho^* \qquad [20]$$

$$y_{\pm} = \frac{\rho^*[1 + \sum_s (M_s/M_A - 1)x_s]f_{\pm}}{\rho[1 + \sum_s (\nu_s - 1)x_s]} = \rho^*(1 + \sum_s M_s m_s)\gamma_{\pm}/\rho \qquad [21]$$

(ii) *Solvent, A*:

The *osmotic coefficient*, ϕ , of a solvent A is defined as (1):

$$\phi = (\mu_A{}^* - \mu_A)/RT\ M_A \sum_s m_s \quad [22]$$

where $\mu_A{}^*$ is the chemical potential of the pure solvent.

The *rational osmotic coefficient*, ϕ_x, is defined as (1):

$$\phi_x = (\mu_A - \mu_A{}^*)/RT\ln x_A = \phi M_A \sum_s m_s/\ln(1 + M_A \sum_s m_s) \quad [23]$$

The activity, a_A, or the activity coefficient, f_A, is sometimes used for the solvent rather than the osmotic coefficient. The activity coefficient is defined relative to pure A, just as for a mixture.

For a mixed solvent, the molar mass in the above equations is replaced by the average molar mass; i.e., for a two-component solvent with components J, K, M_A becomes

$$M_A = M_J + (M_K - M_J)x_{v,K} \quad [24]$$

where $x_{v,K}$ is the solvent mole fraction of component K.

The osmotic coefficient is related directly to the vapor pressure, p, of a solution in equilibrium with vapor containing A only by (14, p.306):

$$\phi M_A \sum_s \nu_s m_s = -\ln(p/p_A{}^*) + (V_{m,A}^* - B_{AA})(p - p_A{}^*)/RT \quad [25]$$

where $p_A{}^*$ is the vapor pressure of pure solvent A, $V_{m,A}^*$ is the molar volume of pure A in the liquid phase, and B_{AA} is the second virial coefficient of the vapor.

The Liquid Phase

A general thermodynamic differential equation which gives solubility as a function of temperature, pressure and composition can be derived. The approach is similar to that of Kirkwood and Oppenheim (9); see also (13, 14). Consider a solid mixture containing c thermodynamic components i. The Gibbs-Duhem equation for this mixture is:

$$\sum_{i=1}^{c} x_i'(S_i'dT - V_i'dp + d\mu_i') = 0 \quad [26]$$

A liquid mixture in equilibrium with this solid phase contains c' thermodynamic components i, where $c' \geqslant c$. The Gibbs-Duhem equation for the liquid mixture is:

$$\sum_{i=1}^{c} x_i(S_idT - V_idp + d\mu_i') + \sum_{i=c+1}^{c'} x_i(S_idT - V_idp + d\mu_i) = 0 \quad [27]$$

Subtract [26] from [27] and use the equation

$$d\mu_i = (d\mu_i)_{T,p} - S_idT + V_idp \quad [28]$$

and the Gibbs-Duhem equation at constant temperature and pressure:

$$\sum_{i=1}^{c} x_i(d\mu_i')_{T,p} + \sum_{i=c+1}^{c'} x_i(d\mu_i)_{T,p} = 0 \quad [29]$$

The resulting equation is:

$$RT\sum_{i=1}^{c} x_i'(d\ln a_i)_{T,p} = \sum_{i=1}^{c} x_i'(H_i - H_i')dT/T - \sum_{i=1}^{c} x_i'(V_i - V_i')dp \quad [30]$$

where

$$H_i - H_i' = T(S_i - S_i') \quad [31]$$

is the enthalpy of transfer of component i from the solid to the liquid phase at a given temperature, pressure and composition, with H_i and S_i the partial molar enthalpy and entropy of component i.

Use of the equations

$$H_i - H_i^{o} = -RT^2(\partial\ln a_i/\partial T)_{x,p} \quad [32]$$

and

$$V_i - V_i^{o} = RT(\partial\ln a_i/\partial p)_{x,T} \quad [33]$$

where superscript o indicates an arbitrary reference state gives:

$$RT\sum_{i=1}^{c} x_i'd\ln a_i = \sum_{i=1}^{c} x_i'(H_i^{o} - H_i')dT/T - \sum_{i=1}^{c} x_i'(V_i^{o} - V_i')dp \quad [34]$$

where

$$d\ln a_i = (d\ln a_i)_{T,p} + (\partial \ln a_i/\partial T)_{x,p} + (\partial \ln a_i/\partial p)_{x,T} \qquad [35]$$

The terms involving enthalpies and volumes in the solid phase can be written as:

$$\sum_{i=1}^{c} x_i' H_i' = H_s^* \qquad \sum_{i=1}^{c} x_i' V_i' = V_s^* \qquad [36]$$

With eqn [36], the final general solubility equation may then be written:

$$R\sum_{i=1}^{c} x_i' d\ln a_i = (H_s^* - \sum_{i=1}^{c} x_i' H_i^0)d(1/T) - (V_s^* - \sum_{i=1}^{c} x_i' V_i^0)dp/T \qquad [37]$$

Note that those components which are not present in both phases do not appear in the solubility equation. However, they do affect the solubility through their effect on the activities of the solutes.

Several applications of eqn [37] (all with pressure held constant) will be discussed below. Other cases will be discussed in individual evaluations.

(a) *Solubility as a function of temperature.*

Consider a binary solid compound A_nB in a single solvent A. There is no fundamental thermodynamic distinction between a binary compound of A and B which dissociates completely or partially on melting and a solid mixture of A and B; the binary compound can be regarded as a solid mixture of constant composition. Thus, with $c = 2$, $x_A' = n/(n + 1)$, $x_B' = 1/(n + 1)$, eqn [37] becomes:

$$d\ln(a_A^n a_B) = -\Delta H_{AB}^0 d(1/RT) \qquad [38]$$

where

$$\Delta H_{AB}^0 = nH_A + H_B - (n + 1)H_s^* \qquad [39]$$

is the molar enthalpy of melting and dissociation of pure solid A_nB to form A and B in their reference states. Integration between T and T_0, the melting point of the pure binary compound A_nB, gives:

$$\ln(a_A^n a_B) = \ln(a_A^n a_B)_{T=T_0} - \int_{T_0}^{T} \Delta H_{AB}^0 d(1/RT) \qquad [40]$$

(i) Non-electrolytes

In eqn [32], introduce the pure liquids as reference states. Then, using a simple first-order dependence of ΔH_{AB}^* on temperature, and assuming that the activitity coefficients conform to those for a simple mixture (8):

$$RT\ln f_A = wx_B^2 \qquad RT\ln f_B = wx_A^2 \qquad [41]$$

then, if w is independent of temperature, eqn [32] and [33] give:

$$\ln\{x_B(1-x_B)^n\} + \ln\left\{\frac{n^n}{(1 + n)^{n+1}}\right\} = G(T) \qquad [42]$$

where

$$G(T) = -\left\{\frac{\Delta H_{AB}^* - T^*\Delta C_p^*}{R}\right\}\left\{\frac{1}{T} - \frac{1}{T^*}\right\} + \frac{\Delta C_p^*}{R}\ln(T/T^*) - \frac{w}{R}\left\{\frac{x_A^2 + nx_B^2}{T} - \frac{n}{(n + 1)T^*}\right\} \qquad [43]$$

where ΔC_p^* is the change in molar heat capacity accompanying fusion plus decomposition of the pure compound to pure liquid A and B at temperature T^*, (assumed here to be independent of temperature and composition), and ΔH_{AB}^* is the corresponding change in enthalpy at $T = T^*$. Equation [42] has the general form:

$$\ln\{x_B(1-x_B)^n\} = A_1 + A_2/(T/K) + A_3\ln(T/K) + A_4(x_A^2 + nx_B^2)/(T/K) \qquad [44]$$

If the solid contains only component B, then $n = 0$ in eqn [42] to [44].

If the infinite dilution reference state is used, then:

$$RT \ln f_{x,B} = w(x_A^2 - 1) \qquad [45]$$

and [39] becomes

$$\Delta H_{AB}^{\infty} = nH_A^* + H_B^{\infty} - (n + 1)H_s^* \qquad [46]$$

where ΔH_{AB}^{∞} is the enthalpy of melting and dissociation of solid compound A_nB to the infinitely dilute reference state of solute B in solvent A; H_A^* and H_B^{∞} are the partial molar enthalpies of the solute and solvent at infinite dilution. Clearly, the integral of eqn [32] will have the same form as eqn [35], with ΔH_{AB}^{∞} replacing ΔH_{AB}^*, ΔC_p^{∞} replacing ΔCp^*, and $x_A^2 - 1$ replacing x_A^2 in the last term.

See (7) and (13) for applications of these equations to experimental data.

(ii) Electrolytes

(a) Mole fraction scale

If the liquid phase is an aqueous electrolyte solution, and the solid is a salt hydrate, the above treatment needs slight modification. Using rational mean activity coefficients, eqn [34] becomes:

$$\ln\left\{\frac{x_B^{\nu}(1 - x_B)^n}{[1 + (\nu - 1)x_B]^{n+\nu}}\right\} - \ln\left\{\frac{n^n}{(n + \nu)^{n+\nu}}\right\} + \ln\left\{\left[\frac{f_B}{f_B^*}\right]^{\nu}\left[\frac{f_A}{f_A^*}\right]^{n}\right\}$$
$$= -\left\{\frac{\Delta H_{AB}^* - T^*\Delta C_p^*}{R}\right\}\left\{\frac{1}{T} - \frac{1}{T^*}\right\} + \frac{\Delta Cp^*}{R}\ln(T/T^*) \qquad [47]$$

where superscript * indicates the pure salt hydrate. If it is assumed that the activity coefficients follow the same temperature dependence as the right-hand side of eqn [47] (15-17), the thermochemical quantities on the right-hand side of eqn [47] are not rigorous thermodynamic enthalpies and heat capacities, but are apparent quantities only. Data on activity coefficients (11) in concentrated solutions indicate that the terms involving these quantities are not negligible, and their dependence on temperature and composition along the solubility-temperature curve is a subject of current research.

A similar equation (with $\nu = 2$ and without the heat capacity terms) or activity coefficients) has been used to fit solubility data for some $MOH-H_2O$ systems, where M is an alkali metal (15); enthalpy values obtained agreed well with known values. The full equation has been deduced by another method in (16) and applied to MCl_2-H_2O systems in (16) and (17). For a summary of the use of equation [47] and similar equations, see (18).

(2) Molality scale

Substitution of the mean activities on the molality scale in eqn [40] gives:

$$\nu\ln\left\{\frac{\gamma_{\pm}m_B}{\gamma_{\pm}^* m_B^*}\right\} - \nu(m_B/m_B^* - 1) - \nu\{m_B(\phi - 1)/m_B^* - \phi^* + 1) = G(T) \qquad [48]$$

where $G(T)$ is the same as in eqn [47], $m_B^* = 1/nM_A$ is the molality of the anhydrous salt in the pure salt hydrate and $\gamma_{\pm}$ and ϕ are the mean activity coefficient and the osmotic coefficient, respectively. Use of the osmotic coefficient for the activity of the solvent leads, therefore, to an equation that has a different appearance to [47]; the content is identical. However, while eqn [47] can be used over the whole range of composition ($0 \leq x_B \leq 1$), the molality in eqn [48] becomes infinite at $x_B = 1$; use of eqn [48] is therefore confined to solutions sufficiently dilute that the molality is a useful measure of composition. The essentials of eqn [48] were deduced by Williamson (19); however, the form used here appears first in the *Solubility Data Series*. For typical applications (where activity and osmotic coefficients are not considered explicitly, so that the enthalpies and heat capacities are apparent values, as explained above), see (20).

The above analysis shows clearly that a rational thermodynamic basis exists for functional representation of solubility-temperature curves in two-component systems, but may be difficult to apply because of lack of experimental or theoretical knowledge of activity coefficients and partial molar enthalpies. Other phenomena which are related ultimately to the stoichiometric activity coefficients and which complicate interpretation include ion pairing, formation of complex ions, and hydrolysis. Similar considerations hold for the variation of solubility with pressure, except that the effects are relatively smaller at the pressures used in many investigations of solubility (7).

(*b*) *Solubility as a function of composition.*

At constant temperature and pressure, the chemical potential of a saturating solid phase is constant:

$$\mu_{A_nB}^* = \mu_{A_nB}(\mathrm{sln}) = n\mu_A + \mu_B \qquad [49]$$
$$= (n\mu_A^* + \nu_+\mu_+^\infty + \nu_-\mu_-^\infty) + nRT\,\ln f_A x_A$$
$$+ \nu RT\,\ln(\gamma_\pm m_\pm Q)$$

for a salt hydrate A_nB which dissociates to water (A), and a salt (B), one mole of which ionizes to give ν_+ cations and ν_- anions in a solution in which other substances (ionized or not) may be present. If the saturated solution is sufficiently dilute, $f_A = x_A = 1$, and the quantity K_s in

$$\Delta G^\infty = (\nu_+\mu_+^\infty + \nu_-\mu_-^\infty + n\mu_A^* - \mu_{AB}^*)$$
$$= -RT\,\ln K_s$$
$$= -\nu RT\,\ln(Q\gamma_\pm m_B) \qquad [50]$$

is called the *solubility product* of the salt. (It should be noted that it is not customary to extend this definition to hydrated salts, but there is no reason why they should be excluded.) Values of the solubility product are often given on mole fraction or concentration scales. In dilute solutions, the theoretical behavior of the activity coefficients as a function of ionic strength is often sufficiently well known that reliable extrapolations to infinite dilution can be made, and values of K_s can be determined. In more concentrated solutions, the same problems with activity coefficients that were outlined in the section on variation of solubility with temperature still occur. If these complications do not arise, the solubility of a hydrate salt $C_\nu A_\nu \cdot nH_2O$ in the presence of other solutes is given by eqn [50] as

$$\nu\,\ln\{m_B/m_B(0)\} = -\nu\ln\{\gamma_\pm/\gamma_\pm(0)\} - n\,\ln\{a_A/a_A(0)\} \qquad [51]$$

where a_A is the activity of water in the saturated solution, m_B is the molality of the salt in the saturated solution, and (0) indicates absence of other solutes. Similar considerations hold for non-electrolytes.

Consideration of *complex mixed ligand equilibria* in the solution phase is also frequently of importance in the interpretation of solubility equilibria. For nomenclature connected with these equilibria (and solubility equilibria as well) see (21, 22).

(*c*) *Alteration of the dissolution medium for pharmaceuticals*

Many substances which are only slightly soluble in water may be made more soluble by the addition of a cosolvent, surface-active agents, or complexing agents.

(i) Addition of a cosolvent. It is frequently necessary to dissolve a quantity of drug in a small volume of liquid so that it may be administered parenterally by injection. If the drug is not sufficiently soluble in water because of its hydrophobicity, the addition of a quantity of water-miscible, but less polar solvent may render the drug soluble in a small quantity of the mixed solvent. Solvents used for this purpose have included propylene glycol, glycerol, ethanol, polyethylene glycol and glycofural. Solubilities of many drug substances in water-organic solvent mixtures have been tabulated by Yalkowsky and Roseman (23).

(ii) Surface-active agents. Another approach to increasing the solubility and rate of dissolution of drug substances is to add a surface-active agent. There is an extensive literature on the application of surfactants and micellar dissolution, which has been summarized recently by Florence (24). Cationic, anionic or neutral surfactants are available. In choosing a surfactant, the possibility of charge-charge interactions between the drug and the surfactant must be considered, as well as the degree of ionization of each species as a function of pH. Micellar dissolution of drugs or additives may protect the dissolved species from hydrolytic degradation by the aqueous solvent. The stability of drugs may therefore be enhanced considerably by the addition of a surfactant. Surfactants may also facilitate the transport of drugs across biological membranes. Examples of substantially improved bioavailability of drugs under the influence of micellar dissolution have been reported (24).

(iii) Other modifications of the dissolution medium. The solubility of weak acid and weak base drugs will usually depend on the pH of the medium. Within reasonable limits for pharmaceutical preparations, pH may be adjusted to obtain the drug in the charged (and usually more soluble) form. The addition of complexing agents such as chelating agents, organic salts, cyclodextrins, or ion-pairing agents may be used to enhance solubility and rate of dissolution. Examples are given in the chapter by A.J. Repta in (3).

The Solid Phase

The definition of solubility permits the occurrence of a single solid phase which may be a pure anhydrous compound, a salt hydrate, a non-stoichiometric compound, or a solid mixture (or solid solution, or "mixed crystals"), and may be stable or metastable. As well, any number of solid phases consistent with the requirements of the phase rule may be present. Metastable solid phases are of widespread occurrence, and may appear as polymorphic (or allotropic) forms or crystal solvates whose rate of transition to more stable forms is very slow. Surface heterogeneity may also give rise to metastability, either when one solid precipitates on the surface of another, or if the size of the solid particles is sufficiently small that surface effects become important. In either case, the solid is not in stable equilibrium with the solution. See (25) for the modern formulation of the effect of particle size on solubility. The stability of a solid may also be affected by the atmosphere in which the system is equilibrated.

Many of these phenomena require very careful, and often prolonged, equilibration for their investigation and elimination. A very general analytical method, the "wet residues" method of Schreinemakers (26), is often used to investigate the composition of solid phases in equilibrium with salt solutions. This method has been reviewed in (27), where [see also (28)] least-squares methods for evaluating the composition of the solid phase from wet residue data (or initial composition data) and solubilities are described. In principle, the same method can be used with systems of other types. Many other techniques for examination of solids, in particular X-ray, optical, and thermal analysis methods, are used in conjunction with chemical analyses (including the wet residues method).

Solid State Manipulation in Pharmaceuticals

(i) Polymorphism. Many drug substances may crystallize in more than one form, a phenomenon called *polymorphism*. The different modifications (polymorphs) arise because of the relative positions and bonding of the molecules in their crystal lattices; true polymorphs do not differ in chemical composition. Polymorphs of the same substance frequently have different physical properties such as solubility and rate of dissolution. Ultimately, the solubility of all forms will revert to that of the form with the lowest Gibbs energy; the solubility of a less-stable form will thus be an initial solubility. The rate of reversion to the most stable form is often very slow, and a form with higher Gibbs energy may exhibit its higher solubility for hours. This phenomenon may be used to advantage by choosing the polymorph with the desired solubility or rate of dissolution. Examples of polymorphism and methods of characterization have been reviewed by Haleblian (29) and Burger (30).

(ii) Crystallinity. In many cases, drug substances may occur in the solid state as amorphous or partly crystalline forms. This is a special case of polymorphism, and may result from rapid precipitation or from freeze-drying. These amorphous or partly crystalline materials are unstable relative to the crystalline form. However, reversion to the crystalline form may be slow, and the less stable forms may be used to enhance solubility and rate of dissolution (31).

(iii) Choice of salt form. Many drug substances are organic salts. In most cases the drug moiety is the organic cation or anion, such as a quaternary ammonium cation or a carboxylate or sulfonate anion. The counterion is frequently an inorganic ion such as sodium or chloride. It is possible to obtain large variations in initial solubility depending on the choice of the salt form of the drug.

COMPILATIONS AND EVALUATIONS

The formats for the compilations and critical evaluations have been standardized for all volumes. A brief description of the data sheets has been given in the FOREWORD; additional explanation is given below.

Guide to the Compilations

The format used for the compilations is, for the most part, self-explanatory. The details presented below are those which are not found in the FOREWORD or which are not self-evident.

Components. Each component is listed according to IUPAC or Chemical Abstracts (CA) name and CA Registry Number. The formula is given either in terms of the IUPAC or Hill (32) system and the choice of formula is governed by what is usual for most current users: i.e., IUPAC for inorganic compounds, and Hill system for organic compounds. Components are ordered according to:

(a) saturating components;
(b) non-saturating components in alphanumerical order;
(c) solvents in alphanumerical order.

The saturating components are arranged in order according to a 18-column periodic table with two additional rows:

Columns 1 and 2: H, alkali elements, ammonium, alkaline earth elements
3 to 12: transition elements
13 to 17: boron, carbon, nitrogen groups; chalcogenides, halogens
18: noble gases
Row 1: Ce to Lu
Row 2: Th to the end of the known elements, in order of atomic number.

Salt hydrates are generally not considered to be saturating components since most solubilities are expressed in terms of the anhydrous salt. The existence of hydrates or solvates is carefully noted in the text, and CA Registry Numbers are given where available, usually in the critical evaluation. Mineralogical names are also quoted, along with their CA Registry Numbers, again usually in the critical evaluation.

Original Measurements. References are abbreviated in the forms given by *Chemical Abstracts Service Source Index* (CASSI). Names originally in other than Roman alphabets are given as transliterated by *Chemical Abstracts*.

Experimental Values. Data are reported in the units used in the original publication, with the exception that modern *names* for units and quantities are used; e.g., mass per cent for weight per cent; mol dm^{-3} for molar; etc. Both mass and molar values are given. Usually, only one type of value (e.g., mass per cent) is found in the original paper, and the compiler has added the other type of value (e.g., mole per cent) from computer calculations based on 1983 atomic weights (33).

Errors in calculations and fitting equations in original papers have been noted and corrected, by computer calculations where necessary.

Method. Source and Purity of Materials. Abbreviations used in *Chemical Abstracts* are often used here to save space.

Estimated Error. If these data were omitted by the original authors, and if relevant information is available, the compilers have attempted to estimate errors from the internal consistency of data and type of apparatus used. Methods used by the compilers for estimating and and reporting errors are based on the papers by Ku and Eisenhart (34).

Comments and/or Additional Data. Many compilations include this section which provides short comments relevant to the general nature of the work or additional experimental and thermodynamic data which are judged by the compiler to be of value to the reader.

References. See the above description for Original Measurements.

Guide to the Evaluations

The evaluator's task is to check whether the compiled data are correct, to assess the reliability and quality of the data, to estimate errors where necessary, and to recommend "best" values. The evaluation takes the form of a summary in which all the data supplied by the compiler have been critically reviewed. A brief description of the evaluation sheets is given below.

Components. See the description for the Compilations.

Evaluator. Name and date up to which the literature was checked.

Critical Evaluation

(a) Critical text. The evaluator produces text evaluating *all* the published data for each given system. Thus, in this section the evaluator reviews the merits or shortcomings of the various data. Only published data are considered; even published data can be considered only if the experimental data permit an assessment of reliability.

(b) Fitting equations. If the use of a smoothing equation is justifiable the evaluator may provide an equation representing the solubility as a function of the variables reported on all the compilation sheets.

(c) Graphical summary. In addition to (b) above, graphical summaries are often given.

(d) Recommended values. Data are *recommended* if the results of at least two independent groups are available and they are in good agreement, and if the evaluator has no doubt as to the adequacy and reliability of the applied experimental and computational procedures. Data are considered as *tentative* if only one set of measurements is available, or if the evaluator considers some aspect of the computational or experimental method as mildly undesirable but estimates that it should cause only minor errors. Data are considered as *doubtful* if the evaluator considers some aspect of the computational or experimental method as undesirable but still considers the data to have some value in those instances where the order of magnitude of the solubility is needed. Data determined by an inadequate method or under ill-defined conditions are *rejected*. However references to these data are included in the evaluation together with a comment by the evaluator as to the reason for their rejection.

(e) References. All pertinent references are given here. References to those data which, by virtue of their poor precision, have been rejected and not compiled are also listed in this section.

(f) Units. While the original data may be reported in the units used by the investigators, the final recommended values are reported in S.I. units (1, 35) when the data can be accurately converted.

References

1. Whiffen, D.H., ed., *Manual of Symbols and Terminology for Physicochemical Quantities and Units*. *Pure Applied Chem.* 1979, 51, No. 1.
2. McGlashan, M.L. *Physicochemical Quantities and Units*. 2nd ed. Royal Institute of Chemistry. London. 1971.
3. Yalkowsky, S.H., ed. *Techniques of Solubilization of Drugs*. Marcel Dekker. New York. 1981.
4. Byron, S.R. *Solid State Chemistry of Drugs*. Academic Press. New York. 1982.
5. Jänecke, E. *Z. Anorg. Chem.* 1906, *51*, 132.
6. Friedman, H.L. *J. Chem. Phys.* 1960, 32, 1351.
7. Prigogine, I.; Defay, R. *Chemical Thermodynamics*. D.H. Everett, transl. Longmans, Green. London, New York, Toronto. 1954.
8. Guggenheim, E.A. *Thermodynamics*. North-Holland. Amsterdam. 1959. 4th ed.
9. Kirkwood, J.G.; Oppenheim, I. *Chemical Thermodynamics*. McGraw-Hill. New York, Toronto, London. 1961.
10. Lewis, G.N.; Randall, M. (rev. Pitzer, K.S.; Brewer, L.). *Thermodynamics*. McGraw Hill. New York, Toronto, London. 1961. 2nd. ed.
11. Robinson, R.A.; Stokes, R.H. *Electrolyte Solutions*. Butterworths. London. 1959. 2nd ed.
12. Harned, H.S.; Owen, B.B. *The Physical Chemistry of Electrolytic Solutions*. Reinhold. New York. 1958. 3rd ed.
13. Haase, R.; Schönert, H. *Solid-Liquid Equilibrium*. E.S. Halberstadt, trans. Pergamon Press, London, 1969.
14. McGlashan, M.L. *Chemical Thermodynamics*. Academic Press. London. 1979.
15. Cohen-Adad, R.; Saugier, M.T.; Said, J. *Rev. Chim. Miner.* 1973, *10*, 631.
16. Counioux, J.-J.; Tenu, R. *J. Chim. Phys.* 1981, 78, 815.
17. Tenu, R.; Counioux, J.-J. *J. Chim. Phys.* 1981, 78, 823.
18. Cohen-Adad, R. *Pure Appl. Chem.* 1985, 57, 255.
19. Williamson, A.T. *Faraday Soc. Trans.* 1944, *40*, 421.
20. Siekierski, S.; Mioduski, T.; Salomon, M. *Solubility Data Series*. Vol. 13. *Scandium, Yttrium, Lanthanum and Lanthanide Nitrates*. Pergamon Press. 1983.
21. Marcus, Y., ed. *Pure Appl. Chem.* 1969, 18, 459.
22. IUPAC Analytical Division. *Proposed Symbols for Metal Complex Mixed Ligand Equilibria* (*Provisional*). *IUPAC Inf. Bull.* 1978, No. 3, 229.

23. Yalkowsky, S.H.; Roseman, T.J. *in* S.H. Yalkowsky, ed. *Techniques of Solubilization of Drugs*. Marcel Dekker. New York. 1981.
24. Florence, A.T. *in* S.H. Yalkowsky, ed. *Techniques of Solubilization of Drugs*. Marcel Dekker. New York. 1981.
25. Enüstün, B.V.; Turkevich, J. *J. Am. Chem. Soc.* 1960, *82*, 4502.
26. Schreinemakers. F.A.H. *Z. Phys. Chem., Stoechiom. Verwandschaftsl.* 1893, 11, 75.
27. Lorimer, J.W. *Can. J. Chem.* 1981, 59, 3076.
28. Lorimer, J.W. *Can. J. Chem.* 1982, *60*, 1978.
29. Haleblian, J.K. *J. Pharm. Sci.* 1975, 64, 1269.
30. Burger, A. *Pharm. Int.* 1982, *3*, 158.
31. Shefter, E. *in* S.H. Yalkowsky, ed. *Techniques of Solubilization of Drugs*. Marcel Dekker. New York. 1981.
32. Hill, E.A. *J. Am. Chem. Soc.* 1900, *22*, 478.
33. IUPAC Commission on Atomic Weights. *Pure Appl. Chem.* 1984, 56, 653.
34. Ku, H.H., p. 73; Eisenhart, C., p. 69; in Ku, H.H., ed. *Precision Measurement and Calibration*. NBS Special Publication 300. Vol. 1. Washington. 1969.
35. *The International System of Units*. Engl. transl. approved by the BIPM of *Le Système International d'Unités*. H.M.S.O. London. 1970.

September, 1986

R. Cohen-Adad,
Villeurbanne, France

S. Lindenbaum,
Lawrence, Kansas, U.S.A.

J.W. Lorimer,
London, Ontario, Canada

A.N. Paruta,
Kingston, R.I., U.S.A.

R. Piekos,
Gdansk, Poland

M. Salomon,
Fair Haven, New Jersey, U.S.A.

Table I-1

Quantities Used as Measures of Solubility
Conversion Table for 2-Component Systems
Containing Solvent A and Solute B

	mole fraction x_B =	mass fraction w_B =	molality m_B =	concentration c_B =
x_B	x_B	$\frac{1}{1 - M_A(1 - 1/x_B)/M_B}$	$\frac{1}{M_A(1/x_B - 1)}$	$\frac{\rho}{M_B + M_A(1/x_B - 1)}$
w_B	$\frac{1}{1 + M_B(1/w_B - 1)}$	w_B	$\frac{1}{M_B(1/w_B - 1)}$	$\rho w_B/M_B$
m_B	$\frac{1}{1 + 1/m_B M_A}$	$\frac{1}{1 + 1/M_B m_B}$	m_B	$\frac{\rho}{M_B + 1/m_B}$
c_B	$\frac{1}{1 + (\rho/c_B - M_B)/M_A}$	$M_B c_B/\rho$	$\frac{1}{\rho/c_B - M_B}$	c_B

ρ = density of solution
M_A, M_B = molar masses of solvent, solute
Formulas are given in forms suitable for rapid computation; all calculations should be made using SI base units.

STRUCTURES

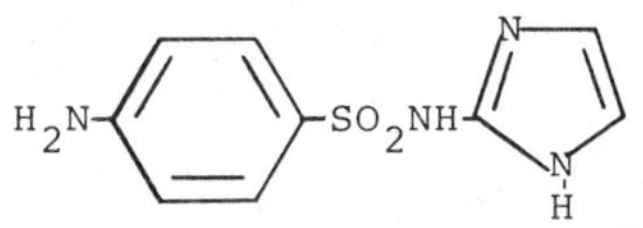

[17103-46-7] M.W.=238.26

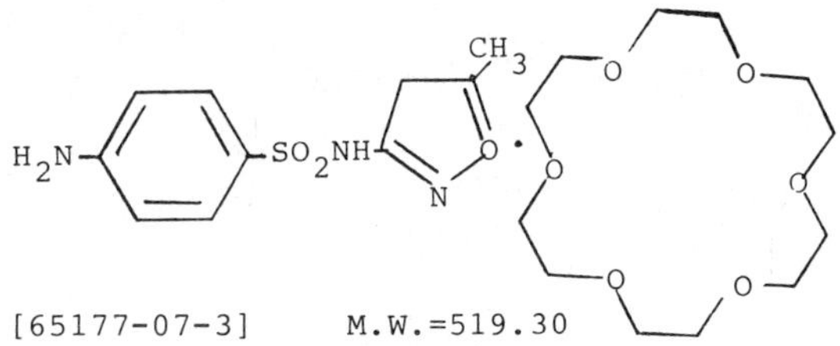

[65177-07-3] M.W.=519.30

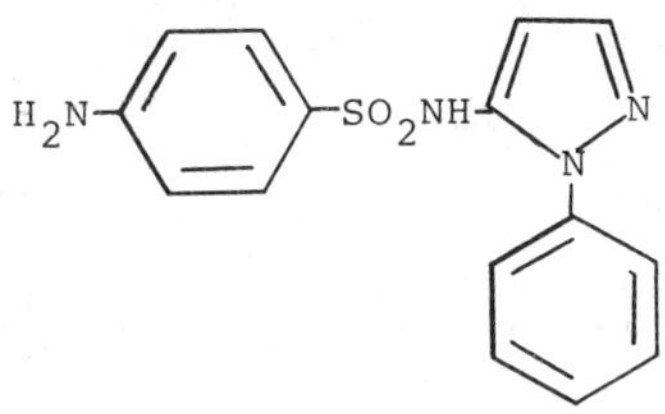

[526-08-9] M.W.=314.36

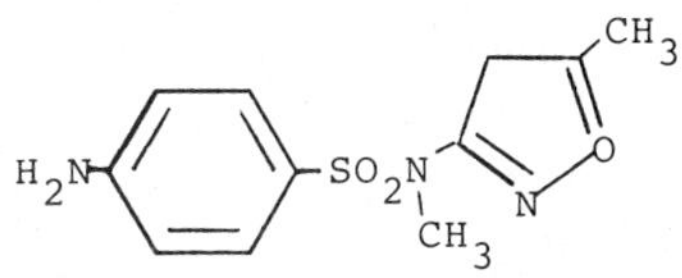

[51543-31-8] M.W.= 267.30

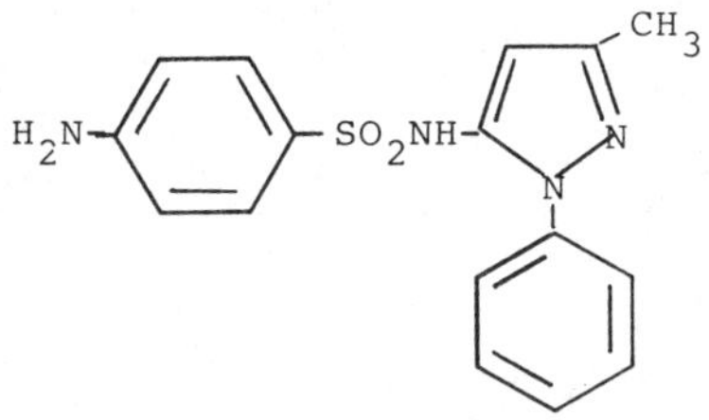

[852-19-7] M.W.=328.39

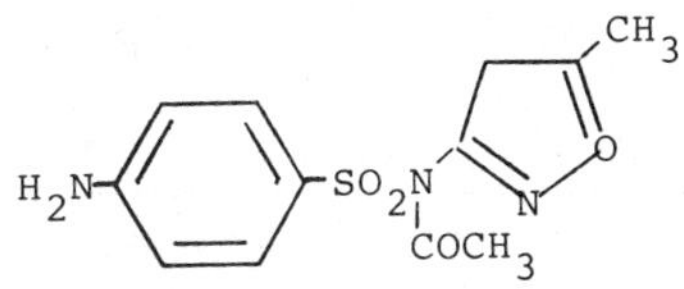

[18607-98-2] M.W.=295.31

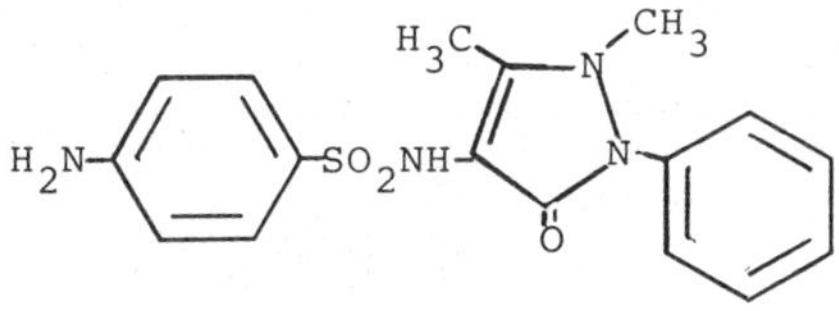

[71119-16-9] M.W.=358.42

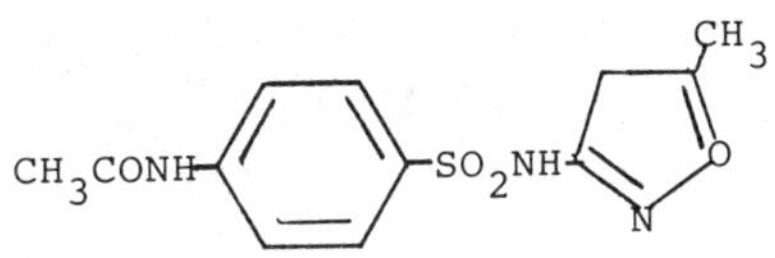

[21312-10-7] M.W.=295.31

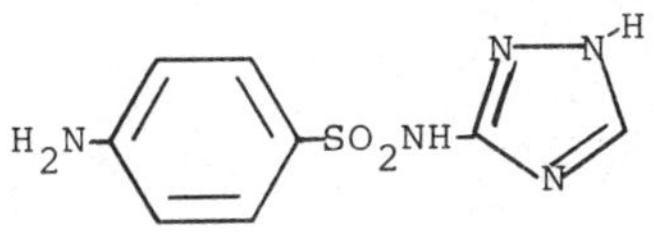

[51732-39-9] M.W.=239.25

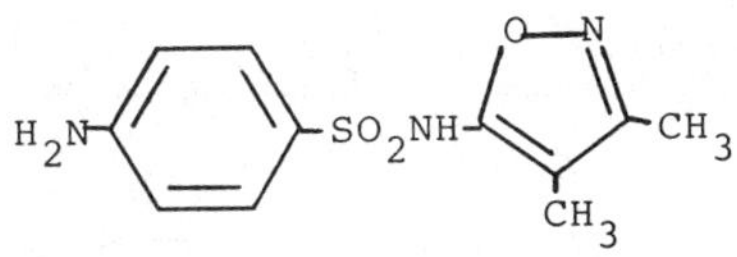

[127-69-5] M.W.=267.30

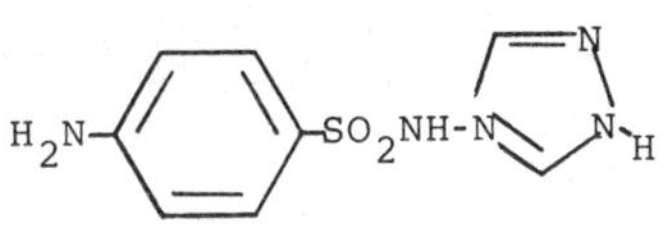

[17103-50-3] M.W.=239.25

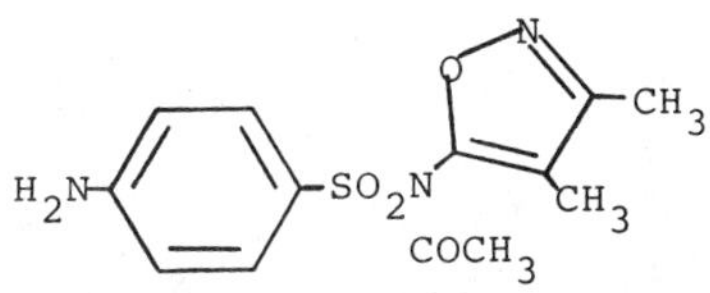

[80-74-0] M.W.=309.34

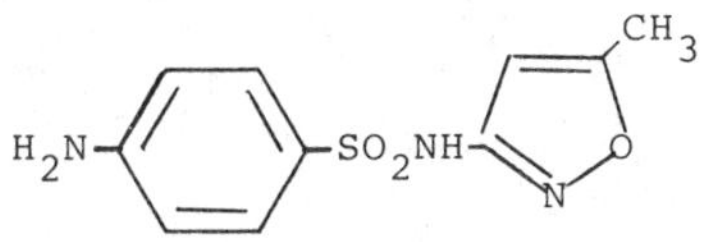

[723-46-6] M.W.=253.27

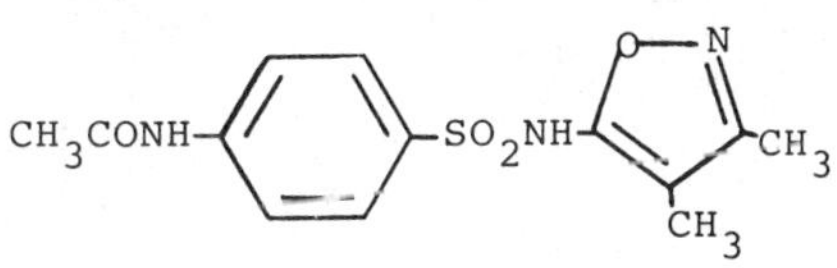

[4206-74-0] M.W.=309.34

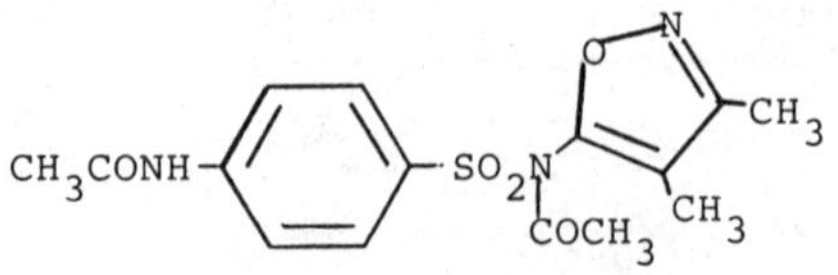

[35943-12-5] M.W.=351.38

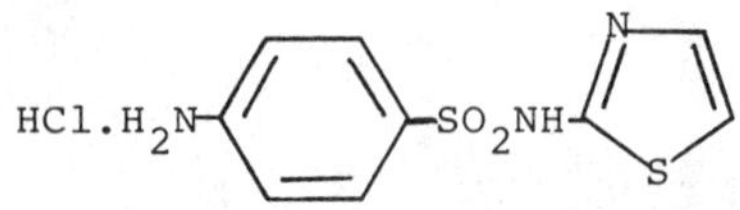

[23325-73-7] M.W.=291.77

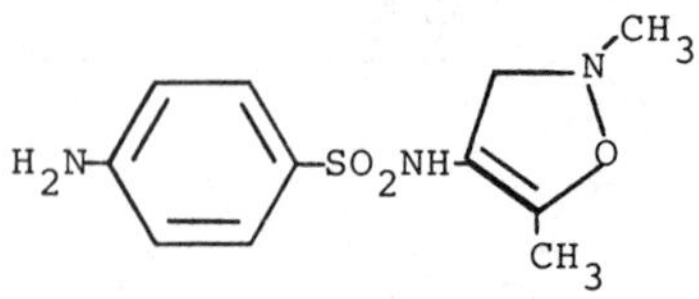

[51543-32-9] M.W.=267.30

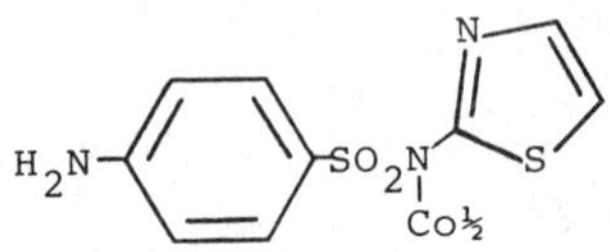

[86729-22-8] M.W.=UNSPECIFIED

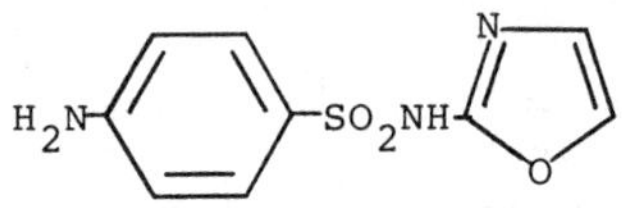

[17103-51-4] M.W.=239.25

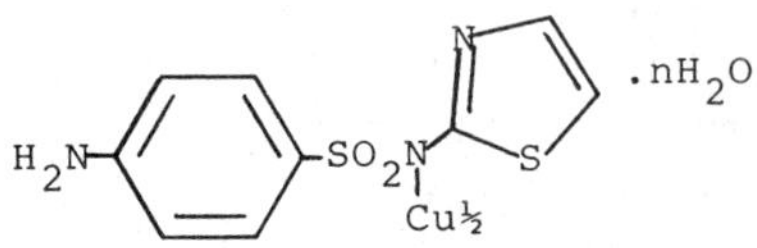

[86729-21-7] M.W.=UNSPECIFIED

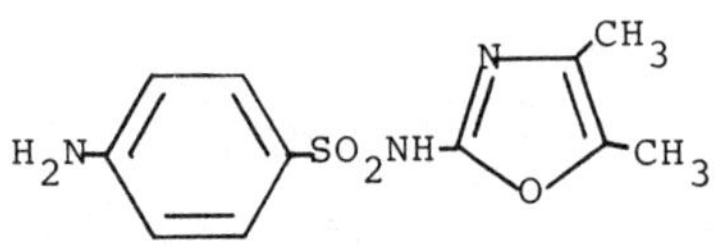

[729-99-7] M.W.=267.30

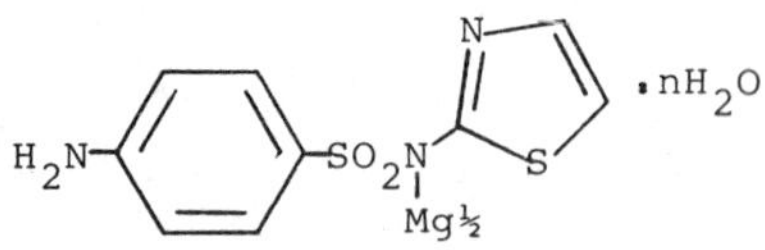

[84812-78-2] M.W.=UNSPECIFIED

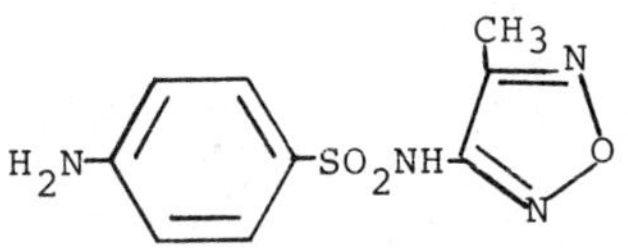

[723-47-7] M.W.=254.26

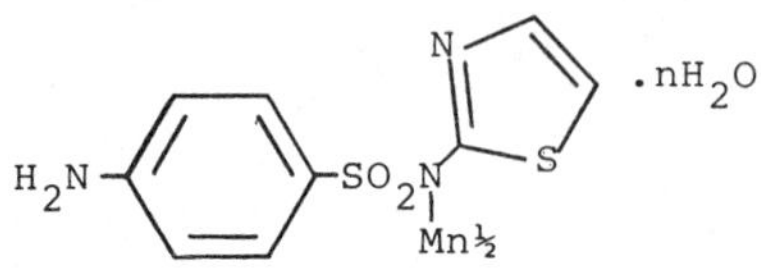

[84812-77-1] M.W.=UNSPECIFIED

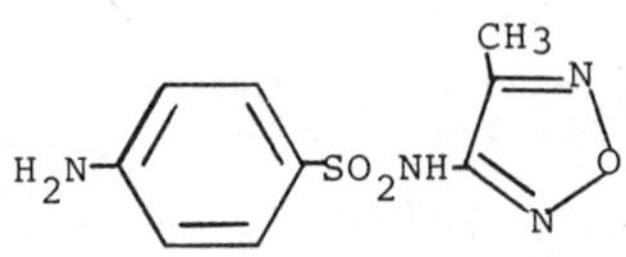

[17103-53-6] M.W.=254.26

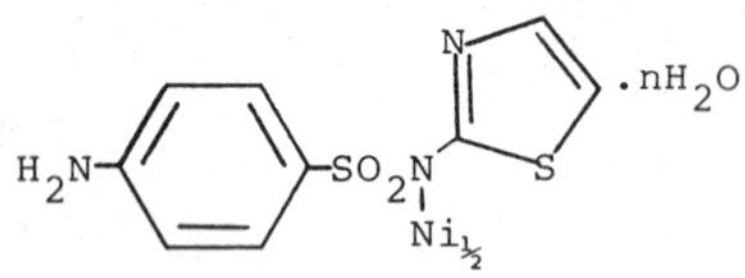

[84812-76-0] M.W.=UNSPECIFIED

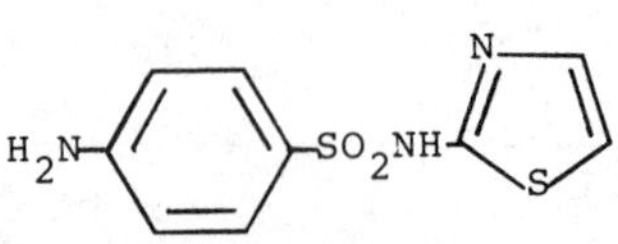

[72-14-0] M.W.=255.31

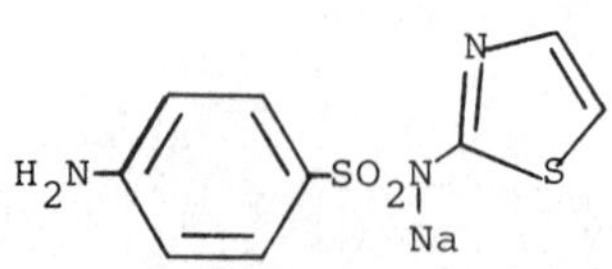

[144-74-1] M.W.=277.29

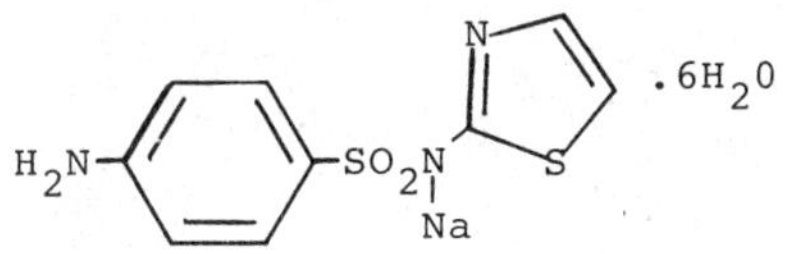 [71119-42-1] M.W.=385.39	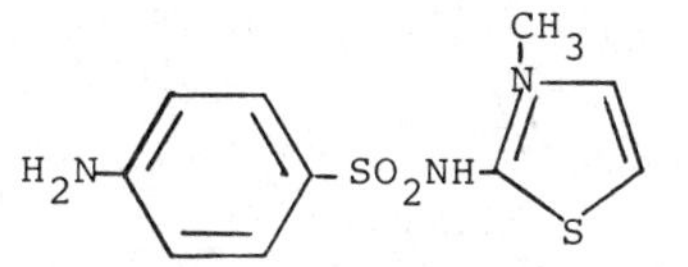 [51203-20-4] M.W.=271.36
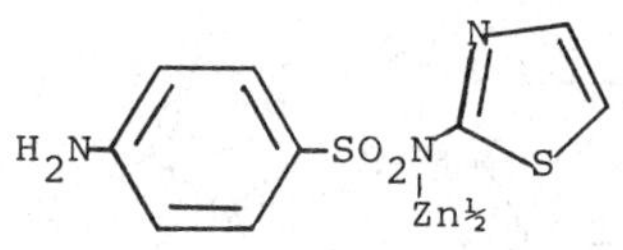 [12286-43-0] M.W.=573.98	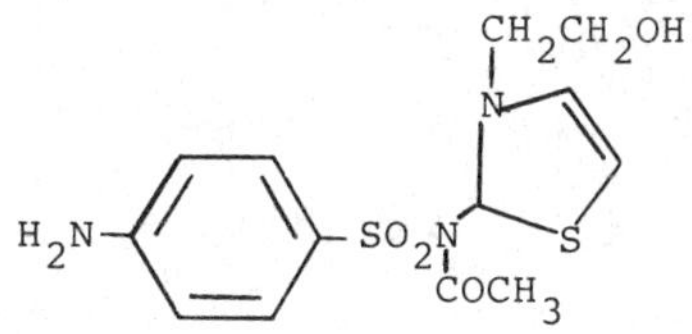 [71119-27-2] M.W.=301.38
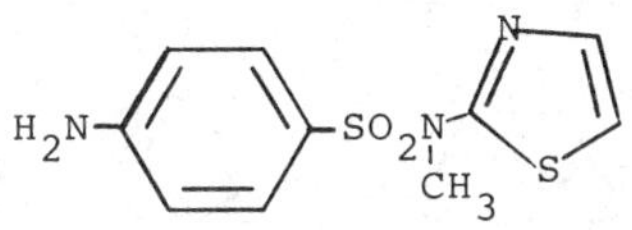 [51203-19-1] M.W.=269.34	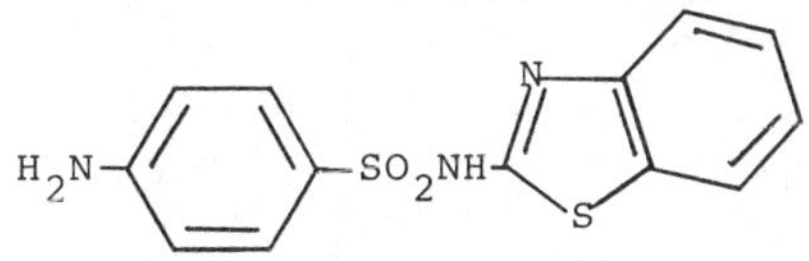 [6138-01-8] M.W.=305.37
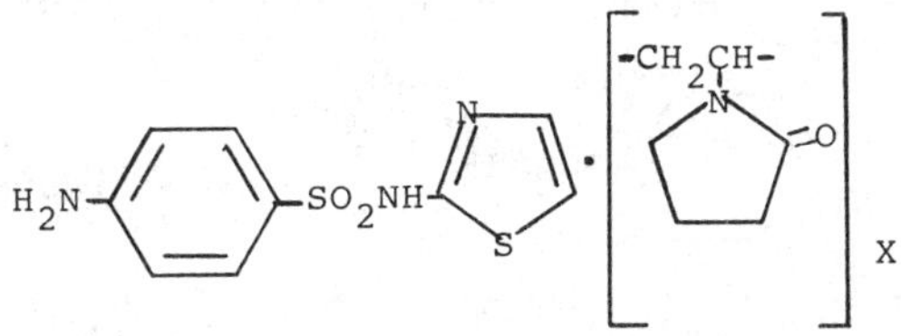 [NOT ASSIGNABLE] M.W.=UNSPECIFIED	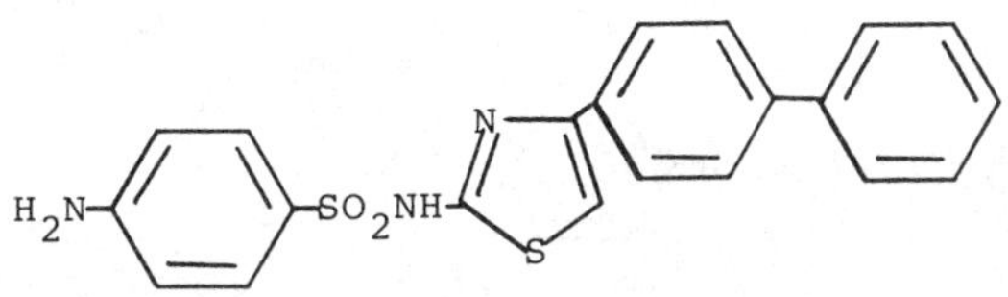 [71119-15-8] M.W.=407.50
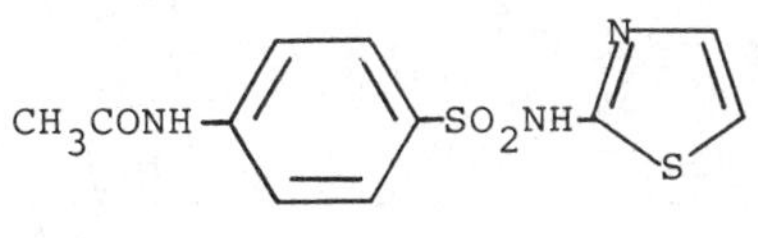 [127-76-4] M.W.=297.35	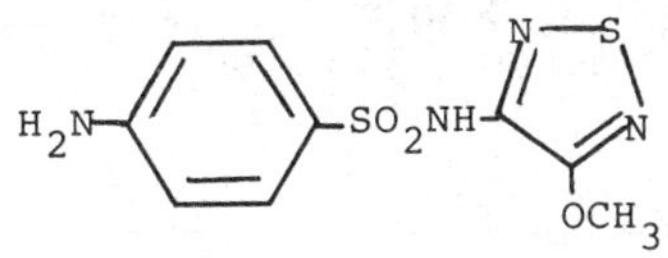 [32909-92-5] M.W.=286.32
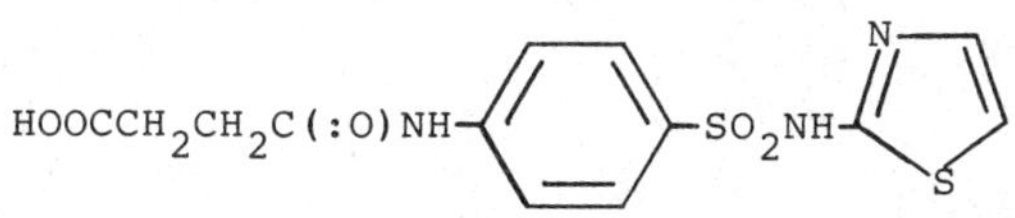 [116-43-8] M.W.=355.38	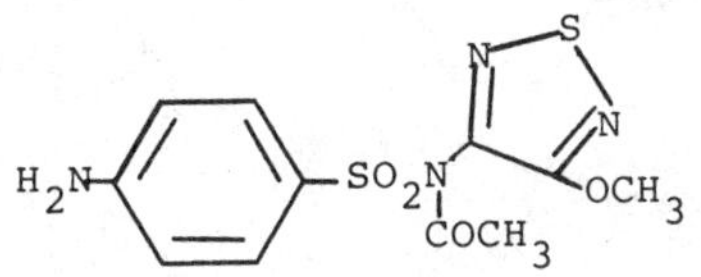 [84930-17-6] M.W.=328.36
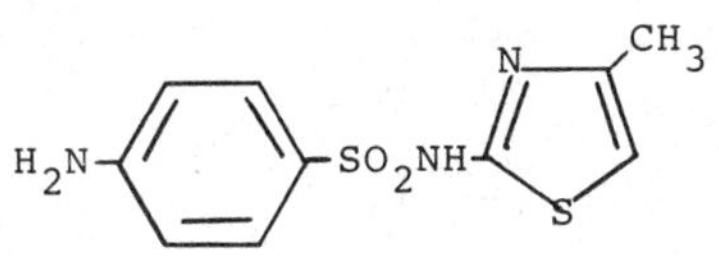 [515-59-3] M.W.=269.34	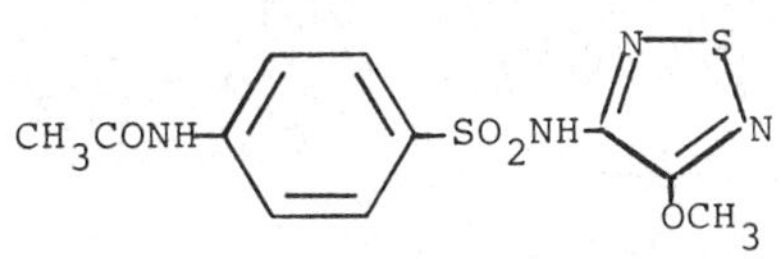 [79962-97-3] M.W.=328.36
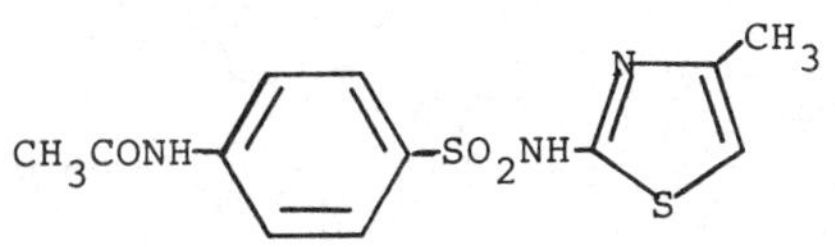 [71119-13-6] M.W.=311.37	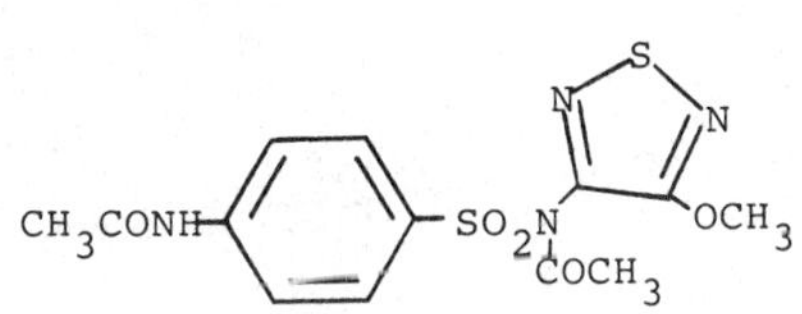 [84930-18-7] M.W.=370.40

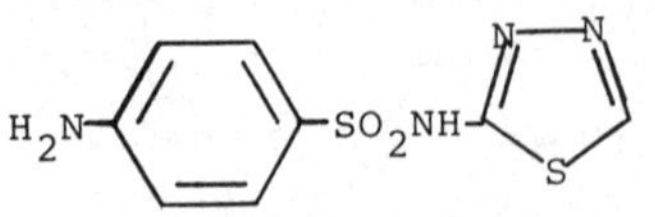

[16808-29-4] M.W.=256.30

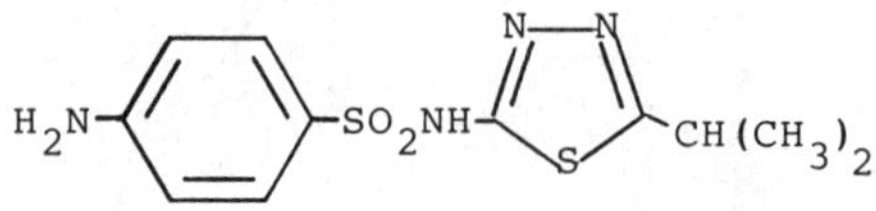

[80-43-2] M.W.=298.38

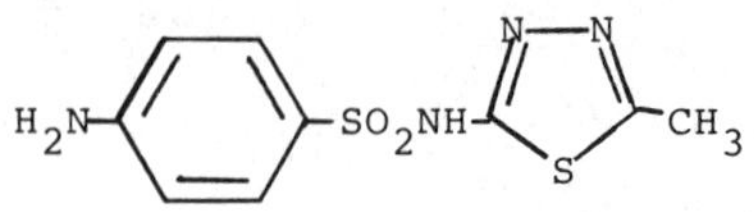

[144-82-1] M.W.=270.32

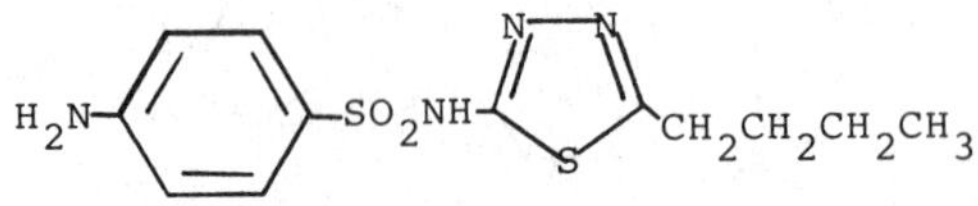

[71119-31-8] M.W.=312.40

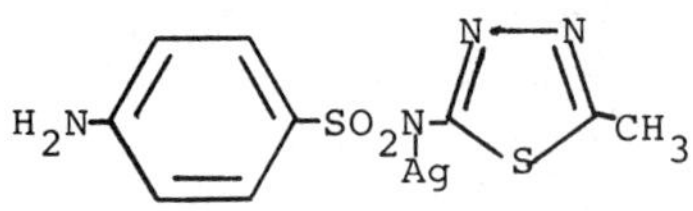

[24342-31-2] M.W.=377.18

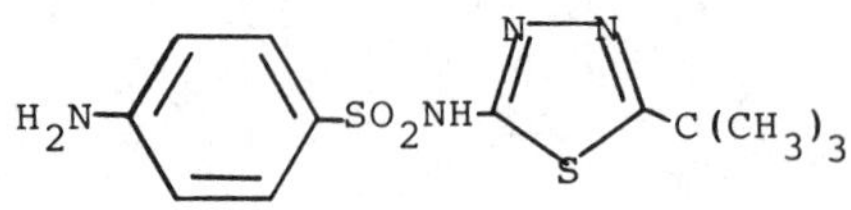

[535-65-9] M.W.=312.40

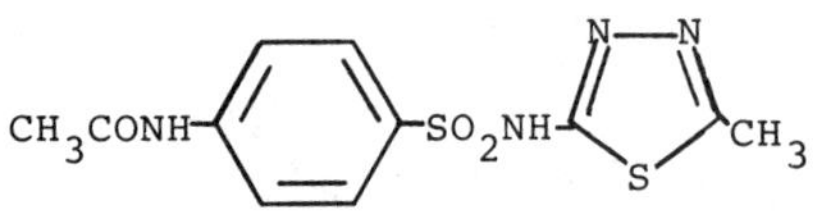

[39719-87-4] M.W.=312.36

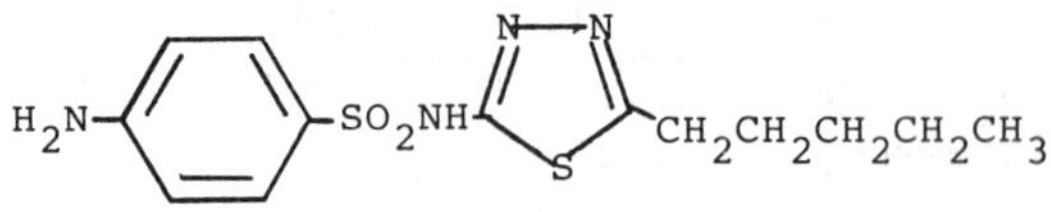

[71119-30-7] M.W.=326.43

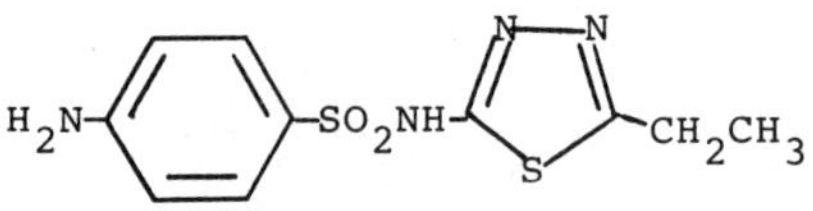

[94-19-9] M.W.=284.35

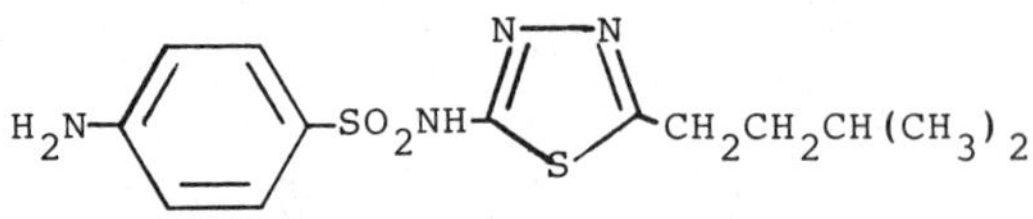

[71119-29-4] M.W.=326.43

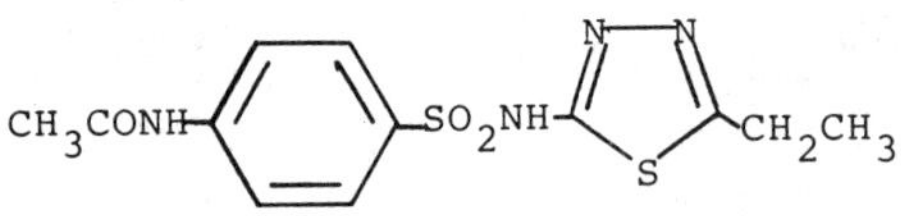

[1037-51-0] M.W.=326.39

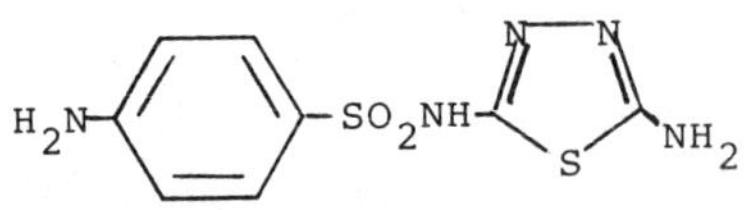

[71119-25-0] M.W.=271.31

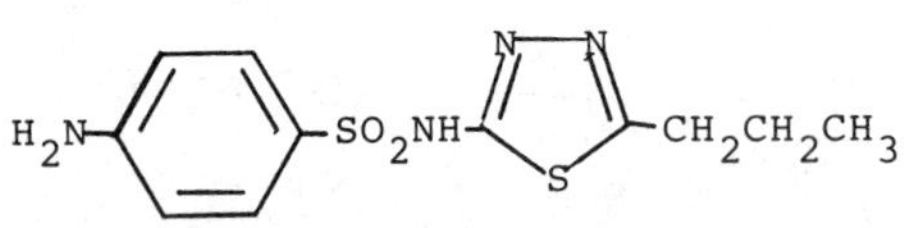

[71119-32-9] M.W.=298.38

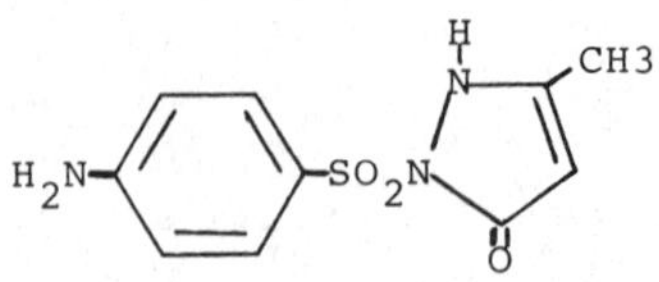

[13269-73-3] M.W.=253.28

COMPONENTS:

(1) Benzenesulfonamide, 4-amino-N-1H-imidazol-2-yl-; $C_9H_{10}N_4O_2S$; [17103-46-7]

(2) Water; H_2O; [7732-18-5]

ORIGINAL MEASUREMENTS:

Anderson, G.W.; Faith, H.E.; Marson, H.W; Winnek, P.S.; Roblin, R.O., Jr.

J. Am. Chem. Soc. 1942, *64*, 2902-5.

VARIABLES:

One temperature: 37°C

PREPARED BY:

R. Piekos

EXPERIMENTAL VALUES:

Solubility of 4-amino-N-1H-imidazol-2-ylbenzenesulfonamide in water at 37°C is 178 mg/100 cm^3 solution (7.47 x 10^{-3} mol dm^{-3}, compiler).

AUXILIARY INFORMATION

METHOD/APPARATUS/PROCEDURE:

Excess sulfonamide in water was heated and stirred on a steam bath for 30 min. The suspension was then agitated for 24 h in a thermostat. A sample of the satd soln was withdrawn through a glass filter, dild, and analyzed by the Marshall method (1) using a General Electric recording spectrophotometer for comparing the colors developed with those of the standards.

SOURCE AND PURITY OF MATERIALS:

The sulfonamide, mp 262°C (cor) was prepd by the authors. Anal: %C 45.8 (calcd 45.4); %H 4.6 (4.2); %N 23.7 (23.5).

Purity of the water was not specified.

ESTIMATED ERROR:

Nothing specified.

REFERENCES:

1. Bratton, A.C.; Marshall, E.K., Jr. *J. Pharmacol.* 1939, *66*, 4.

COMPONENTS:

(1) Benzenesulfonamide, 4-amino-N-(1-phenyl-1H-pyrazol-5-yl)- (sulfaphenazole); $C_{15}H_{14}N_4O_2S$; [526-08-9]

(2) Water; H_2O; [7732-18-5]

ORIGINAL MEASUREMENTS:

Yamasaki, M.; Aoki, M.; Kamada, A.; Yata, N. *Yakuzaigaku* 1967, *27(1)*, 37-40.

VARIABLES:

One temperature: 30°C

PREPARED BY:

R. Piekos

EXPERIMENTAL VALUES:

Solubility of sulfaphenazole in water at 30°C is 0.48 mmol/L (0.15 g dm^{-3}, compiler).

AUXILIARY INFORMATION

METHOD/APPARATUS/PROCEDURE:

Sulfaphenazole (0.5 g) was placed in an L-shaped tube together with 20 ml of water. The mixt was shaken in a thermostat until equilibrium was attained. The sulfaphenazole was assayed in the supernatant spectrophotometrically at 545 nm on a Beckman DU spectrophotometer. The results were taken from a calibration graph.

SOURCE AND PURITY OF MATERIALS:

Nothing specified

ESTIMATED ERROR:

Soly: not specified

Temp: ±1°C (authors)

REFERENCES:

COMPONENTS:	ORIGINAL MEASUREMENTS:
(1) Benzenesulfonamide, 4-amino-N-(1-phenyl-1H-pyrazol-5-yl)- (sulfaphenazole); $C_{15}H_{14}N_4O_2S$; [526-08-9] (2) Hydrochloric acid; HCl; [7647-01-0] (3) Water; H_2O; [7732-18-5]	Ogata, H.; Shibazaki, T.; Inoue, T.; Ejima, A. *Chem. Pharm. Bull.* 1979, *27(6)*, 1281-6.
VARIABLES:	PREPARED BY:
One temperature: 37°C	R. Piekos

EXPERIMENTAL VALUES:

Solubility of sulfaphenazole in 0.1N HCl at 37°C is 1.199 mg/ml (3.814×10^{-3} mol dm^{-3}, compiler).

AUXILIARY INFORMATION

METHOD/APPARATUS/PROCEDURE:

A centrifuge tube contg 30 ml of 0.1N HCl and 0.5-3.0 g of the sulfaphenazole powder was tightly sealed and shaken at 37°C. The concn of the dissolved drug was detd spectrophotometrically following filtration (type EH, pore size 0.5 μm), and the procedure was repeated every 24 h until a const concn was obtained. A Millipore filter was used for filtration.

SOURCE AND PURITY OF MATERIALS:

Comm available 500-mg uncoated tablets of sulfaphenazole were used. Hydrochloric acid was of reagent grade.

ESTIMATED ERROR:

Nothing specified

REFERENCES:

COMPONENTS:

(1) Benzenesulfonamide, 4-amino-N-(1-phenyl-1H-pyrazol-5-yl)- (sulfaphenazole); $C_{15}H_{14}N_4O_2S$; [526-08-9]

(2) Phosphoric acid, disodium salts; Na_2HPO_4; [7558-94-4]

(3) Phosphoric acid, monopotassium salt; KH_2PO_4; [7778-77-0]

(4) Water; H_2O; [7732-18-5]

ORIGINAL MEASUREMENTS:

Riess, W.

Intern. Congr. Chemotherapy, Proc. 3rd, Stuttgart 1963, *7*, 627-32.

VARIABLES:

One temperature: 20°C; one pH: 7.4

PREPARED BY:

R. Piekos

EXPERIMENTAL VALUES:

Solubility of sulfaphenazole in a M/15 Sörensen buffer solution (pH 7.4) at 20°C is 130 mg% (4.14×10^{-3} mol dm^{-3} solution, compiler).

AUXILIARY INFORMATION

METHOD/APPARATUS/PROCEDURE:

Sörensen buffer solns of pH varying between 7 and 8 were prepd, satd with sulfaphenazole at 20°C, their pH was measured at equilibrium, and the sulfaphenazole was assayed colorimetrically. The measured pH values were plotted against concn, and the soly at pH 7.4 was detd by interpolation (personal communication).

SOURCE AND PURITY OF MATERIALS:

Nothing specified

ESTIMATED ERROR:

Nothing specified

REFERENCES:

COMPONENTS:

(1) Benzenesulfonamide, 4-amino-N-(1-phenyl-1H-pyrazol-5-yl)- (sulfaphenazole); $C_{15}H_{14}N_4O_2S$; [526-08-9]

(2) Phosporic acid, disodium salt; Na_2HPO_4; [7558-94-4]

(3) Phosporic acid, monopotassium salt; KH_2PO_4; [7778-77-0]

(4) Water; H_2O; [7732-18-5]

ORIGINAL MEASUREMENTS:

Yamazaki, M.; Aoki, M.; Kamada, A.; Yata, N, *Yakuzaigaku* 1967, *27(1)*, 37-40.

PREPARED BY:

R. Piekos

VARIABLES:

One temperature: 30°C; one pH. 7.4

EXPERIMENTAL VALUES:

Solubility of sulfaphenazole in a phosphate buffer solution of pH 7.4[a] (μ = 0.17) at 30°C is 6.63 mmol/L (2.01 g dm^{-3}, compiler).

[a]At the end of experiment the pH was 7.1

AUXILIARY INFORMATION

METHOD/APPARATUS/PROCEDURE:

Sulfaphenazole (0.5 g) was placed in an L-shaped tube together with 20 ml of the buffer soln. The mixt was shaken in a thermostat until equilibrium was attained. The sulfaphenazole was assayed in the supernatant spectrophotometrically at 545 nm on a Beckman DU spectrophotometer. The results were taken from a calibration graph.

SOURCE AND PURITY OF MATERIALS:

Nothing specified

ESTIMATED ERROR:

Soly and pH: not specified

Temp: ±1°C (authors)

REFERENCES:

COMPONENTS:

(1) Benzenesulfonamide, 4-amino-N-(1-phenyl-1H-pyrazol-5-yl)- (sulfaphenazole); $C_{15}H_{14}N_4O_2S$; [526-08-9]

(2) Phosphoric acid, disodium salt; Na_2HPO_4; [7558-94-4]

(3) 1,2,3-Propanetricarboxylic acid, 2-hydroxy- (citric acid); $C_6H_8O_7$; [77-92-9]

(4) Water; H_2O; [7732-18-5]

ORIGINAL MEASUREMENTS:

Bertazzoli, C.; Buogo, A.; Ciceri, C. Ghione, M.; Turolla, E.; Zavaglio, V. *Minerva Med.* 1961, *52(40)*, 1789-96.

PREPARED BY:

R. Piekos

VARIABLES:

pH

EXPERIMENTAL VALUES:

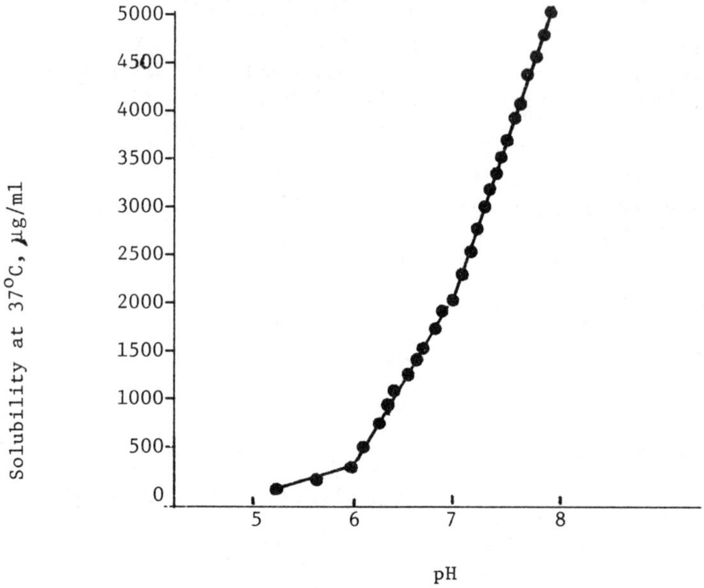

AUXILIARY INFORMATION

METHOD/APPARATUS/PROCEDURE:

The soly of sulfaphenazole in McIlvaine's Na_2HPO_4 - citric acid buffer solns was detd under agitation at 37°C. No details were given.

SOURCE AND PURITY OF MATERIALS:

Nothing specified

ESTIMATED ERROR:

Nothing specified

REFERENCES:

COMPONENTS:	ORIGINAL MEASUREMENTS:
(1) Benzenesulfonamide, 4-amino-N-(1-phenyl-1H-pyrazol-5-yl)- (sulfaphenazole); $C_{15}H_{14}N_4O_2S$; [526-08-9] (2) Methane, trichloro- (chloroform); $CHCl_3$; [67-66-3]	Riess, W. *Intern. Congr. Chemotherapy, Proc., 3rd, Stuttgart* 1963, *1*, 627-32.
VARIABLES:	**PREPARED BY:**
One temperature: 20°C	R. Piekos

EXPERIMENTAL VALUES:

Solubility of sulfaphenazole in chloroform at 20°C is 247 mg% (7.86 x 10^{-3} mol dm^{-3} solution, compiler).

AUXILIARY INFORMATION

METHOD/APPARATUS/PROCEDURE:

Nothing specified

SOURCE AND PURITY OF MATERIALS:

Nothing specified

ESTIMATED ERROR:

Nothing specified

REFERENCES:

COMPONENTS:	ORIGINAL MEASUREMENTS:
(1) Benzenesulfonamide, 4-amino-N-(1-phenyl-1H-pyrazol-5-yl)- (sulfaphenazole); $C_{15}H_{14}N_4O_2S$; [526-08-9] (2) Methane, trichloro- (chloroform); $CHCl_3$; [67-66-3]	Yamazaki, M.; Aoki, M.; Kamada, A.; Yata, N. *Yakuzaigaku* 1967, *27(1)*, 37-40.
VARIABLES:	PREPARED BY:
One temperature: 30°C	R. Piekos

EXPERIMENTAL VALUES:

Solubility of sulfaphenazole in chloroform at 30°C is 9.97 mmol/L (3.01 g dm^{-3}, compiler).

AUXILIARY INFORMATION

METHOD/APPARATUS/PROCEDURE:

Sulfaphenazole (0.5 g) was placed in an L-shaped tube together with 20 ml of chloroform. The mixt was shaken in a thermostat until equilibrium was attained. The sulfaphenazole was assayed in the supernatant spectrophotometrically at 545 nm on a Beckman DU spectrophotometer. The results were taken from a calibration graph.

SOURCE AND PURITY OF MATERIALS:

Nothing specified

ESTIMATED ERROR:

Soly: not specified

Temp: ±1°C (authors)

REFERENCES:

COMPONENTS:

(1) Benzenesulfonamide, 4-amino-N-(3-methyl-1-phenyl-1H-pyrazol-5-yl)-(sulfamethylphenazole); $C_{16}H_{16}N_4O_2S$; [852-19-7]

(2) Phosphoric acid, disodium salt; Na_2HPO_4; [7558-94-4]

(3) Phosphoric acid, monopotassium salt; KH_2PO_4; [7778-77-0]

(4) Water; H_2O; [7732-18-5]

ORIGINAL MEASUREMENTS:

Riess, W.

Intern. Congr. Chemotherapy, Proc. 3rd, Stuttgart 1963, *1*, 627-32

PREPARED BY:

R. Piekos

VARIABLES:

One temperature: 20°C; one pH: 7.4

EXPERIMENTAL VALUES:

Solubility of sulfamethylphenazole in a M/15 Sörensen buffer solution (pH 7.4) at 20°C is 63 mg% (2.0 x 10^{-3} mol dm^{-3} solution, compiler).

AUXILIARY INFORMATION

METHOD/APPARATUS/PROCEDURE:

Sörensen buffer solns of pH varying between 7 and 8 were prepd, satd with sulfamethylphenazole at 20°C, their pH was measured at equilibrium, and the sulfamethylphenazole was assayed colorimetrically. The measured pH values were plotted against concn, and the soly at pH 7.4 was detd by interpolation (personal communication).

SOURCE AND PURITY OF MATERIALS:

Nothing specified

ESTIMATED ERROR:

Nothing specified

REFERENCES:

COMPONENTS:	ORIGINAL MEASUREMENTS:
(1) Benzenesulfonamide, 4-amino-N-(3-methyl-1-phenyl- 1H-pyrazol-5-yl)- (sulfamethylphenazole); $C_{16}H_{16}N_4O_2S$ [852-19-7] (2) Methane, trichloro- (chloroform); $CHCl_3$; [67-66-3]	Riess, W. *Intern. Congr. Chemotherapy, Proc., 3rd, Stuttgart* 1963, *1*, 627-32.
VARIABLES: One temperature: 20°C	PREPARED BY: R. Piekos

EXPERIMENTAL VALUES:

Solubility of sulfamethylphenazole in chloroform at 20°C is 363 mg% (1.15×10^{-2} mol dm^{-3} solution, compiler).

AUXILIARY INFORMATION

METHOD/APPARATUS/PROCEDURE:	SOURCE AND PURITY OF MATERIALS:
Nothing specified	Nothing specified
	ESTIMATED ERROR: Nothing specified
	REFERENCES:

COMPONENTS:

(1) Benzenesulfonamide, 4-amino-N-(1-phenyl-2,3-dimethyl-5-oxopyrazol-4-yl)-; $C_{17}H_{18}N_4O_3S$; [71119-16-9]

(2) Water; H_2O; [7732-18-5]

ORIGINAL MEASUREMENTS:

Roblin, R.O., Jr.; Williams, J.H.; Winnek, P.S.; English, J. P.

J. Am. Chem. Soc. 1940, *62*, 2002-5.

VARIABLES:

One temperature: 37°C

PREPARED BY:

R. Piekos

EXPERIMENTAL VALUES:

Solubility of 4-amino-N-(1-phenyl-2,3-dimethyl-5-oxopyrazol-4-yl)benzenesulfonamide in water at 37°C is 15.6 mg/100 cm^3 solution (4.35×10^{-4} mol dm^{-3}, compiler).

AUXILIARY INFORMATION

METHOD/APPARATUS/PROCEDURE:

Excess sulfonamide in water was heated and stirred on a steam bath for 30 min. The suspension was then agitated for 24 h in a thermostat at 37°C. A sample of the satd soln was withdrawn through a glass filter, dild, and analyzed by the Marshall method (1) using a General Electric recording spectrophotometer for comparing the colors developed with those of the standards.

SOURCE AND PURITY OF MATERIALS:

The sulfonamide, mp 260-1°C (dec, cor), was prepd by the authors. Anal: %C 57.5 (calcd 57.0); %H 5.1 (5.0); %N 16.1 (15.6). Purity of the water was not specified.

ESTIMATED ERROR:

Nothing specified

REFERENCES:

1. Bratton, A.C.; Marshall, E.K., Jr. *J. Pharmacol.* 1939, *66*, 4.

COMPONENTS:	ORIGINAL MEASUREMENTS:
(1) Benzenesulfonamide, 4-amino-N-1H-1,2,4-triazol-3-yl-; $C_8H_9N_5O_2S$; [51732-39-9] (2) Water; H_2O; [7732-18-5]	Anderson, G. W.; Faith, H.E.; Marson, H. W.; Winnek, P. S.; Roblin, R. O., Jr. *J. Am. Chem. Soc.* <u>1942</u>, *64*, 2902-5.
VARIABLES:	PREPARED BY:
One temperature: 37°C	R. Piekos

EXPERIMENTAL VALUES:

Solubility of 4-amino-N-1H-1,2,4-triazol-3-ylbenzenesulfonamide in water at 37°C is 60 mg/100 cm^3 solution (2.5 x 10^{-3} mol dm^{-3}, compiler).

AUXILIARY INFORMATION

METHOD/APPARATUS/PROCEDURE:

Excess sulfonamide in water was heated and stirred on a steam bath for 30 min. The suspension was then agitated for 24 h in a thermostat. A sample of the satd soln was withdrawn through a glass filter, dild, and analyzed by the Marshall method (1) using a General Electric recording spectrophotometer for comparing the colors developed with those of the standards.

SOURCE AND PURITY OF MATERIALS:

The sulfonamide, mp 195-6°C (cor), was prepd by the authors. Anal: %C 40.6 (calcd 40.2); %H 3.8 (3.8); %N 29.1 (29.3).
Purity of the water was not specified.

ESTIMATED ERROR:

Nothing specified.

REFERENCES:

1. Bratton, A. C.; Marshall, E. K., Jr. *J. Pharmacol.* <u>1939</u>, *66*, 4.

COMPONENTS:

(1) Benzenesulfonamide, 4-amino-N-4H-1,2,4-triazol-4-yl-; $C_8H_9N_5O_2S$; [17103-50-3]

(2) Water; H_2O; [7732-18-5]

ORIGINAL MEASUREMENTS:

Anderson, G. W.; Faith, H. E.; Marson, H.W.; Winnek, P. S.; Roblin, R. O., Jr.

J. Am. Chem. Soc. 1942, *64*, 2902-5.

VARIABLES:

One temperature: 37°C

PREPARED BY:

R. Piekos

EXPERIMENTAL VALUES:

Solubility of 4-amino-N-4H-1,2,4-triazole-4-ylbenzenesulfonamide in water at 37°C is 216 mg/100 cm^3 solution (9.03×10^{-3} mol dm^{-3}, compiler).

AUXILIARY INFORMATION

METHOD/APPARATUS/PROCEDURE:

Excess sulfonamide in water was heated and stirred on a steam bath for 30 min. The suspension was then agitated for 24 h in a thermostat. A sample of the satd soln was withdrawn through a glass filter, dild, and analyzed by the Marshall method (1) using a General Electric recording spectrophotometer for comparing the colors developed with those of the standards.

SOURCE AND PURITY OF MATERIALS:

The sulfonamide, mp 237°C (cor) was prepd by the authors. Anal: %C 40.1 (calcd 40.2); %H 3.8 (3.8); %N 29.4 (29.3). Purity of the water was not specified.

ESTIMATED ERROR:

Nothing specified.

REFERENCES:

1. Bratton, A. C.; Marshall, E. K., Jr. *J. Pharmacol.* 1939, *66*, 4.

COMPONENTS:	EVALUATOR:
(1) Benzenesulfonamide, 4-amino-N-(5-methyl-3-isoxazolyl)- (sulfamethoxazole) $C_{10}H_{11}N_3O_3S$; [723-46-6] (2) Water (3) Aqueous HCl; (4) Aqueous NaOH (5) Aqueous ethanol (6) Methanol	Anthony N. Paruta Department of Pharmaceutics University of Rhode Island Kingston, Rhode Island, USA and Ryszard Piekos Faculty of Pharmacy, University of Gdansk Gdansk, Poland 1986

CRITICAL EVALUATION:

The aqueous solubility data on the above compound are summarized in Table I. Yamazaki's (2) value was the only value available at 303K, and is not considered further. It is lower than those at 298K (4,6), thus probably unreasonable. Rudy and Senkowski (4) and Shah et al. (6) give identical values for the aqueous solubility at 289K. The solubility values can thus be given as 2 x 10^{-3} mol dm^{-3} in water at 298K. Kitao et. al. (3) determined the solubility at 310K at a pH value of 4. Since there are no concurring values (1,5) no recommended value can be given for this temperature. The value of Kitao et al. (3) is somewhat similar to that of Ghanem et al. (5), which is interesting since it would be expected that a broad invarient solubility isotherm over a span of pH values should exist. Thus, even though these values are similar, the solubility suggested by Ghanem et al. (5) is probably valid, and can be proposed as the tentative value.

Table I: Solubility of Sulfamethoxazole in water at various temperatures

	10^3 mol dm^{-3} (*indicates mol kg^{-1})		
Reference	298K	303K	310K
1	-	-	4.11
2	-	1.59	-
3	-	-	2.48 (pH=4)
4	2.0	-	-
5	-	-	2.37
6	2	-	-

For ampholytes of this type, solubility can be enhanced by the addition of either acids or bases. The condition produce a more water soluble cationic species (protonation) under acidic conditions, and the more water soluble anionic form under basic conditions at high pH. In two reports (7,8), the solubility was determined in 0.1N HCl both at 298K and 310K. Ogata et al. (7) records a value of 1.24 x 10^{-2} mol dm^{-3} in 0.1N HCl at 310K which is 6.2 times the solubility in water. Shah et al. (8) give a value of 1 x 10^{-4} mol dm^{-3} at a pH = 1, which is clearly incorrect being only a small fraction of the solubility in water (about 5%). However, at a concentration of 0.84N HCl (pH = 0.076) at 298K, a value of 1.12 x 10^{-2} mol dm^{-3} is reported which is in line with the value of Ogata et al. (7) being about 5.6 times the solubility in water. The value of Shah et al. (8) in 0.84N HCl is some 95 times greater than that in 0.1N HCl. In this context it might be instructive to point out the trend (magnitude enhancement) by comparing the solubility of sulfamethoxazole in different systems. The recommended values at 298K are 2 x 10^{-3} mol dm^{-3} in water, 63 x 10^{-3} mol dm^{-3} in 0.1N NaOH, 149 x 10^{-3} mol dm^{-3} in 95% ethanol in water and 350 x 10^{-3} mol dm^{-3} in methanol. There is a 31 fold increase in solubility in 0.1N NaOH no doubt due to the formation of the anionic form of the compound which has a much higher aqueous solubility. There is a dramatic shift in pH from near neutrality to pH = 13, a strong alkaline solution that forms a water soluble sodium salt of this compound. In methanol, there is a 175 fold increase in solubility due to the semipolar nature of solute and solvent. In 95% ethanol in water (10-12) there is about a 75 fold increase in solubility. The enhancements are quite striking and illustrate the significant latitude that can be used. The solubility of this compound was given by Rudy and Senkowski (9) and Shah et al. (8) in 1973 and 1981 respectively are in excellent agreement and a recommended value of 6.3 x 10^{-2} mol dm^{-3} can be given in aqueous 0.1N NaOH solution at 298K.

Further, the values of solubility in methanol were given by these workers (8,9) and were also in excellent agreement and is given as 0.35 mol dm^{-3} at 298K. The recommended value in 95% ethanol in water is 0.13 mol dm^{-3} at 298K.

REFERENCES:

(1) Anderson, G.W.; Faith, H.E.; Marson, H.W.; Winnek, P.S.; Roblin, R.O., Jr. *J. Am. Chem. Soc.* 1942, *64*, 2902-5.
(2) Yamazaki, M.; Aoki, M.; Kamada, A.; Yata, N.; *Yakuzaigaku* 1967, *27(1)*, 37-40.
(3) Kitao, K,; Kubo, K.; Morishita, T.; Yata, N.; Kamada, A. *Chem. Pharm. Bull.* 1973, *21*, 2417-26.
(4) Rudy, B.C.; Senkowski, B.Z. *Anal. Profiles Drug Subst.* 1973, *2*, 467-86.
(5) Ghanem, A.; Meshali, M.; Ibraheem, Y. *J. Pharm. Pharmacol.* 1980, *32*, 675-7.
(6) Shah, N.H.; Lazarus, J.H.; Sheth, P.R.; Jarowski, C.I. *J. Pharm. Sci.* 1981, *70(6)*, 611-13.

REFERENCES: Continuation

(7) Rudy, B.C.; Senkowski, B.Z.; *Anal. Profiles Drug Subst.* 1973, *2*, 467-86.

(8) Shah, N.H.; Lazarus, J.H.; Sheth, P.R.; Jarowski, C.I. *J. Pharm. Sci.* 1981, *70(6)*, 611-13.

(9) Rudy, B.C.; Senkowski, B.Z. *Anal. Profiles Drug Subts.* 1973, *2*, 467-86.

(10) Shah, N.H.; Lazarus, J.H.; Sheth, P.R.; Jarowski, C.I. *J. Pharm. Sci.* 1981, *70(6)*, 611-13.

(11) Rudy, B.C.; Senkowski, B.Z., *Anal. Profiles Drug Subst.* 1973, *2*, 467-86.

(12) Shah, N.H.; Lazarus, J.H.; Sheth, P.R.; Jarowski, C.I. *J. Pharm. Sci.* 1981, *70(6)*, 611-13.

COMPONENTS:	ORIGINAL MEASUREMENTS:
(1) Benzenesulfonamide, 4-amino-N-(5-methyl-3-isoxazolyl)- (sulfamethoxazole); $C_{10}H_{11}N_3O_3S$; [723-46-6] (2) Water; H_2O; [7732-18-5]	Anderson, G.W.; Faith, H.E.; Marson, H.W.; Winnek, P.S.; Roblin, R.O.,Jr. *J. Am. Chem. Soc.* 1942, *64*, 2902-5.
VARIABLES:	PREPARED BY:
One temperature: 37°C	R. Piekos

EXPERIMENTAL VALUES:

Solubility of sulfamethoxazole in water at 37°C is 104 mg/100 cm^3 solution (4.11×10^{-3} mol dm^{-3}, compiler).

AUXILIARY INFORMATION

METHOD/APPARATUS/PROCEDURE:

Excess sulfonamide in water was heated and stirred on a steam bath for 30 min. The suspension was then agitated for 24 h in a thermostat. A sample of the satd soln was withdrawn through a glass filter, dild, and analyzed by the Marshall method (1) using a General Electric recording spectrophotometer for comparing the colors developed with those of the standards.

SOURCE AND PURITY OF MATERIALS:

The sulfonamide, mp 169-70°C (cor) was prepd by the authors. Anal: %C 47.4 (calcd 47.4); %H 4.2 (4.4); %N 16.5 (16.6).

Purity of the water was not specified.

ESTIMATED ERROR:

Nothing specified

REFERENCES:

1. Bratton, A.C.; Marshall, E.K., Jr. *J. Pharmacol.* 1939, *66*, 4.

COMPONENTS:	ORIGINAL MEASUREMENTS:
(1) Benzenesulfonamide, 4-amino-N-(5-methyl-3-isoxazolyl)- (sulfisomezole)*; $C_{10}H_{11}N_3O_3S$; [723-46-6] (2) Water; H_2O; [7732-18-5]	Yamazaki, M.; Aoki, M.; Kamada, A.; Yata, N. *Yakuzaigaku* 1967, *27(1)*, 37-40.
VARIABLES:	**PREPARED BY:**
One temperature: 30°C	R. Piekos

EXPERIMENTAL VALUES:

Solubility of sulfisomezole* in water at 30°C is 1.59 mmol/L (0.403 g dm^{-3}, compiler).

*Another common trivial name is sulfamethoxazole.

AUXILIARY INFORMATION

METHOD/APPARATUS/PROCEDURE:

Sulfisomezole* (0.5 g) was placed in an L-shaped tube together with 20 ml of water. The mixt was shaken in a thermostat until equilibrium was attained. The sulfisomezole* was assayed in the supernatant spectrophotometrically at 545 nm on a Beckman DU spectrophotometer. The results were taken from a calibration graph.

SOURCE AND PURITY OF MATERIALS:

Nothing specified

ESTIMATED ERROR:

Soly: not specified

Temp: ±1°C (authors)

REFERENCES:

COMPONENTS:	ORIGINAL MEASUREMENTS:
(1) Benzenesulfonamide, 4-amino-N-(5-methyl-3-isoxazolyl)- (sulfamethoxazole); $C_{10}H_{11}N_3O_3S$; [723-46-6] (2) Water; H_2O; [7732-18-5]	Kitao, K.; Kubo, K.; Morishita, T.; Yata, N.; Kamada, A. *Chem. Pharm. Bull.* 1973, *21*, 2417-26.
VARIABLES:	**PREPARED BY:**
One temperature: 37°C	R. Piekos

EXPERIMENTAL VALUES:

Solubility of sulfamethoxazole in water at 37°C is 2.48 mmol dm^{-3} solution.

AUXILIARY INFORMATION

METHOD/APPARATUS/PROCEDURE:	SOURCE AND PURITY OF MATERIALS:
Soly was detd by continuously adjusting the pH of the aq soln to 4 with 0.05N NaOH. The concn. of sulfamethoxazole was detd by diazotization.	Comm available sulfamethoxazole (source not specified) was used as supplied. Deionized water was used.
	ESTIMATED ERROR: Soly: not specified Temp: ±1°C (authors).
	REFERENCES:

COMPONENTS:	ORIGINAL MEASUREMENTS:
(1) Benzenesulfonamide, 4-amino-N-(5-methyl-3-isoxazolyl)- (sulfamethoxazole); $C_{10}H_{11}N_3O_3S$; [723-46-6] (2) Water; H_2O; [7732-18-5]	Rudy, B.C.; Senkowski, B.Z. *Anal. Profiles Drug Subst.* 1973, *2*, 467-86.
VARIABLES:	**PREPARED BY:**
One temperature" 25°C	R. Piekos

EXPERIMENTAL VALUES:

Solubility of sulfamethoxazole in water at 25°C is 0.5 mg/ml (2.0 x 10^{-3} mol dm^{-3}, compiler). [a]

[a]The temperature and all auxilliary information was given by Edward A. MacMullan from Roche Products Inc., Manati, P.R., in a personal communication.

AUXILIARY INFORMATION

METHOD/APPARATUS/PROCEDURE:

An excess of solute was equilibrated with the solvent overnight at const temp. (25°C) while being agitated with a 60 cycle vibrator (VIBROMIXER). A portion of the clear supernatant soln was then taken and its concn was detd by uv spectrophotometry after suitable diln.

SOURCE AND PURITY OF MATERIALS:

The sulfamethoxazole was of reference standard quality equivalent to USP. The distd and deionized water of high resistivity was used.

ESTIMATED ERROR:

Soly: precision ±1% (MacMullan)
Temp: not specified

REFERENCES:

COMPONENTS:	ORIGINAL MEASUREMENTS:
(1) Benzenesulfonamide, 4-amino-N-(5-methyl-3-isoxazolyl)- (sulfamethoxazole); $C_{10}H_{11}N_3O_3S$; [723-46-6] (2) Water; H_2O; [7732-18-5]	Ghanem, A.; Meshali, M.; Ibraheem, Y. *J. Pharm. Pharmacol.* 1980, *32*, 675-7.
VARIABLES:	**PREPARED BY:**
One temperature: 37°C	R. Piekos

EXPERIMENTAL VALUES:

Solubility of sulfamethoxazole in water at 37°C is 0.6 g $litre^{-1}$ (2 x 10^{-3} mol dm^{-3}, compiler).

AUXILIARY INFORMATION

METHOD/APPARATUS/PROCEDURE:

An excess of sulfamethoxazole was added to 15 ml of water in a 30-ml glass stoppered bottle which was rotated on a water bath at 37°C until equilibrium was attained. The sample was filtered and the sulfonamide was assayed spectrophotometrically at 265 nm after dilg with 0.1M HCl. A coulometric assay gave similar results.

SOURCE AND PURITY OF MATERIALS:

Sulfamethoxazole was from Kahira Pharm and Chem Ind Co. Egypt.

Purity of the water was not specified.

ESTIMATED ERROR:

Soly: detns were carried out at least in duplicate (authors).

Temp: ± 1°C (authors).

REFERENCES:

COMPONENTS:	ORIGINAL MEASUREMENTS:
(1) Benzenesulfonamide, 4-amino-N-(5-methyl-3-isoxazolyl)- (sulfamethoxazole); $C_{10}H_{11}N_3O_3S$; [723-46-6] (2) Water; H_2O; [7732-18-5]	Shah, N.H.; Lazarus, J.H.; Sheth, P.R.; Jarowski, C.I. *J. Pharm. Sci.* 1981, *70(6)*, 611-13.
VARIABLES:	**PREPARED BY:**
One temperature: 25°C	R. Piekos

EXPERIMENTAL VALUES:

Solubility of sulfamethoxazole in water at 25°C is 0.5 mg/ml (2 x 10^{-3} mol dm^{-3}, compiler).

AUXILIARY INFORMATION

METHOD/APPARATUS/PROCEDURE:

The solubility of sulfamethoxazole was determined by the method specified in USP XX (1).

SOURCE AND PURITY OF MATERIALS:

Sulfamethoxazole was a research compd purchased from Hoffman - LaRoche, Nutley, N.J. Its purity was not specified.
The purity of water was not specified.

ESTIMATED ERROR:

Nothing specified

REFERENCES:

1. *"The United States Pharmacopeia"*, 20th rev., U.S. Pharmacopeial Convention, Rockville, Md., 1980, p. 120

COMPONENTS:	ORIGINAL MEASUREMENTS:
(1) Benzenesulfonamide, 4-amino-N-(5-methyl-3-isoxazolyl)- (sulfamethoxazole); $C_{10}H_{11}N_3O_3S$; [723-46-6] (2) Hydrochloric acid; HCl; [7647-01-0] (3) Water; H_2O; [7732-18-5]	Ogata, H.; Shibazaki, T.; Inoue, T.; Ejima, A. *Chem. Pharm. Bull.* 1979 *27(6)*, 1281-6.
VARIABLES:	**PREPARED BY:**
One temperature: 37°C	R. Piekos

EXPERIMENTAL VALUES:

Solubility of sulfamethoxazole in 0.1N HCl at 37°C is 3.140 mg/ml (1.240×10^{-2} mol dm^{-3}, compiler).

AUXILIARY INFORMATION

METHOD/APPARATUS/PROCEDURE:

A centrifuge tube contg 30 ml of 0.1N HCl and 0.5-3.0 g of the sulfamethoxazole powder was tightly sealed and shaken at 37°C. The concn of the dissolved drug was detd spectrophotometrically following filtration through a Millipore filter (type EH, pore size 0.5 μm), and the procedure was repeated every 24 h until a const concn was obtained.

SOURCE AND PURITY OF MATERIALS:

Comm available 500-mg uncoated tablets of sulfamethoxazole were used. Hydrochloric acid was of reagent grade.

ESTIMATED ERROR:

Nothing specified

REFERENCES:

COMPONENTS:

(1) Benzenesulfonamide, 4-amino-N-(5-methyl-3-isoxazolyl)- (sulfamethoxazole); $C_{10}H_{11}N_3O_3S$; [723-46-6]

(2) Hydrochloric acid; HCl; [7647-01-0]

(3) Water; H_2O; [7732-18-5]

ORIGINAL MEASUREMENTS:

Shah, N.H.; Lazarus, J.H.; Sheth, P.R.; Jarowski, C.I.

J. Pharm. Sci. 1981, *70(6)*, 611-13.

VARIABLES:

Concentration of HCl

PREPARED BY:

R.Piekos

EXPERIMENTAL VALUES:

Concentration of HCl, N	Solubility at 25°C mg/ml	Solubility at 25°C mol dm^{-3} [a]
0.1	0.03	1×10^{-4}
0.84	2.85	1.12×10^{-2}

[a]Calculated by compiler

AUXILIARY INFORMATION

METHOD/APPARATUS/PROCEDURE:

The solubility of sulfamethoxazole was determined by the method specified in USP XX (1).

SOURCE AND PURITY OF MATERIALS:

Sulfamethoxazole was a research compd purchased from Hoffman - LaRoche, Nutley, N.J. Its purity was not specified.

The source and purity of hydrochloric acid was not specified.

ESTIMATED ERROR:

Nothing specified

REFERENCES:

1. *"The United States Pharmacopeia"*, 20th rev., U.S. Pharmacopeial Convention, Rockville, Md., 1980, p. 120.

COMPONENTS:
(1) Benzenesulfonamide, 4-amino-N-(5-methyl-3-isoxazolyl)- (sulfamethoxazole); $C_{10}H_{11}N_3O_3S$; [723-46-6]
(2) Sodium hydroxide; NaOH; [1310-73-2]
(3) Water; H_2O; [7732-18-5]

ORIGINAL MEASUREMENTS:
Rudy, B.C.; Senkowski, B.Z.
Anal. Profiles Drug Subst. 1973, *2*, 467-86.

VARIABLES:
One temperature: 25°C

PREPARED BY:
R. Piekos

EXPERIMENTAL VALUES:

Solubility of sulfamethoxazole in a 0.1N NaOH solution at 25°C is 16.0 mg/ml (6.32×10^{-2} mol dm^{-3}, compiler).[a]

[a]The temperature and all auxiliary information was given by Edward A. MacMullan from Roche Products Inc., Manati, P.R., in a personal communication.

AUXILIARY INFORMATION

METHOD/APPARATUS/PROCEDURE:
An excess of solute was equilibrated with the solvent overnight at const temp (25°) while being agitated with a 60 cycle vibrator (VIBROMIXER). A portion of the clear supernatant solution was then taken and its concn was detd by uv spectrophotometry after suitable diln.

SOURCE AND PURITY OF MATERIALS:
The sulfamethoxazole was of reference standard quality equivalent to USP. Reagent grade NaOH was used. The distd and deionized water of high resistivity was used.

ESTIMATED ERROR:
Soly: precision ±1% (MacMullan)
Temp: not specified

REFERENCES:

COMPONENTS:	ORIGINAL MEASUREMENTS:
(1) Benzenesulfonamide, 4-amino-N-(5-methyl-3-isoxazolyl)- (sulfamethoxazole); $C_{10}H_{11}N_3O_3S$; [723-46-6] (2) Sodium hydroxide; NaOH; [1310-73-2] (3) Water; H_2O; [7732-18-5]	Shah, N.H.; Lazarus, J.H.; Sheth, P.R.; Jarowski, C.I. *J. Pharm. Sci.* 1981, *70(6)* 611-13.
VARIABLES:	**PREPARED BY:**
One temperature: 25°C	R. Piekos

EXPERIMENTAL VALUES:

Solubility of sulfamethoxazole in a 0.1N NaOH solution at 25°C is 16 mg/ml (6.3×10^{-2} dm^{-3}, compiler).

AUXILIARY INFORMATION

METHOD/APPARATUS/PROCEDURE:

The solubility of sulfamethoxazole was determined by the method specified in USP XX (1).

SOURCE AND PURITY OF MATERIALS:

Sulfamethoxazole was a research compd purchased from Hoffman - LaRoche, Nutley, N.J. Its purity was not specified.

The source and purity of NaOH and water was not specified.

ESTIMATED ERROR:

Nothing specified

REFERENCES:

1. *"The United States Pharmacopeia"*, 20th rev., U.S. Pharmacopeial Convention, Rockville, Md., 1980 p. 120.

COMPONENTS:	ORIGINAL MEASUREMENTS:
(1) Benzenesulfonamide, 4-amino-N-(5-methyl-3-isoxazolyl)- (sulfamethoxazole); $C_{10}H_{11}N_3O_3S$; [723-46-6] (2) Sodium chloride; NaCl; [7647-14-5] (3) Water; H_2O; [7732-18-5]	Hirano, K.; Ichihashi, T.; Yamada, H. *Chem. Pharm.Bull.* 1981, *29(3)*, 817-27.
VARIABLES:	**PREPARED BY:**
One temperature: 37°C	R. Piekos

EXPERIMENTAL VALUES:

Solubility of sulfamethoxazole in a 0.9% NaCl solution at 37°C is 0.61 mg/ml (2.4×10^{-3} mol dm^{-3}, compiler).

AUXILIARY INFORMATION

METHOD/APPARATUS/PROCEDURE:	SOURCE AND PURITY OF MATERIALS:
An excess amt of powdered sulfamethoxazole was shaken well at 37°C with a 0.9% NaCl soln until attaining satn. The undissolved crystals were removed by filtration through a G5 glass filter or by centrifugation, and the concn of solute in the filtrate or supernatant was assayed spectrophotometrically at 267 nm, after diln with EtOH - H_2O (1:1, v/v), using a Perkin Elmer UV-VIS spectrophotometer (Hitachi Co., Ltd., Tokyo)	Sulfamethoxazole was synthesized by the authors and was of medicinal grade. The remaining materials were of anal or reagent grade. **ESTIMATED ERROR:** Nothing specified **REFERENCES:**

COMPONENTS:

(1) Benzenesulfonamide, 4-amino-N-(5-methyl-3-isoxazolyl)- (sulfamethoxazole); $C_{10}H_{11}N_3O_3S$; [723-46-6]

(2) Phosphoric acid, disodium salt Na_2HPO_4; [7558-94-4]

(3) Phosphoric acid, monopotassium salt; KH_2PO_4; [7778-77-0]

(4) Water; H_2O; [7732-18-5]

ORIGINAL MEASUREMENTS:

Riess, W.

Intern. Congr. Chemotherapy, Proc., 3rd, Stuttgart 1963, *1*, 627-32.

PREPARED BY:

R. Piekos

VARIABLES:

One temperature: 20°C; one pH: 7.4

EXPERIMENTAL VALUES:

Solubility of sulfamethoxazole in a M/15 Sörensen buffer solution (pH 7.4) at 20°C is 930 mg% (3.67×10^{-2} mol dm^{-3} solution, compiler).

AUXILIARY INFORMATION

METHOD/APPARATUS/PROCEDURE:

Sörensen buffer solns of pH varying between 7 and 8 were prepd, satd with sulfamethoxazole at 20°C, their pH was measured at equilibrium, and the sulfamethoxazole was assayed colorimetrically. The measured pH values were then plotted against concn, and the soly at pH 7.4 was detd by interpolation (personal communication).

SOURCE AND PURITY OF MATERIALS:

Nothing specified

ESTIMATED ERROR:

Nothing specified

REFERENCES:

COMPONENTS:

(1) Benzenesulfonamide, 4-amino-N-(5-methyl-3-isoxazolyl)- (sulfisomezole)*; $C_{10}H_{11}N_3O_3S$; [723-46-6]

(2) Phosphoric acid, disodium salt; Na_2HPO_4; [7558-94-4]

(3) Phosphoric acid, monopotassium salt; KH_2PO_4; [7778-77-0]

(4) Water; H_2O; [7732-18-5]

ORIGINAL MEASUREMENTS:

Yamazaki, M; Aoki, M.; Kamada, A.; Yata, N. *Yakuzaigaku* 1967, *27(1)*, 37-40.

PREPARED BY:

R. Piekos

VARIABLES:

One temperature: 30°C; one pH: 7.4

EXPERIMENTAL VALUES:

Solubility of sulfisomezole* in a phosphate buffer solution of pH 7.4[a] (μ = 0.17) at 30°C is 20.7 mmol/L (5.24 g dm^{-3}, compiler).

[a]At the end of experiment the pH was 6.9

*Another common trivial name is sulfamethoxazole.

AUXILIARY INFORMATION

METHOD/APPARATUS/PROCEDURE:

Sulfisomezole* (0.5 g) was placed in an L-shaped tube together with 20 ml of the buffer soln. The mixt was shaken in a thermostat until equilibrium was attained. The sulfisomezole* was assayed in the supernatant spectrophotometrically at 545 nm on a Beckmann DU spectrophotometer. The results were taken from a calibration graph.

SOURCE AND PURITY OF MATERIALS:

Nothing specified

ESTIMATED ERROR:

Soly and pH: not specified

Temp: ±1°C (authors)

REFERENCES:

COMPONENTS:

(1) Benzenesulfonamide, 4-amino-N-(5-methyl-3-isoxazolyl)- (sulfamethoxazole); $C_{10}H_{11}N_3O_3S$; [723-46-6]

(2) Phosphoric acid, disodium salt; Na_2HPO_4; [7558-94-4]

(3) Phosphoric acid, monopotassium salt; KH_2PO_4; [7778-77-0]

(4) Water; H_2O; [7732-18-5]

ORIGINAL MEASUREMENTS:

Hekster, Y.A.; Vree, T.B.; Damsma, J.E.; Friesen, W.T.
J. Antimicrob. Chemother. 1981, 8, 133-44.

PREPARED BY:

R. Piekos

VARIABLES:

pH

EXPERIMENTAL VALUES:

pH	Solubility at 25°C	
	mg/l	10^3 mol dm^{-3} [a]
5.5	300	1.18
7.5	1900	7.50

[a]Calculated by compiler

AUXILIARY INFORMATION

METHOD/APPARATUS/PROCEDURE:

Satd solns of sulfamethoxazole were prepd in phosphate buffers of pH 5.5 and 7.5 at room temp (25°C). The concn of the solute was measured by means of a Spectra Physics 3500B high-performance liquid chromatograph equipped with a column oven (Model 748) and a Pye-Unicam LC-UV spectrophotometric detector. The detector was connected to a 1-mV recorder. A stainless steel column (10 cm x 4.6 mm i.d.) was packed with Lichrosorb RPS, 5 µm, obtained from Chrompack. An injection loop of 100 µl was used. The oven temp was 40°C. Detection of sulfamethoxazole was performed at 260 nm.

SOURCE AND PURITY OF MATERIALS:

The source and purity of the materials was not specified.

ESTIMATED ERROR:

The detection limit of the solute by HPLC was 0.5 mg/l (authors). The error in temperature and pH was not specified.

REFERENCES:

COMPONENTS:

(1) Benzenesulfonamide, 4-amino-N-(5-methyl-3-isoxazolyl)- (sulfamethoxazole); $C_{10}H_{11}N_3O_3S$; [723-46-6]

(2) Phosphoric acid, disodium salt; Na_2HPO_4; [7558-94-4]

(3) Phosphoric acid, monopotassium salt; KH_2PO_4; [7778-77-0]

(4) Water; H_2O; [7732-18-5]

ORIGINAL MEASUREMENTS:

Meshali, M.; El Sabbagh, H.; Ghanem, A.; Foda, A.

Pharmazie 1983, *38(6)*, 403-6.

VARIABLES:

One temperature: 37°C

PREPARED BY:

R. Piekos

EXPERIMENTAL VALUES:

Equilibrium solubility of sulfamethoxazole in a phosphate buffer solution of pH 7.2, at 37°C, is 0.6% (2×10^{-2} mol kg^{-1} solution, compiler).

AUXILIARY INFORMATION

METHOD/APPARATUS/PROCEDURE:

A tablet of sulfamethoxazole was placed in 500 ml of a phosphate buffer of pH 7.2 and stirred at 37°C. Samples were taken at time intervals and the solute concn was detd by the method reported by the authors (1).

SOURCE AND PURITY OF MATERIALS:

Sulfamethoxazole tablets were picked up from the market. They satisfied the USP requirements for uniformity of wt and the BP requirements for uniformity of content. The source and purity of the remaining materials were not specified.

ESTIMATED ERROR:

Nothing specified.

REFERENCES:

1. Ghanem, A.; Meshali, M.; Foda, A. *J. Pharm. Pharmacol* 1979, *31*, 122.

COMPONENTS:

(1) Benzenesulfonamide, 4-amino-N-(5-methyl-3-isoxazolyl)- (sulfamethoxazole); $C_{10}H_{11}N_3O_3S$; [723-46-6]

(2) Aminoacetic acid (glycine); $C_2H_5NO_2$; [56-40-6]

(3) Hydrochloric acid; HCl; [7647-01-0]

(4) Sodium chloride; NaCl; [7647-14-5]

(5) Water; H_2O; [7732-18-5]

ORIGINAL MEASUREMENTS:

Meshali, M.; El Sabbagh, H.; Ghanem, A.; Foda, A.

Pharmazie, 1983, *38(6)*, 403-6.

VARIABLES:

One temperature: 37°C

PREPARED BY:

R. Piekos

EXPERIMENTAL VALUES:

Equilibrium solubility of sulfamethoxazole in artificial gastric juice (0.5 g glycine, 0.35 g NaCl and 9.4 ml HCl per liter of solution; pH 1.1) at 37°C is 0.338% (1.33×10^{-2} mol kg^{-1} solution - compiler).

AUXILIARY INFORMATION

METHOD/APPARATUS/PROCEDURE:

A tablet of sulfamethoxazole was placed in 500 ml of artificial gastric juice of pH 1.1, and the suspension was stirred at 37°C. Samples were taken at time intervals and the solute concn was detd by the method reported by the authors.

SOURCE AND PURITY OF MATERIALS:

Sulfamethoxazole tablets were picked up from the market. They satisfied the USP requirements for uniformity of wt and the BP requirements for uniformity of content. The source and purity of the remaining materials were not specified.

ESTIMATED ERROR:

Nothing specified.

REFERENCES:

1. Ghanem. A.; Meshali, M.; Foda, A. *J. Pharm. Pharmacol.* 1979, *31*, 122.

COMPONENTS:

(1) Benzenesulfonamide, 4-amino-N-(5-methyl-3-isoxazolyl)- (sulfamethoxazole); $C_{10}H_{11}N_3O_3S$; [723-46-6]

(2) Phosphoric acid, disodium salt; Na_2HPO_4; [7558-94-4]

(3) Phosphoric acid, monopotassium salt KH_2PO_4; [7778-77-0]

(4) Sodium chloride; NaCl; [7647-14-5]

(5) Water; H_2O; [7732-18-5]

ORIGINAL MEASUREMENTS:

Hirano, K.; Ichihashi, T.; Yamada, H.; *Chem. Pharm. Bull.* 1981, *29(3)*, 817-27.

VARIABLES:

One temperature: 37°C

PREPARED BY:

R. Piekos

EXPERIMENTAL VALUES:

Solubility of sulfamethoxazole in a 1/15M phosphate buffer solution of pH 7.25, isotonized with NaCl, at 37°C, is 5.7 mg/ml (2.2×10^{-2} mol dm^{-3}, compiler).

AUXILIARY INFORMATION

METHOD/APPARATUS/PROCEDURE:

An excess amt of powdered sulfamethoxazole was shaken well at 37°C with 1/15M phosphate buffer of pH 7.25, isotonized with NaCl, until attaining satn. The undissolved crystals were removed by filtration through a G5 glass filter or by centrifugation, and the concn of solute in the filtrate or supernatant was assayed spectrophotometrically at 267 nm. after diln with EtOH - H_2O (1:1, v/v), using a Perkin Elmer UV-VIS spectrophotometer (Hitachi Co., Ltd., Tokyo).

SOURCE AND PURITY OF MATERIALS:

Sulfamethoxazole was synthesized by the authors and was of medicinal grade. The remaining materials were of anal or reagent grade.

ESTIMATED ERROR:

Nothing specified

REFERENCES:

COMPONENTS:	ORIGINAL MEASUREMENTS:
(1) Benzenesulfonamide, 4-amino-N-(5-methyl-3-isoxazolyl)- (sulfamethoxazole); $C_{10}H_{11}N_3O_3S$; [723-46-6] (2) Ethanol; C_2H_6O; [64-17-5] (3) Water; H_2O; [7732-18-6]	Rudy, B.C.; Senkowski, B.Z. *Anal. Profiles Drug Subst.* 1973, *2*, 467-86
VARIABLES:	**PREPARED BY:**
One temperature: 25°C	R. Piekos

EXPERIMENTAL VALUES:

Solubility of sulfamethoxazole in 95% ethanol at 25°C is 37.8 mg/ml (0.149 mol dm^{-3}, compiler).[a]

[a]The temperature and all auxiliary information was given by Edward A. MacMullan from Roche Products Inc., Manati, P.R., in a personal communication.

AUXILIARY INFORMATION

METHOD/APPARATUS/PROCEDURE:

An excess of solute was equilibrated with the solvent overnight at const temp (25°C) while being agitated with a 60 cycle vibrator (VIBROMIXER). A portion of the clear supernatant soln was then taken, weighed and the solvent removed in a vacuum oven. The wt of solute was detd after the residue had been dried to const wt. All weighing was done on a Mettler microbalance using microanal techniques.

SOURCE AND PURITY OF MATERIALS:

The sulfamethoxazole was of reference standard quality equivalent to USP. The solvent was purchased Reagent Grade and used without further purification.

ESTIMATED ERROR:

Soly: precision ±1% (MacMullan)

Temp: not specified

REFERENCES:

COMPONENTS:

(1) Benzenesulfonamide, 4-amino-N-(5-methyl-3-isoxazolyl)- (sulfamethoxazole); $C_{10}H_{11}N_3O_3S$; [723-46-6]

(2) Ethanol; C_2H_6O; [64-17-5]

(3) Water; H_2O; [7732-18-5]

ORIGINAL MEASUREMENTS:

Shah, N.H.; Lazarus, J.H.; Sheth, P.R.; Jarowski, C.I.

J. Pharm. Sci. 1981, *70(6)*, 611-13.

VARIABLES:

One temperature: 25°C

PREPARED BY:

R. Piekos

EXPERIMENTAL VALUES:

Solubility of sulfamethoxazole in 95% ethanol at 25°C is 30 mg/ml (0.12 mol dm^{-3}, compiler).

AUXILIARY INFORMATION

METHOD/APPARATUS/PROCEDURE:

The solubility of sulfamethoxazole was determined by the method specified in USP XX (1).

SOURCE AND PURITY OF MATERIALS:

Sulfamethoxazole was a research compd purchased from Hoffman - LaRoche, Nutley, N.J. Its purity was not specified.

The source and purity of the 95% EtOH was not specified.

ESTIMATED ERROR:

Nothing specified

REFERENCES:

1. *"The United States Pharmacopeia"*, 20th rev., U.S. Pharmacopeial Convention, Rockville, Md., 1980, p. 120.

COMPONENTS:

(1) Benzenesulfonamide, 4-amino-N-(5-methyl-3-isoxazolyl)- (sulfamethoxazole); $C_{10}H_{11}N_3O_3S$; [723-46-6]

(2) Bovine serum albumin

(3) Phosphoric acid, disodium salt; Na_2HPO_4; [7558-94-4]

(4) Phosphoric acid, monopotassium salt; KH_2PO_4; [7778-77-0]

(5) Sodium chloride; NaCl; [7647-14-5]

(6) Water; H_2O; [7732-18-5]

ORIGINAL MEASUREMENTS:

Hirano, K.; Yamada, H.

J. Pharm. Sci. 1982, *71(5)*, 500-5.

VARIABLES:

One temperature: 37°C; one pH: 7.25

PREPARED BY:

R. Piekos

EXPERIMENTAL VALUES:

Solubility of sulfamethoxazole in a 2% (w/v) bovine serum albumin in pH 7.25 phosphate buffer (0.067 M Na_2HPO_4-KH_2PO_4) isotonized with NaCl, at 37°C, is 7.2 mg/ml (2.8×10^{-2} mol dm^{-3}, compiler).

AUXILIARY INFORMATION

METHOD/APPARATUS/PROCEDURE:

The previously developed method was employed (1). An excess of powd sulfamethoxazole was shaken well at 37°C with the 2% bovine serum albumin in pH 7.25 phosphate buffer isotonized with NaCl until attaining satn. The undissolved crystals were removed by filtration through a G5 glass filter or by centrifugation, and the concn of solute in the filtrate or supernatant was assayed spectrophotometrically at 267 nm using a Perkin Elmer UV-VIS spectrophotometer (Hitachi Co., Ltd., Tokyo).

SOURCE AND PURITY OF MATERIALS:

Sulfamethoxazole was synthesized by the authors and was of medicinal grade. Bovine serum albumin (purity not specified) was from Sigma Chemical Co., St. Louis, Mo. The remaining materials were of anal or reagent grade.

ESTIMATED ERROR:

Nothing specified

REFERENCES:

1. Hirano, K.; Ichihashi, T.; Yamada, H. *Chem. Pharm. Bull.* 1981, *29(3)*, 817.

COMPONENTS:

(1) Benzenesulfonamide, 4-amino-N-(5-methyl-3-isoxazolyl)- (sulfamethoxazole); $C_{10}H_{11}N_3O_3S$; [723-46-6]

(2) 1,4,7,10,13,16-Hexaoxacyclooctadecane (18-C-6); $C_{12}H_{24}O_6$; [17455-13-9]

(3) Hydrochloric acid; HCl; [7647-01-0]

(4) Water; H_2O; [7732-18-5]

ORIGINAL MEASUREMENTS:

Takayama, K; Nambu, N.; Nagai, T. *Chem. Pharm. Bull.* 1978, *26(10)*, 2965-70.

PREPARED BY:

R. Piekos

VARIABLES:

Temperature

EXPERIMENTAL VALUES:

t/°C	Saturated concentration of sulfamethoxazole after decomplexation of its 1:1 complex with 18-C-6 in 0.2N HCl 10^2M
30	1.11
35	1.31
40	1.64

AUXILIARY INFORMATION

METHOD/APPARATUS/PROCEDURE:

An excess of the complex was dissolved in 50 ml of 0.2N HCl. The sampling was done by a 1-ml pipet fitted with a G-4 glass filter. The concentration of the sulfonamide was detd by uv spectrophotometry after dilg with 0.2N HCl.

SOURCE AND PURITY OF MATERIALS:

Sulfamethoxazole (Shionogi Pharmaceutical Co.) was recrystd from a 30% (V/V) Me_2CO-H_2O soln. 18-C-6 was of the reagent grade. The 1:1 complex was prepd by the authors. Purity of the HCl soln was not specified.

ESTIMATED ERROR:

Nothing specified

REFERENCES:

COMPONENTS:

(1) Benzenesulfonamide,4-amino-N-(5-methyl-3-isoxazolyl)- (sulfamethoxazole); $C_{10}H_{11}N_3O_3S$; [723-46-6]

(2) D-Glucitol (sorbitol); $C_6H_{14}O_6$; [50-70-4]

(3) Water; H_2O; [7732-18-5]

ORIGINAL MEASUREMENTS:

Ghanem, A.; Meshali, M.; Ibraheem, Y. *J. Pharm. Pharmacol.* 1980, *32*, 675-7.

VARIABLES:

Concentration of sorbitol

PREPARED BY:

R. Piekos

EXPERIMENTAL VALUES:

Concentration of sorbitol Weight%	Solubility at 37°C g $litre^{-1}$	10^3 mol dm^{-3} [a]
0.5	0.60	2.37
1.0	0.595	2.35
1.5	0.60	2.37

[a]Calculated by compiler

AUXILIARY INFORMATION

METHOD/APPARATUS/PROCEDURE:

An excess of sulfamethoxazole was added to 15 ml of sorbitol soln in 30-ml glass-stoppered bottles which were rotated on a water bath at 37°C until equilibrium was attained. Samples were filtered and the sulfonamide was assayed spectrophotometrically at 265 nm after dilg with 0.1M HCl. Coulometric assays gave similar results.

SOURCE AND PURITY OF MATERIALS:

Sulfamethoxazole was from Kahira Pharm and Chem Ind Co, Egypt. Sorbitol was purchased from El-Nasr Chem Co, Egypt. Distd waster was used.

ESTIMATED ERROR:

Soly: detns were carried out in duplicate (authors)

Temp: ±1°C (authors)

REFERENCES:

COMPONENTS:

(1) Benzenesulfonamide, 4-amino-N-(5-methyl-3-isoxazolyl)- (sulfamethoxazole); $C_{10}H_{11}N_3O_3S$; [723-46-6]

(2) Mannitol; $C_6H_{14}O_6$; [87-78-5]

(3) Water; H_2O; [7732-18-5]

ORIGINAL MEASUREMENTS:

Ghanem, A.; Meshali, M.; Ibraheem, Y. *J. Pharm. Pharmacol.* 1980, *32*, 675-7.

VARIABLES:

Concentration of mannitol

PREPARED BY:

R. Piekos

EXPERIMENTAL VALUES:

Concentration of mannitol Weight%	Solubility at 37°C g $litre^{-1}$	$10^3 mol\ dm^{-3}$ [a]
0.5	0.615	2.43
1.0	0.605	2.39
1.5	0.603	2.38

[a] Calculated by compiler

AUXILIARY INFORMATION

METHOD/APPARATUS/PROCEDURE:

An excess of sulfamethoxazole was added to 15 ml of mannitol soln in 30-ml glass-stoppered bottles which were rotated on a water bath at 37°C until equilibrium was attained. Samples were filtered and the sulfonamide was assayed spectrophotometrically at 265 nm after dilg with 0.1M HCl. Coulometric assays gave similar results.

SOURCE AND PURITY OF MATERIALS:

Sulfamethoxazole was from Kahira Pharm and Chem Ind Co, Egypt. Mannitol was purchased from El-Nasr Chem Co, Egypt. Distd water was used.

ESTIMATED ERROR:

Soly: detns were carried out in duplicate (authors).

Temp: ±1°C (authors).

REFERENCES:

COMPONENTS:

(1) Benzenesulfonamide, 4-amino-N-(5-methyl-3-isoxazolyl)- (sulfamethoxazole); $C_{10}H_{11}N_3O_3S$; [723-46-6]

(2) Glucose; $C_6H_{12}O_6$; [50-99-7]

(3) Water; H_2O; [7732-18-5]

ORIGINAL MEASUREMENTS:

Ghanem, A.; Meshali, M.; Ibraheem, Y.

J. Pharm. Pharmacol. 1980, *32*, 675-7.

VARIABLES:

Concentration of glucose

PREPARED BY:

R. Piekos

EXPERIMENTAL VALUES:

Concentration of glucose	Solubility at 37°C	
Weight%	g $litre^{-1}$	10^3 mol dm^{-3} [a]
0.5	0.67	2.6
1.0	0.755	3.0
1.5	0.76	3.0

[a] Calculated by compiler

AUXILIARY INFORMATION

METHOD/APPARATUS/PROCEDURE:

An excess of sulfamethoxazole was added to 15 ml of glucose soln in 30-ml glass-stoppered bottles which were rotated on a water bath at 37°C until equilibrium was attained. Samples were filtered and the sulfonamide was assayed spectrophotometrically at 265 nm after dilg with 0.1M HCl. Coulometric assays gave similar results.

SOURCE AND PURITY OF MATERIALS:

Sulfamethoxazole was from Kahira Pharm and Chem Ind Co, Egypt. Glucose was purchased from El-Nasr Chem Co, Egypt. Distd water was used.

ESTIMATED ERROR:

Soly: detns were carried out in duplicate (authors).

Temp: ±1°C

REFERENCES:

COMPONENTS:	ORIGINAL MEASUREMENTS:
(1) Benzenesulfonamide, 4-amino-N-(5-methyl-3-isoxazolyl)- (sulfamethoxazole); $C_{10}H_{11}N_3O_3S$; [723-46-6] (2) Galactose; $C_6H_{12}O_6$; [26566-61-0] (3) Water; H_2O; [7732-18-5]	Ghanem, A.; Meshali, M.; Ibraheem, Y. *J. Pharm. Pharmacol*, 1980, *32*, 675-7.

VARIABLES:	PREPARED BY:
Concentration of galactose	R. Piekos

EXPERIMENTAL VALUES:

Concentration of galactose Weight%	Solubility at 37°C g $litre^{-1}$	Solubility at 37°C 10^3 mol dm^{-3} [a]
0.5	0.67	2.64
1.0	0.775	3.06
1.5	0.80	3.16

[a]Calculated by compiler

AUXILIARY INFORMATION

METHOD/APPARATUS/PROCEDURE:

An excess of sulfamethoxazole was added to 15 ml of galactose soln in 30-ml glass-stoppered bottles which were rotated on a water bath at 37°C until equilibrium was attained. Samples were filtered and the sulfonamide was assayed spectrophotometrically at 265 nm after dilg with 0.1M HCl. Coulometric assays gave similar results.

SOURCE AND PURITY OF MATERIALS:

Sulfamethoxazole was from Kahira Pharm and Chem Ind Co, Egypt. Galactose was purchased from E. Merck.
Distd water was used.

ESTIMATED ERROR:

Soly: detns were carried out in duplicate (authors).
Temp: ±1°C (authors).

REFERENCES:

COMPONENTS:

(1) Benzenesulfonamide, 4-amino-N-(5-methyl-3-isoxazolyl)- (sulfamethoxazole); $C_{10}H_{11}N_3O_3S$; [723-46-6]

(2) α-D-Glucopyranoside, β-D-fructofuranosyl- (sucrose); $C_{12}H_{22}O_{11}$; [57-60-1]

(3) Water; H_2O; [7732-18-5]

ORIGINAL MEASUREMENTS:

Ghanem, A.; Meshali, M.; Ibraheem, Y.

J. Pharm. Pharmacol. 1980, *32*, 675-7.

VARIABLES:

Concentration of sucrose

PREPARED BY:

R. Piekos

EXPERIMENTAL VALUES:

Concentration of sucrose Weight%	Solubility at 37°C g $litre^{-1}$	Solubility at 37°C 10^3 mol dm^{-3} [a]
0.5	0.615	2.43
1.0	0.605	2.39
1.5	0.615	2.43

[a]Calculated by compiler

AUXILIARY INFORMATION

METHOD/APPARATUS/PROCEDURE:

A excess of sulfamethoxazole was added to 15 ml of sucrose soln in 30-ml glass-stoppered bottles which were rotated on a water bath at 37°C until equilibrium was attained. Samples were filtered and the sulfonamide was assayed spectrophotometrically at 265 nm after dilg with 0.1M HCl. Coulometric assays gave similar results.

SOURCE AND PURITY OF MATERIALS:

Sulfamethoxazole was from Kahira Pharm and Chem Ind Co, Egypt. Sucrose was purchased.

Purity of the water was not specified.

ESTIMATED ERROR:

Soly: detns were carried out in duplicate (authors).

Temp: ±1°C (authors)

REFERENCES:

COMPONENTS:

(1) Benzenesulfonamide, 4-amino-N-(5-methyl-3-isoxazolyl)- (sulfamethoxazole); $C_{10}H_{11}N_3O_3S$; [723-46-6]

(2) D-Glucose, 4-O- α-D-glucopyranosyl- (maltose); $C_{12}H_{22}O_{11}$; [69-79-4]

(3) Water; H_2O; [7732-18-5]

ORIGINAL MEASUREMENTS:

Ghanem, A.; Meshali, M.; Ibraheem, Y.

J. Pharm. Pharmacol. 1980, *32*, 675-7.

VARIABLES:

Concentration of maltose

PREPARED BY:

R. Piekos

EXPERIMENTAL VALUES:

Concentration of maltose Weight%	Solubility at $37^{\circ}C$ g $litre^{-1}$	Solubility at $37^{\circ}C$ 10^3 mol dm^{-3} [a]
0.5	0.66	2.6
1.0	0.79	3.1
1.5	0.83	3.3

[a] Calculated by compiler

AUXILIARY INFORMATION

METHOD/APPARATUS/PROCEDURE:

An excess of sulfamethoxazole was added to 15 ml of maltose soln in 30-ml glass-stoppered bottles which were rotated on a water bath at $37^{\circ}C$ until equilibrium was attained. Samples were filtered and the sulfonamide was assayed spectrophotometrically at 265 nm after dilg with 0.1M HCl. Coulometric assays gave similar results.

SOURCE AND PURITY OF MATERIALS:

Sulfamethoxazole was from Kahira Pharm and Chem Ind Co, Egypt. Maltose was purchased from Spolek, Czechoslovakia. Distd water was used.

ESTIMATED ERROR:

Soly: detns were carried out in duplicate (authors).

Temp: $\pm 1^{\circ}C$

REFERENCES:

COMPONENTS:

(1) Benzenesulfonamide, 4-amino-N-(5-methyl-3-isoxazolyl)- (sulfamethoxazole); $C_{10}H_{11}N_3O_3S$; [723-46-6]

(2) Methanol; CH_4O; [67-56-1]

ORIGINAL MEASUREMENTS:

Rudy, B.C.; Senkowski, B.Z.
Anal. Profiles Drug Subst, 1973, *2*, 467-86.

VARIABLES:

One temperature: 25°C

PREPARED BY:

R. Piekos

EXPERIMENTAL VALUES:

Solubility of sulfamethoxazole in methanol at 25°C is 90.3 mg/ml (0.356 mol dm^{-3}, compiler).[a]

[a]The temperature and all auxiliary information was given by Edward A. MacMullan from Roche Products Inc., Manati, P.R., in personal communication.

AUXILIARY INFORMATION

METHOD/APPARATUS/PROCEDURE:

An excess of solute was equilibrated with the solvent overnight at const temp (25°C) while being agitated with a 60 cycle vibrator (VIBROMIXER). A portion of the clear supernatant soln was then taken, weighed and the solvent removed in a vacuum oven. The wt. of solute was detd after the residue had been dried to const wt. All weighing was done on a Mettler microbalance using microanal techniques.

SOURCE AND PURITY OF MATERIALS:

The sulfamethoxazole was of reference standard quality equivalent to USP.
The solvent was purchased Reagent Grade and used without further purification.

ESTIMATED ERROR:

Soly: precision ±1% (MacMullan)

Temp: not specified

REFERENCES:

COMPONENTS:	ORIGINAL MEASUREMENTS:
(1) Benzenesulfonamide, 4-amino-N-(5-methyl-3-isoxazolyl)- (sulfamethoxazole); $C_{10}H_{11}N_3O_3S$; [723-46-6] (2) Methanol; CH_4O; [67-56-1]	Shah, N.H.; Lazarus, J.H.; Sheth, P.R.; Jarowski, C.I. *J. Pharm. Sci.* 1981, *70(6)*, 611-13.
VARIABLES:	**PREPARED BY:**
One temperature: 25°C	R. Piekos

EXPERIMENTAL VALUES:

Solubility of sulfamethoxazole in methanol at 25°C is 90 mg/ml (0.35 mol dm^{-3}, compiler).

AUXILIARY INFORMATION

METHOD/APPARATUS/PROCEDURE:

The solubility of sulfamethoxazole was determined by the method specified in USP XX (1).

SOURCE AND PURITY OF MATERIALS:

Sulfamethoxazole was a research compd purchased from Hoffmann - LaRoche, Nutley, N.J. Its purity was not specified.
The source and purity of methanol was not specified.

ESTIMATED ERROR:

Nothing specified

REFERENCES:

1. *"The United States Pharmacopeia"*, 20th rev. U.S. Pharmacopeial Convention, Rockville, Md., 1980, p. 120.

COMPONENTS:	ORIGINAL MEASUREMENTS:
(1) Benzenesulfonamide, 4-amino-N-(5-methyl-3-isoxazolyl)- (sulfamethoxazole); $C_{10}H_{11}N_3O_3S$; [723-46-6] (2) 2-Propanol; C_3H_8O; [67-63-0]	Rudy, B.C.; Senkowski, B.Z. *Anal. Profiles Drug Subst.* 1973, *2*, 467-86.
VARIABLES:	PREPARED BY:
One temperature: 25°C	R. Piekos

EXPERIMENTAL VALUES:

Solubility of sulfamethoxazole in 2-propanol at 25°C is 8.8 mg/ml (3.5×10^{-2} mol dm^{-3}, compiler). [a]

[a]The temperature and all auxiliary information was given by Edward A. MacMullan from Roche Products Inc., Manati, P.R., in a personal communication.

AUXILIARY INFORMATION

METHOD/APPARATUS/PROCEDURE:

An excess of solute was equilibrated with the solvent overnight at const temp (25°C) while being agitated with a 60 cycle vibrator (VIBROMIXER). A portion of the clear supernatant soln was then taken, weighed and the solvent removed in a vacuum oven. The wt of solute was detd after the residue had been dried to const wt. All weighing was done on a Mettler microbalance using microanal techniques.

SOURCE AND PURITY OF MATERIALS:

The sulfamethoxazole was of reference standard quality equivalent to USP. The solvent was purchased Reagent Grade and used without further purification.

ESTIMATED ERROR:

Soly: precision ±1% (MacMullan)
Temp: not specified

REFERENCES:

COMPONENTS:	ORIGINAL MEASUREMENTS:
(1) Benzenesulfonamide,4-amino-N-(5-methyl-3-isoxazolyl)- (sulfamethoxazole); $C_{10}H_{11}N_3O_3S$; [723-46-6] (2) Ethanol, 2-ethoxy-; $C_4H_{10}O_2$; [110-80-5]	Sunwoo, C.; Eisen, H. *J. Pharm. Sci.*, 1971, *60*, 238-44.
VARIABLES:	**PREPARED BY:**
One temperature: 25°C	R. Piekos

EXPERIMENTAL VALUES:

The mole fraction solubility of sulfamethoxazole in 2-ethoxyethanol at 25°C is 0.0911 (22.0 g/100 g solution, compiler).

AUXILIARY INFORMATION

METHOD/APPARATUS/PROCEDURE:	SOURCE AND PURITY OF MATERIALS:
Soly was detd by the method reported by Restaino and Martin. Sulfamethoxazole was assayed on a Coleman-Hitachi 124 double-beam spectrophotometer at 271 nm after diln of a sample with 95% alcohol or water.	Sulfamethoxazole (Hoffmann - La Roche, Inc., Nutley, N.J.) was recrystd from warm alcohol. 2-Ethoxyethanol (Cellosolve solvent, Union Carbide, New York, N.Y.) was of industrial grade.
	ESTIMATED ERROR: Soly: the mean of 3 runs was given (authors). Temp: ±1.0°C (authors).
	REFERENCES: 1. Restaino, F.A.; Martin, A.N. *J. Pharm. Sci.* 1964, *53*, 636.

COMPONENTS:	ORIGINAL MEASUREMENTS:
(1) Benzenesulfonamide, 4-amino-N-(5-methyl-3-isoxazolyl)- (sulfamethoxazole); $C_{10}H_{11}N_3O_3S$; [723-46-6] (2) Ethanol; C_2H_6O; [64-17-5] (3) Methanol; CH_4O; [67-56-1]	Rudy, B.C.; Senkowski, B.Z. *Anal. Profiles Drug Subst.* 1973, *2*, 467-86.

VARIABLES:	PREPARED BY:
One temperature: 25°C	R. Piekos

EXPERIMENTAL VALUES:

Solubility of sulfamethoxazole in 3A alcohol (ethanol containing approximately 5% methanol) at 25°C is 30.6 mg/ml (0.121 mol dm^{-3}, compiler).[a]

[a]The temperature, the composition of 3A alcohol and all auxiliary information was given by Edward A. MacMullan from Roche Products Inc., Manati, P.R., in a personal communication.

AUXILIARY INFORMATION

METHOD/APPARATUS/PROCEDURE:

An excess of solute was equilibrated with the solvent overnight at const temp (25°) while being agitated with a 60 cycle vibrator (VIBROMIXER). A portion of the clear supernatant soln was then taken, weighed and the solvent removed in a vacuum oven. The wt of solute was detd after the residue had been dried to const wt. All weighing was done on a Mettler microbalance using microanal techniques.

SOURCE AND PURITY OF MATERIALS:

The sulfamethoxazole was of reference standard quality equivalent to USP. The solvent was purchased Reagent Grade and used without further purification.

ESTIMATED ERROR:

Soly: precision ±1% (MacMullan)

Temp: not specified

REFERENCES:

COMPONENTS:	ORIGINAL MEASUREMENTS:
(1) Benzenesulfonamide, 4-amino-N-(5-methyl-3-isoxazolyl)- (sulfamethoxazole); $C_{10}H_{11}N_3O_3S$; [723-46-6] (2) Ethane, 1,1'-oxybis- (ethyl ether); $C_4H_{10}O$; [60-29-7]	Rudy, B.C.; Senkowski, B.Z. *Anal. Profiles Drug Subst.* 1973, *2*, 467-86.
VARIABLES:	**PREPARED BY:**
One temperature: 25°C	R. Piekos

EXPERIMENTAL VALUES:

Solubility of sulfamethoxazole in ethyl ether at 25°C is 2.7 mg/ml (1.1×10^{-2} mol dm^{-3}, compiler).[a]

[a]The temperature and all auxiliary information was given by Edward A. MacMullan from Roche Products Inc., Manati, P.R., in a personal communication.

AUXILIARY INFORMATION

METHOD/APPARATUS/PROCEDURE:

An excess of solute was equilibrated with the solvent overnight at const temp (25°C) while being agitated with a 60 cycle vibrator (VIBROMIXER). A portion of the clear supernatant soln was then taken, weighed and the solvent removed in a vacuum oven. The wt of solute was detd after the residue had been dried to const wt. All weighing was done on a Mettler microbalance using microanal techniques.

SOURCE AND PURITY OF MATERIALS:

The sulfamethoxazole was of reference standard quality equivalent to USP. The solvent was purchased Reagent Grade and used without further purification.

ESTIMATED ERROR:

Soly: precision ±1% (MacMullan)

Temp: not specified

REFERENCES:

COMPONENTS:	ORIGINAL MEASUREMENTS:
(1) Benzenesulfonamide, 4-amino-N-(5-methyl-3-isoxazolyl)- (sulfamethoxazole); $C_{10}H_{11}N_3O_3S$; [723-46-6] (2) Petroleum ether; [8032-32-4]	Rudy, B.C.; Senkowski, B.Z. *Anal. Profiles Drug Subst.* 1973, *2*, 467-86.
VARIABLES: One temperature: 25°C	**PREPARED BY:** R. Piekos

EXPERIMENTAL VALUES:

Solubility of sulfamethoxazole in petroleum ether (boiling range 30-60°C) at 25°C is 0.2 mg/ml (8×10^{-4} mol dm^{-3}, compiler). [a]

[a]The temperature and all auxiliary information was given by Edward A. MacMullan from Roche Products Inc., Manati, P.R., in a personal communication.

AUXILIARY INFORMATION

METHOD/APPARATUS/PROCEDURE:

An excess of solute was equilibrated with the solvent overnight at const temp (25°C) while being agitated with a 60 cycle vibrator (VIBROMIXER). A portion of the clear supernatant soln was then taken, weighed and the solvent removed in a vacuum oven. The wt of solute was detd after the residue had been dried to const wt. All weighing was done on a Mettler microbalance using microanal techniques.

SOURCE AND PURITY OF MATERIALS:

The sulfamethoxazole was of reference standard quality equivalent to USP. The solvent was purchased Reagent Grade and used without further purification.

ESTIMATED ERROR:

Soly: precision ±1% (MacMullan)
Temp: not specified

REFERENCES:

COMPONENTS:	ORIGINAL MEASUREMENTS:
(1) Benzenesulfonamide, 4-amino-N-(5-methyl-3-isoxazolyl)- (sulfamethoxazole); $C_{10}H_{11}N_3O_3S$; [723-46-6] (2) Benzene; C_6H_6; [71-43-2]	Rudy, B.C.; Senkowski, B.Z. *Anal. Profiles Drug Subst.* 1973, *2*, 467-86.
VARIABLES:	PREPARED BY:
One temperature: 25°C	R. Piekos

EXPERIMENTAL VALUES:

Solubility of sulfamethoxazole in benzene at 25°C is 0.5 mg/ml (2.0×10^{-3} mol dm^{-3}, compiler). [a]

[a] The temperature and all auxiliary information was given by Edward A. MacMullan from Roche Products Inc., Manati, P.R., in a personal communication.

AUXILIARY INFORMATION

METHOD/APPARATUS/PROCEDURE:	SOURCE AND PURITY OF MATERIALS:
An excess of solute was equilibrated with benzene overnight at const temp (25°C) while being agitated with a 60 cycle vibrator (VIBROMIXER). A portion of the clear supernatant soln was then taken, weighed and the solvent removed in a vacuum oven. The wt of solute was detd after the residue had been dried to const wt. All weighing was done on a Mettler microbalance using microanal techniques.	The sulfamethoxazole was of reference standard quality equivalent to USP. The solvent was purchased Reagent Grade and used without further purification. ESTIMATED ERROR: Soly: precision ±1% (MacMullan) Temp: not specified REFERENCES:

COMPONENTS:

(1) Benzenesulfonamide, 4-amino-N-(5-methyl-3-isoxazolyl)- (sulfamethoxazole); $C_{10}H_{11}N_3O_3S$; [723-46-6]

(2) Benzene; C_6H_6; [71-43-2]

ORIGINAL MEASUREMENTS:

Takayama, K.; Nambu, N.; Nagai, T., *Chem. Pharm. Bull.* 1977, *25*, 2608-12.

VARIABLES:

One temperature: 10°C

PREPARED BY:

R. Piekos

EXPERIMENTAL VALUES:

Solubility of sulfamethoxazole in benzene at 10°C is 3.60×10^{-4} mol dm^{-3} [a].

[a] Numerical value supplied by the authors.

AUXILIARY INFORMATION

METHOD/APPARATUS/PROCEDURE:

The system was equilibrated in a sealed vial for 72 h at 10°C. The satd soln was rapidly filtered through a Toyo filter paper No. 5B, 1 cm^3 of the filtrate was evapd at 40°C and the residue was dissolved in $CHCl_3$ to det the concn in the UV region using a Hitachi 124 spectrophotometer.

SOURCE AND PURITY OF MATERIALS:

Sulfamethoxazole, m.p. 167°C, was a very pure compd supplied by Shionogi Pharmaceutical Co., Ltd.

Purity of the benzene was not specified.

ESTIMATED ERROR:

None specified

REFERENCES:

COMPONENTS:	ORIGINAL MEASUREMENTS:
(1) Benzenesulfonamide, 4-amino-N-(5-methyl-3-isoxazolyl)- (sulfamethoxazole); $C_{10}H_{11}N_3O_3S$; [723-46-6] (2) 1,4,7,10,13,16-Hexaoxacyclooctadecane (18-Crown-6); $C_{12}H_{24}O_6$; [17455-13-9] (3) Benzene; C_6H_6; [71-43-2]	Takayama, K.; Nambu, N.; Nagai, T. *Chem. Pharm. Bull.* 1977, *25*, 2608-12.
VARIABLES:	**PREPARED BY:**
Concentration of 18-Crown-6	R. Piekos

EXPERIMENTAL VALUES:

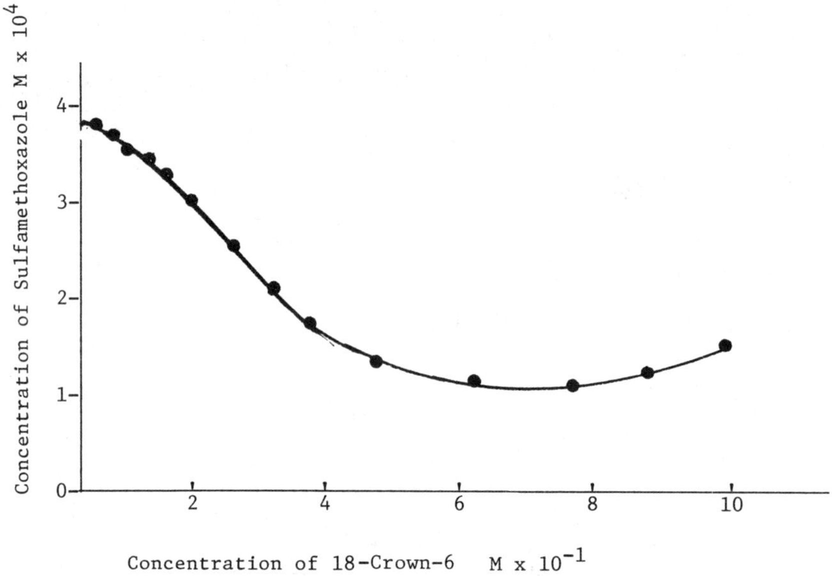

AUXILIARY INFORMATION

METHOD/APPARATUS/PROCEDURE:	SOURCE AND PURITY OF MATERIALS:
The system was equilibrated in a sealed vial for 72 h at 10°C. The satd. soln. was rapidly filtered through a Toyo filter paper No. 5B, 1 cm^3 of the filtrate was evapd. at 40°C and the residue was dissolved in $CHCl_3$ to det. the concn. in the UV region using a Hitachi 124 spectrophotometer.	Sulfamethoxazole, m.p. 167°C, was a very pure compd. supplied by Shionogi Pharmaceutical Co., Ltd. Purity of the benzene was not specified. **ESTIMATED ERROR:** None specified **REFERENCES:**

COMPONENTS:

(1) Benzenesulfonamide, 4-amino-N-(5-methyl-3-isoxazolyl)- (sulfamethoxazole); $C_{10}H_{11}N_3O_3S$; [723-46-6]

(2) 1,4,7,10,13,16-Hexaoxacyclooctadecane (18-Crown-6); $C_{12}H_{24}O_6$; [17455-13-9]

(3) Benzene; C_6H_6; [71-43-2]

ORIGINAL MEASUREMENTS:

Takayama, K.; Nambu, N.; Nagai, T., *Chem. Pharm. Bull.* 1977, *25*, 2608-12.

VARIABLES:

Concentration of 18-Crown-6

PREPARED BY:

R. Piekos

EXPERIMENTAL VALUES:

Concentration of 18-Crown-6 10^3 mol dm^{-3}	Solubility at 10^oC [a] 10^4 mol dm^{-3}
0.30	3.70
0.60	3.42
0.90	3.56
1.20	3.30
1.50	3.10
1.80	2.92
2.10	2.82
2.40	2.66
2.70	2.50
3.00	2.44
4.00	2.25
5.00	1.54
6.00	1.40
7.00	1.30
8.00	0.95
9.00	1.25
10.00	1.28

[a]Numerical values supplied by the authors

AUXILIARY INFORMATION

METHOD/APPARATUS/PROCEDURE:

The system was equilibrated in a sealed vial for 72 h at 10^oC. The satd soln was rapidly filtered through a Toyo filter paper No. 5B, 1 cm^3 of the filtrate was evapd at 40^oC and the residue was dissolved in $CHCl_3$ to det the concn in the UV region using a Hitachi 124 spectrophotometer .

SOURCE AND PURITY OF MATERIALS:

Sulfamethoxazole, m.p. 167^oC, was a very pure compd supplied by Shionogi Pharmaceutical Co., Ltd.

18-Crown-6 was of the reagent grade.

Purity of the benzene was not specified.

ESTIMATED ERROR:

None specified

REFERENCES:

COMPONENTS:	ORIGINAL MEASUREMENTS:
(1) Benzenesulfonamide, 4-amino-N-(5-methyl-3-isoxazolyl)- (sulfamethoxazole); $C_{10}H_{11}N_3O_3S$; [723-46-6] (2) Methane, trichloro- (chloroform); $CHCl_3$; [67-66-3]	Riess, W. *Intern. Congr. Chemotherapy, Proc., 3rd, Stuttgart* 1963, *1*, 627-32.
VARIABLES:	**PREPARED BY:**
One temperature: 20°C	R. Piekos

EXPERIMENTAL VALUES:

Solubility of sulfamethoxazole in chloroform at 20°C is 206 mg% (8.13×10^{-3} mol dm^{-3} solution, compiler).

AUXILIARY INFORMATION

METHOD/APPARATUS/PROCEDURE:	SOURCE AND PURITY OF MATERIALS:
Nothing specified	Nothing specified
	ESTIMATED ERROR:
	Nothing specified
	REFERENCES:

COMPONENTS:	ORIGINAL MEASUREMENTS:
(1) Benzenesulfonamide, 4-amino-N-(5-methyl-3-isoxazolyl)- (sulfisomezole); $C_{10}H_{11}N_3O_3S$; [723-46-6] (2) Methane, trichloro- (chloroform); $CHCl_3$; [67-66-3]	Yamazaki, M.; Aoki, M.; Kamada, A.; Yata, N. *Yakuzaigaku,* <u>1967</u>, *27(1)*, 37-40.
VARIABLES:	**PREPARED BY:**
One temperature: 30°C	R. Piekos

EXPERIMENTAL VALUES:

Solubility of sulfisomezole*in chloroform at 30°C is 6.75 mmol/L (1.71 g dm^{-3}, compiler).

* Another common trivial name is sulfamethoxazole.

AUXILIARY INFORMATION

METHOD/APPARATUS/PROCEDURE:

Sulfisomezole (0.5 g) was placed in an L-shaped tube together with 20 ml of chloroform. The mixt was shaken in a thermostat until equilibrium was attained. The sulfisomezole was assayed in the supernatant spectrophotometrically at 545 nm on a Beckmann DU spectrophotometer. The results were taken from a calibration graph.

SOURCE AND PURITY OF MATERIALS:

Nothing specified

ESTIMATED ERROR:

Soly: not specified

Temp: ±1°C (authors)

REFERENCES:

COMPONENTS:	ORIGINAL MEASUREMENTS:
(1) Benzenesulfonamide, 4-amino-N-(5-methyl-3-isoxazolyl)- (sulfamethoxazole); $C_{10}H_{11}N_3O_3S$; [723-46-6] (2) Methane, trichloro-; $CHCl_3$; [67-66-3]	Kitao, K.; Kubo, K.; Morishita, T.; Yata, N.; Kamada, A. *Chem. Pharm. Bull.* 1973, *21*, 2417-26.
VARIABLES:	**PREPARED BY:**
One temperature: 37°C	R. Piekos

EXPERIMENTAL VALUES:

Solubility of sulfamethoxazole in $CHCl_3$ at 37°C is 13.1 mmol dm^{-3} solution.

AUXILIARY INFORMATION

METHOD/APPARATUS/PROCEDURE:

One ml of the $CHCl_3$ soln of sulfamethoxazole at equilibrium was taken into a test tube. After evapn of the solvent, the residue was dissolved in 1N NaOH, the soln was properly dild with deionized water, and the concn of sulfamethoxazole was detd by diazotization.

SOURCE AND PURITY OF MATERIALS:

Comm available sulfamethoxazole (source not specified) was used as supplied. Neither source nor the purity of the $CHCl_3$ was specified.

ESTIMATED ERROR:

Soly: not specified

Temp: ±1°C (authors)

REFERENCES:

COMPONENTS:	ORIGINAL MEASUREMENTS:
(1) Benzenesulfonamide, 4-amino-N-(5-methyl-3-isoxazolyl)- (sulfamethoxazole); $C_{10}H_{11}N_3O_3S$; [723-46-6] (2) Methane, trichloro-; $CHCl_3$ [67-66-3]	Rudy, B.C.; Senkowski, B.Z. *Anal. Profiles Drug Subst.* 1973, *2*, 467-86.

VARIABLES:	PREPARED BY:
One temperature: 25°C	R. Piekos

EXPERIMENTAL VALUES:

Solubility of sulfamethoxazole in trichloromethane at 25°C is 2.3 mg/ml (9.1×10^{-3} mol dm^{-3}, compiler). [a]

[a]The temperature and all auxiliary information was given by Edward A. MacMullan from Roche Products Inc., Manati, P.R., in a personal communication.

AUXILIARY INFORMATION

METHOD/APPARATUS/PROCEDURE:

An excess of solute was equilibrated with trichloromethane overnight at const temp (25°C) while being agitated with a 60 cycle vibrator (VIBROMIXER). A portion of the clear supernatant soln was then taken, weighed and the solvent removed in a vacuum oven. The wt of solute was detd after the residue had been dried to const wt. All weighing was done on a Mettler microbalance using microanal techniques.

SOURCE AND PURITY OF MATERIALS:

The sulfamethoxazole was of reference standard quality equivalent to USP. The solvent was purchased Reagent Grade and used without further purification.

ESTIMATED ERROR:

Soly: precision ±1% (MacMullan)

Temp: not specified

REFERENCES:

COMPONENTS:

(1) Benzenesulfonamide, 4-amino-N-(5-methyl-3-isoxazolyl)- - 1,4,7,10,13,16-hexaoxacyclooctadecane complex (1:1); $C_{10}H_{11}N_3O_3S \cdot C_{12}H_{24}O_6$; [65177-07-3]

(2) Hydrochloric acid; HCl; [7647-01-0]

(3) Water; H_2O; [7732-18-5]

ORIGINAL MEASUREMENTS:

Takayama, K.; Nambu, N.; Nagai, T.
Chem. Pharm. Bull. 1978, *26(10)*, 2965-70.

VARIABLES:

Temperature

PREPARED BY:

R. Piekos

EXPERIMENTAL VALUES:

t/°C	Saturated concentration of the complex in 0.2N HCl 10^2M
30	2.23
35	2.43
40	2.64

AUXILIARY INFORMATION

METHOD/APPARATUS/PROCEDURE:

An excess of the complex was dissolved in 50 ml of 0.2N HCl. Sampling was done by a 1-ml pipet fitted with a G-4 glass filter. The concn of the sulfonamide in the complex was detd by uv spectrophotometry after dilg with 0.2N HCl.

SOURCE AND PURITY OF MATERIALS:

The complex was prepd by sealing 5 g of the sulfonamide (Shionogi Pharmaceutical Co.) with 5.2 g of the reagent grade crown ether in a flask and stirring well for 10 days at 10°C. The complex was filtered off, washed with benzene and dried under vacuum for 24 h. Purity of the HCl soln was not specified.

ESTIMATED ERROR:

Nothing specified

REFERENCES:

COMPONENTS:	ORIGINAL MEASUREMENTS:
(1) Benzenesulfonamide, 4-amino-N-methyl-N-(5-methyl-3-isoxazolyl)-; $C_{11}H_{13}N_3O_3S$; [51543-31-8] (2) Water; H_2O; [7732-18-5]	Kitao, K.; Kubo, K.; Morishita, T.; Yata, N.; Kamada, A. *Chem. Pharm. Bull.* 1973, *21*, 2417-26.
VARIABLES:	PREPARED BY:
One temperature: 37°C	R. Piekos

EXPERIMENTAL VALUES:

Solubility of 4-amino-N-methyl-N-(5-methyl-3-isoxazolyl)-benzenesulfonamide in water at 37°C is 0.628 mmol dm^{-3} solution.

AUXILIARY INFORMATION

METHOD/APPARATUS/PROCEDURE:	SOURCE AND PURITY OF MATERIALS:
The sulfonamide was detd in the aq soln (pH 6) by diazotization . No details were given.	The sulfonamide was synthesized by the authors. Its purity was not specified. Deionized water was used.
	ESTIMATED ERROR: Soly: not specified. Temp: ±1°C (authors).
	REFERENCES:

COMPONENTS:	ORIGINAL MEASUREMENTS:
(1) Benzenesulfonamide, 4-amino-N-methyl-N-(5-methyl-3-isoxazolyl)-; $C_{11}H_{13}N_3O_3S$; [51543-31-8] (2) Methane, trichloro-; $CHCl_3$; [67-66-3]	Kitao, K.; Kubo, K.; Morishita, T.; Yata, N.; Kamada, A. *Chem. Pharm. Bull.* 1973, *21*, 2417-26.
VARIABLES:	PREPARED BY:
One temperature: 37°C	R. Piekos

EXPERIMENTAL VALUES:

Solublity of 4-amino-N-methyl-N-(5-methyl-3-isoxazolyl)-benzenesulfonamide in $CHCl_3$ at 37°C is 1000 mmol dm^{-3} solution.

AUXILIARY INFORMATION

METHOD/APPARATUS/PROCEDURE:

One ml of the $CHCl_3$ soln of the sulfonamide at equilibrium was taken into a test tube. After evapn of the solvent, the residue was dissolved in EtOH, the soln was properly dild with deionized water, and the concn of the sulfonamide was detd by diazotization.

SOURCE AND PURITY OF MATERIALS:

The sulfonamide was synthesized by the authors. Its purity was not specified. Neither source nor purity of the $CHCl_3$ was specified.

ESTIMATED ERROR:

Soly: not specified

Temp: ±1°C (authors)

REFERENCES:

COMPONENTS:

(1) Acetamide, N-[(4-aminophenyl)sulfonyl]-N-(5-methyl-3-isoxazolyl)-(N^1-acetylsulfamethoxazole); $C_{12}H_{13}N_3O_4S$; [18607-98-2]

(2) Sodium chloride; NaCl; [7647-14-5]

(3) Water; H_2O ; [7732-18-5]

ORIGINAL MEASUREMENTS:

Hirano, H.; Ichihashi, T.; Yamada, H.
Chem. Pharm. Bull. 1981, *29(3)*, 817-27.

VARIABLES:

One temperature: 37°C

PREPARED BY:

R. Piekos

EXPERIMENTAL VALUES:

Solubility of N^1 -acetylsulfamethoxazole in a 0.9% NaCl solution at 37°C is 0.076 mg/ml (2.6×10^{-4} mol dm^{-3}, compiler).

AUXILIARY INFORMATION

METHOD/APPARATUS/PROCEDURE:

An excess of powdered N^1 -acetylsulfamethoxazole was shaken well at 37°C with a 0.9% NaCl soln until attaining satn. The undissolved crystals were removed by filtration through a G5 glass filter or by centrifugation, and the concn of the solute was assayed spectrophotometrically at 289 nm, after diln with EtOH - H_2O (1:1, v/v) using a Perkin Elmer UV-VIS spectrophotometer (Hitachi Co., Ltd., Tokyo).

SOURCE AND PURITY OF MATERIALS:

N^1 -Acetylsulfamethoxazole was synthesized by the authors and was of medical grade. The remaining materials were of anal or reagent grade.

ESTIMATED ERROR:

Nothing specified

REFERENCES:

COMPONENTS:

(1) Acetamide, N-[(4-aminophenyl)sulfonyl]-N-(5-methyl-3-isoxazolyl)-(N^1-acetylsulfamethoxazole); $C_{12}H_{13}N_3O_4S$; [18607-98-2]

(2) Phosphoric acid, disodium salt; Na_2HPO_4; [7558-94-4]

(3) Phosphoric acid, monopotassium salt; KH_2PO_4; [7778-77-0]

(4) Water; H_2O; [7732-18-5]

ORIGINAL MEASUREMENTS:

Hekster, Ch. A.; Vree, T. B.

Antibiotics Chemother. 1982, *31*, 22-118.

VARIABLES:

pH

PREPARED BY:

R. Piekos

EXPERIMENTAL VALUES:

pH	Solubility at 25°C	
	mg/l	10^4 mol dm^{-3} [a]
5.5	66	2.2
7.5[b]	66	2.2

[a]Calculated by compiler

[b]Erroneous pH value of 7.0 is given in the article

AUXILIARY INFORMATION

METHOD/APPARATUS/PROCEDURE:

The earlier developed method (1) was used (personal communication). Satd solns of N^1-acetylsulfamethoxazole were prepd in phosphate buffers of pH 5.5 and 7.5 at 25°C. The concn of the solute was measured by means of a Spectra Physics 3500B high-performance liquid chromatograph equipped with a Model 748 column oven and a Pye-Unicam LC-UV spectrophotometric detector.

SOURCE AND PURITY OF MATERIALS:

Neither source nor the purity of the materials was specified.

ESTIMATED ERROR:

Soly: the detection limit of the solute by HPLC was 0.5 mg/l (authors). The errors in temp and pH were not specified.

REFERENCES:

Hekster, Y. A.; Vree, T.B.; Damsma, J. E.; Friesen, W. T. *J. Antimicrob. Chemother.* 1981, *8*, 133.

COMPONENTS:

(1) Acetamide, N-[4[[(5-methyl-3-isoxazolyl) -amino]sulfonyl]phenyl]- (N^4-acetylsulfamethoxazole); $C_{12}H_{13}N_3O_4S$; [21312-10-7]
(2) Phosphoric acid, disodium salt; Na_2HPO_4; [7558-94-4]
(3) Phosphoric acid, monopotassium salt; KH_2PO_4; [7778-77-0]
(4) Water; H_2O; [7732-18-5]

ORIGINAL MEASUREMENTS:

Hekster, Y.A.; Vree, T.B.; Damsma, J.E.; Friesen, W. T.

J. Antimicrob. Chemother, 1981, *8*, 133-44.

VARIABLES:

pH

PREPARED BY:

R. Piekos

EXPERIMENTAL VALUES:

pH	Solubility at 25°C	
	mg/l	10^3 mol dm^{-3} [a]
5.5	115	0.389
7.5	1000	3.386

[a]Calculated by compiler

AUXILIARY INFORMATION

METHOD/APPARATUS/PROCEDURE:

Satd solns of N^4-acetylsulfamethoxazole were prepd in phosphate buffers of pH 5.5 and 7.5 at room temp (25°C). The concn of the solute was measured by means of a Spectra Physics 3500B high-performance liquid chromatograph equipped with a column oven (Model 748) and a Pye-Unicam LC-UV spectrophotometric detector. The detector was connected to a 1-mV recorder. A stainless steel column (10 cm x 4.6 mm i.d.) was packed with Lichrosorb RPS, 5 μm, obtained from Chrompack. An injection loop of 100 μl was used. The oven temp was 40°C. Detection of the solute was performed at 260 nm.

SOURCE AND PURITY OF MATERIALS:

The source and purity of the materials was not specified.

ESTIMATED ERROR:

The detection limit of solute by HPLC was 0.5 mg/l (authors). The error in temperature and pH was not specified.

REFERENCES:

COMPONENTS:	EVALUATOR:
(1) Benzenesulfonamide, 4-amino-N-(3-4-dimethyl-5-isoxazolyl)-(sulfisoxazole) $C_{11}H_{13}N_3O_3S$; [127-69-5] (2) Water (3) Ethanol	Anthony N. Paruta Department of Pharmaceutics University of Rhode Island Kingston, Rhode Island, USA and Ryszard Piekos Faculty of Pharmacy, University of Gdansk Gdansk, Poland 1986

CRITICAL EVALUATION:

Aqueous solubilities of the compound at 310K as determined twice, in 1978 and 1980, by the same laboratory (1,2) using virtually identical methods and procedures and are the same. Assuming that the values were independently determined, the recommended value is 1.09 x 10^{-3} mol dm^{-3} in water at 298K.

Ethanolic solubilities were determined at 303K by two independent groups (3,4). The results are only within 10%, and the equilibrium time unclear (4). The tentative average value of sulfisoxazole in ethanol at 303K is given as 81.6 x 10^{-3} mol dm^{-3}. This value is about 75 times higher than that of water.

REFERENCES:

(1) Kaneniwa, N.; Watari, N. *Chem. Pharm. Bull.* 1978, *26(3)*, 813-26.
(2) Watari, N.; Kaneniwa, N.; Hanano, M. *Int. J. Pharm.* 1980, *6(2)*, 155-66.
(3) Mauger, J.W.; Petersen, H., Jr.; Alexander, K.S.; Paruta, A.N. *Drug Dev. Ind. Pharm.* 1977, *3(2)*, 163-83.
(4) Sekikawa, H.; Nakano, M.; Arita, T. *Chem. Pharm. Bull.* 1978, *26(1)*, 118-26.

COMPONENTS:	ORIGINAL MEASUREMENTS:
(1) Benzenesulfonamide, 4-amino-N-(3,4-dimethyl-5-isoxazolyl)- (sulfisoxazole); $C_{11}H_{13}N_3O_3S$; [127-69-5] (2) Water; H_2O; [7732-18-5]	Yamazaki, M; Aoki, M.; Kamada, A.; Yata, N.; *Yakuzaigaku* 1967, *27(1)*, 37-40.
VARIABLES:	**PREPARED BY:**
One temperature: 30°C	R. Piekos

EXPERIMENTAL VALUES:

Solubility of sulfisoxazole in water at 30°C is 0.83 mmol/L (0.22 g dm^{-3}, compiler).

AUXILIARY INFORMATION

METHOD/APPARATUS/PROCEDURE:

Sulfisoxazole (0.5 g) was placed in an L-shaped tube together with 20 ml of water. The mixt was shaken in a thermostat until equilibrium was attained. The sulfisoxazole was assayed in the supernatant spectrophotometrically at 545 nm on a Beckman DU spectrophotometer. The results were taken from a calibration graph.

SOURCE AND PURITY OF MATERIALS:

Nothing specified

ESTIMATED ERROR:

Soly: not specified

Temp: ±1°C (authors)

REFERENCES:

COMPONENTS:

(1) Benzenesulfonamide, 4-amino-N-(3,4-dimethyl-5-isoxazolyl)- (sulfisoxazole); $C_{11}H_{13}N_3O_3S$; [127-69-5]

(2) Water; H_2O; [7732-18-5]

ORIGINAL MEASUREMENTS:

Kaneniwa, N.; Watari, N.

Chem. Pharm. Bull. 1978, *26(3)*, 813-26.

VARIABLES:

One temperature: 37°C

PREPARED BY:

R. Piekos

EXPERIMENTAL VALUES:

Solubility of sulfisoxazole in water at 37°C is 0.292 mg/ml solution (1.09×10^{-3} mol dm^{-3}, compiler).

AUXILIARY INFORMATION

METHOD/APPARATUS/PROCEDURE:

An excess of sulfisoxazole was placed in a flask contg 25 ml of water. The flask was shaken (2 strokes/s at the amplitude of 3 cm) in a thermostatically controlled water bath at 37°C. One-ml sample was withdrawn every 6 h (total equilibration period was 3-5 days) using a warmed Millipore filter syringe with a filter pore size of 0.45 μ (Millipore HAWP 01300) and the filtrate was dild with water and assayed spectrophotometrically (1).

SOURCE AND PURITY OF MATERIALS:

Commercial sulfisoxazole of the Japanese Pharmacopeia grade and distd water were used.

ESTIMATED ERROR:

Soly: not specified.

Temp: ±0.05°C (authors).

REFERENCES:

1. Kaneniwa, N.; Watari, N. *Chem. Pharm. Bull.* 1974, *22*, 1699.

COMPONENTS:

(1) Benzenesulfonamide, 4-amino-N-(3,4-dimethyl-5-isoxazolyl)- (sulfisoxazole); $C_{11}H_{13}N_3O_3S$; [127-69-5]

(2) Water; H_2O; [7732-18-5]

ORIGINAL MEASUREMENTS:

Watari, N; Kaneniwa, N.; Hanano, M.
Int. J. Pharm. 1980, *6(2)*, 155-66.

VARIABLES:

One temperature: 37°C

PREPARED BY:

R. Piekos

EXPERIMENTAL VALUES:

Solubility of sulfisoxazole in water at 37°C is 29.2 mg/100 ml (1.09×10^{-3} mol dm^{-3}, compiler).

AUXILIARY INFORMATION

METHOD/APPARATUS/PROCEDURE:

The earlier developed method was employed (1), whereby an excess of sulfisoxazole, required to saturate medium, was placed in a flask contg 25 ml of water. The flask was shaken (2 strokes/s) at an amplitude of 3 cm, in a thermostatically controlled bath. One-ml sample was removed every 6 h (total equilibration time was 3-5 days) using a warmed Millipore filter syringe with a filter pore size of 0.45 μ (Millipore HAWP 01300) and the filtrate was dild with water and assayed spectrophotometrically.

SOURCE AND PURITY OF MATERIALS:

Sulfisoxazole was of the Japanese Pharmacopeia grade.
Distilled water was used.

ESTIMATED ERROR:

Soly: not specified
Temp: ±0.05°C (authors)

REFERENCES:

1. Kaneniwa, N.; Watari, N.
Chem. Pharm. Bull. 1974, *22*, 1699.

COMPONENTS:

(1) Benzenesulfonamide, 4-amino-N-(3,4-dimethyl-5-isoxazolyl)- (sulfisoxazole); $C_{11}H_{13}N_3O_3S$; [127-69-5]

(2) Hydrochloric acid; HCl; [7647-01-0]

(3) Water; H_2O; [7732-18-5]

ORIGINAL MEASUREMENTS:

Ogata, H.; Shibazaki, T.; Inoue, T.; Ejima, A. *Chem. Pharm. Bull.* 1979, *27(6)*, 1281-6.

VARIABLES:

One temperature: 37°C

PREPARED BY:

R. Piekos

EXPERIMENTAL VALUES:

Solubility of sulfisoxazole in 0.1N HCl at 37°C is 1.440 mg/ml (5.387×10^{-3} mol dm^{-3}, compiler).

AUXILIARY INFORMATION

METHOD/APPARATUS/PROCEDURE:

A centrifuge tube contg 30 ml of 0.1N HCl and 0.5-3.0 g of the sulfisoxazole powder was tightly sealed and shaken at 37°C. The concn of the dissolved drug was detd spectrophotometrically following filtration through a Millipore filter (type EH, pore size 0.5 μm), and the procedure was repeated every 24 h until a const concn was obtained.

SOURCE AND PURITY OF MATERIALS:

Comm available 500-mg uncoated tablets of sulfisoxazole were used. Hydrochloric acid was of reagent grade.

ESTIMATED ERROR:

Nothing specified

REFERENCES:

COMPONENTS:

(1) Benzenesulfonamide, 4-amino-N-(3,4-dimethyl-5-isoxazolyl)-(sulfisoxazole); $C_{11}H_{13}N_3O_3S$; [127-69-5]

(2) Carbonic acid, monosodium salt; $NaHCO_3$; [144-55-8]

(3) Water; H_2O; [7732-18-5]

ORIGINAL MEASUREMENTS:

Takubo, T.; Matsumaru, H.; Tsuchiya, S.; Hiura, M. *Chem. Pharm. Bull.* 1973, *21(7)*, 1440-5.

VARIABLES:

One temperature: 37°C; one pH: 8.4

PREPARED BY:

R. Piekos

EXPERIMENTAL VALUES:

Solubility of sulfisoxazole in a $NaHCO_3$ solution (1.680 g $NaHCO_3$/100 ml water) of pH 8.4 at 37°C is 31.25 mg/ml solution[a] (1.169×10^{-1} mol dm^{-3} solution, compiler).

[a]Numerical value to the graphical data was given by one of the authors (S. T.) in personal communication.

AUXILIARY INFORMATION

METHOD/APPARATUS/PROCEDURE:

Aliquots of the $NaHCO_3$ soln were placed in glass-stoppered flasks with excess of sulfisoxazole. The flasks were allowed to stand at 37±1°C and shaken vigorously for 4 h until equilibrium was attained. One ml of the supernatant was removed by means of a filter pipet and sulfisoxazole was assayed by the previously reported method (1).

SOURCE AND PURITY OF MATERIALS:

The sulfisoxazole was of the pharmaceutical grade. The source and purity of $NaHCO_3$ was not specified.
Distd was used.

ESTIMATED ERROR:

Soly and pH: not specified.
Temp: ±1°C (authors).

REFERENCES:

1. Takubo, T.; Tsuchiya, S.; Hiura, M. *Yakuzaigaku* 1971, *31*, 298.

COMPONENTS:

(1) Benzenesulfonamide, 4-amino-N-(3,4-dimethyl-5-isoxazolyl)-(sulfisoxazole); $C_{11}H_{13}N_3O_3S$; [127-69-5]

(2) Carbonic acid; disodium salt; Na_2CO_3; [497-19-8]

(3) Water; H_2O; [7732-18-5]

ORIGINAL MEASUREMENTS:

Takubo, T. ; Matsumaru, H.; Tsuchiya, S.; Hiura, M. *Chem. Pharm. Bull.* 1973, *21(7)*, 1440-5.

VARIABLES:

One temperature: 37°C; one pH: 11.3

PREPARED BY:

R. Piekos

EXPERIMENTAL VALUES:

Solubility of sulfisoxazole in a Na_2CO_3 solution (2.120 g Na_2CO_3/100 ml water) of pH 11.3 at 37°C is 54.12 mg/ml solution[a] (2.025×10^{-1} mol dm^{-3} solution, compiler).

[a]Numerical value for the graphical data was given by one of the authors (S. T.) in personal communication.

AUXILIARY INFORMATION

METHOD/APPARATUS/PROCEDURE:

Aliquots of the Na_2CO_3 solution were placed in glass-stoppered flasks with excess of sulfisoxazole. The flasks were allowed to stand at 37±1°C and shaken vigorously for 4 h until equilibrium was established. One ml of the supernatant was removed by means of a filter pipet and sulfisoxazole was assayed by the previously reported method (1).

SOURCE AND PURITY OF MATERIALS:

The sulfisoxazole was of pharmaceutical grade. The source and purity of Na_2CO_3 was not specified.

Distd water was used.

ESTIMATED ERROR:

Soly and pH: not specified.

Temp: ±1°C.

REFERENCES:

1. Takubo, T.; Tsuchiya, S.; Hiura, M. *Yakuzaigaku* 1971, *31*, 298.

COMPONENTS:

(1) Benzenesulfonamide, 4-amino-N-(3,4-dimethyl-5-isoxazolyl)-(sulfisoxazole); $C_{11}H_{13}N_3O_3S$; [127-69-5]

(2) Carbonic acid; disodium salt; Na_2CO_3; [497-19-8]

(3) Carbonic acid; monosodium salt; $NaHCO_3$; [144-55-8]

(4) Water; H_2O; [7732-18-5]

ORIGINAL MEASUREMENTS:

Takubo, T.; Matsumaru; H. ; Tsuchiya, S.; Hiura, M. *Chem. Pharm. Bull.* 1973, *21(7)*, 1440-5.

VARIABLES:

pH

PREPARED BY:

R. Piekos

EXPERIMENTAL VALUES:

Na_2CO_3 g/100 ml water	$NaHCO_3$ g/100 ml water	pH	Solubility at $37^{\circ}C$ mg/ml soln[a]	Solubility at $37^{\circ}C$ 10 mol dm^{-3} soln[b]
0.212	1.512	9.1	35.84	1.341
0.848	1.008	9.8	48.97	1.832
1.908	0.168	10.7	54.12	2.025

[a]Numerical values to the graphical data were given by one of the authors (S.T.) in personal communication.

[b]Calculated by compiler.

AUXILIARY INFORMATION

METHOD/APPARATUS/PROCEDURE:

Aliquots of carbonate buffer solns were placed in glass-stoppered flasks with excess of sulfisoxazole. The flasks were allowed to stand at 37±1°C and shaken vigorously for 4 h until equilibrium was established. One ml of the supernatant was removed by means of a filter pipet and sulfisoxazole was assayed by the previously reported method (1).

SOURCE AND PURITY OF MATERIALS:

The sulfisoxazole was of pharmaceutical grade. The source and purity of Na_2CO_3 and $NaHCO_3$ were not specified. Distd water was used.

ESTIMATED ERROR:

Soly and pH: not specified.

Temp: ±1°C (authors).

REFERENCES:

1. Takubo, T.; Tsuchiya, S.; Hiura, M. *Yakuzaigaku*, 1971, *31*, 298.

COMPONENTS:

(1) Benzenesulfonamide, 4-amino-N-(3,4-dimethyl-5-isoxazolyl)- (sulfisoxazole); $C_{11}H_{13}N_3O_3S$; [127-69-5]

(2) Phosphoric acid, disodium salt; Na_2HPO_4; [7558-94-4]

(3) Phosphoric acid, monopotassium salt; KH_2PO_4; [7778-77-0]

(4) Water; H_2O; [7732-18-5]

ORIGINAL MEASUREMENTS:

Bandelin, F.J.; Malesh, W.

J. Am.Pharm. Assoc., Sci. Ed. 1959, *48*, 177-81.

VARIABLE:

pH

PREPARED BY:

R. Piekos

EXPERIMENTAL VALUES:

Solubility of sulfisoxazole in buffers of varying mixtures of $Na_2HPO_4 \cdot 7H_2O$ (71.6 g/1; distilled water; 0.27 mol dm^{-3}, compiler) and KH_2PO_4 (36.3 g/1 distilled water; 0.27 mol dm^{-3}, compiler) at 37°C

Initial pH	Solubility	
	mg/100 ml	10^2 mol dm^{-3} [a]
4.5	33	0.12
5.0	45	0.16
5.5	70	0.26
6.0	175	0.65
6.5	405	1.51
7.0	1360	5.08
7.5	2870	10.73

[a] calculated by compiler

AUXILIARY INFORMATION

METHOD/APPARATUS/PROCEDURE:

Solns were prepd by adding an excess of sulfisoxazole to 10 ml of buffer soln at each pH level in 18 x 150-mm test tubes, stoppering the tubes and placing them in a water bath at 37°C with gentle agitation for 24 h. The mixt was then filtered and a 1-ml aliquot was accurately pipetted into a volumetric flask for diln and analysis. The balance was retained for pH detn to ascertain any change in pH value. The sulfonamide was assayed colorimetrically by the method of Bratton and Marshall as described in detail by Biamonte and Schneller (1). A standard curve was prepd using accurately prepd standard solutions.

SOURCE AND PURITY OF MATERIALS:

Neither source nor purity of the reagents were specified.

Distilled water was used.

ESTIMATED ERROR:

Soly: av values of duplicate runs are reported (authors).

Temp and pH: not specified.

REFERENCES:

1. Biamonte, A.R.; Schneller, G.E. *J. Am. Pharm. Assoc., Sci. Ed.*, 1952, *41*, 341.

COMPONENTS:

(1) Benzenesulfonamide, 4-amino-N-(3,4-dimethyl-5-isoxazolyl)- (sulfisoxazole); $C_{11}H_{13}N_3O_3S$; [127-69-5]

(2) Phosphoric acid, disodium salt; Na_2HPO_4; [7558-94-4]

(3) Phosphoric acid, monopotassium salt; KH_2PO_4 [7778-77-0]

(4) Water; H_2O; [7732-18-5]

ORIGINAL MEASUREMENTS:

Riess, W.

Intern. Congr. Chemotherapy, Proc. 3rd, Stuttgart 1963, *1*, 627-32.

VARIABLE:

One temperature: 20°C; one pH: 7.4

PREPARED BY:

R. Piekos

EXPERIMENTAL VALUES:

Solubility of sulfisoxazole in a M/15 Sörensen buffer solution (pH 7.4) at 20°C is 4000 mg% (0.1496 mol dm^{-3} solution, compiler).

AUXILIARY INFORMATION

METHOD/APPARATUS/PROCEDURE:

Sörensen buffer solns of pH varying between 7 and 8 were prepd, satd with sulfisoxazole at 20°C, their pH was measured at equilibrium and the sulfisoxazole was assayed colorimetrically. The measured pH values were then plotted against concn., and the soly at pH 7.4 was detd by interpolation (personal communication).

SOURCE AND PURITY OF MATERIALS:

Nothing specified.

ESTIMATED ERROR:

Nothing specified.

REFERENCES:

COMPONENTS:

(1) Benzenesulfonamide, 4-amino-N-(3,4-dimethyl-5-isoxazolyl)- (sulfisoxazole); $C_{11}H_{13}N_3O_3S$; [127-69-5]

(2) Phosphoric acid, disodium salt; Na_2HPO_4; [7558-94-4]

(3) Phosphoric acid, monopotassium salt; KH_2PO_4; [7778-77-0]

(4) Water; H_2O; [7732-18-5]

ORIGINAL MEASUREMENTS:

Yamazaki, M.; Aoki, M.; Kamada, A.; Yata, N. *Yakuzaigaku,* 1967, *27(1),* 37-40.

VARIABLES:

One temperature: 30°C; one pH: 7.4

PREPARED BY:

R. Piekos

EXPERIMENTAL VALUES:

Solubility of sulfisoxazole in a phosphate buffer solution of pH 7.4[a] ($\mu = 0.17$) at 30°C is 32.1 mmol/L (8.580 g dm^{-3}, compiler).

[a]At the end of experiment the pH was 6.5

AUXILIARY INFORMATION

METHOD/APPARATUS/PROCEDURE:

Sulfisoxazole (0.5 g) was placed in an L-shaped tube together with 20 ml of the buffer soln. The mixt was shaken in a thermostat until equilibrium was attained. The sulfisoxazole was assayed in the supernatant spectrophotometrically at 545 nm on a Beckmann DU spectrophotometer. The results were taken from a calibration graph.

SOURCE AND PURITY OF MATERIALS:

Nothing specified

ESTIMATED ERROR:

Soly and pH: not specified

Temp: ±1°C (authors)

REFERENCES:

COMPONENTS:

(1) Benzenesulfonamide, 4-amino-N-(3,4-dimethyl-5-isoxazolyl)- (sulfafurazole)*; $C_{11}H_{13}N_3O_3S$; [127-69-5]

(2) Phosphoric acid, disodium salt; Na_2HPO_4; [7558-94-4]

(3) Phosphoric acid, monopotassium salt; KH_2PO_4; [7778-77-0]

(4) Water; H_2O; [7732-18-5]

ORIGINAL MEASUREMENTS:

Hekster, Ch. A.; Vree, T.B.

Antibiotics Chemother. 1982, *31*, 22-118.

VARIABLES:

pH

PREPARED BY:

R. Piekos

EXPERIMENTAL VALUES:

pH	Solubility at 25°C mg/l	Solubility at 25°C 10^3 mol dm^{-3}
5.5	1,533	5.735
7.5[b]	4,724	17.670

[a]Calculated by compiler

[b]Erroneous pH value of 7.0 is given in the article

*Another common trivial name is sulfisoxazole.

AUXILIARY INFORMATION

METHOD/APPARATUS/PROCEDURE:

The earlier developed method (1) was used (personal communication). Satd solns of sulfafurazole* were prepd in phosphate buffers of pH 5.5 and 7.5 at 25°C. The concn of the solute was measured by means of a Spectra Physics 3500B high-performance liquid chromatograph equipped with a Model 748 column oven and a Pye-Unicam LC-UV spectrophotometric detector.

SOURCE AND PURITY OF MATERIALS:

Neither source nor the purity of the materials was specified.

ESTIMATED ERROR:

Soly: the detection limit of the solute by HPLC was 0.5 mg/l (authors). The errors in temp and pH were not specified.

REFERENCES:

1. Hekster, Y.A.; Vree, T.B.; Damsma, J.E.; Friesen, W.T. *J. Antimicrob. Chemother.* 1981, *8*, 133.

COMPONENTS:

(1) Benzenesulfonamide, 4-amino-N-(3,4-dimethyl-5-isoxazolyl)- (sulfisoxazole); $C_{11}H_{13}N_3O_3S$; [127-69-5]

(2) Calcium chloride; $CaCl_2$; [10043-52-4]

(3) Magnesium chloride; $MgCl_2$; [7786-30-3]

(4) Phosphoric acid, monoammonium salt; $NH_4H_2PO_4$; [7722-76-1]

(5) Potassium chloride; KCl; [7447-40-7]

(6) Sodium chloride; NaCl; [7647-14-5]

(7) Urea; CH_4N_2O; [57-13-6]

(8) Water; H_2O; [7732-18-5]

ORIGINAL MEASUREMENTS:

Bandelin, F. J.; Malesh, W.

J. Am. Pharm. Assoc., Sci. Ed. <u>1959</u>, *48*, 177-81.

VARIABLES:

pH at 37°

PREPARED BY:

R. Piekos

EXPERIMENTAL VALUES:

Solubility of sulfisoxazole in a solution containing $CaCl_2$ 0.143, $MgCl_2$ 0.121, $NH_4H_2PO_4$ 0.300, KCl 1.660, NaCl 2.950 and urea 20 g/dm^3 (synthetic urine, Mosher Vehicle) at 37°C.

Equilibrium pH	Solubility	
	mg/100 ml	10^2 mol/dm^3 [a]
4.5	36	0.13
5.0	51	0.19
5.5	80	0.29
6.0	220	0.82
6.4	710	2.66
6.7	2600	9.73

[a]calculated by compiler

AUXILIARY INFORMATION

METHOD/APPARATUS/PROCEDURE:

Excess sulfisoxazole was added to aliquots of synthetic urine solns and 1% H_3PO_4 or 1% NaOH solns were used to adjust the pH to the required value. The solns were agitated for 24 h with addn of acid or base to keep them at the desired pH level until equilibrium was attained. Then the solns were filtered and in aliquots the sulfonamide was assayed spectrophotometrically by the method described by Biamonte and Schneller (1).

SOURCE AND PURITY OF MATERIALS:

Nothing specified

ESTIMATED ERROR:

Soly: average values of 2 detns were given.

Temp: not specified

pH : not specified

REFERENCES:

1. Biamonte, A.R.; Schneller, G. E. *J. Am. Pharm. Assoc., Sci. Ed.* <u>1952</u>, *41*, 341.

COMPONENTS:

(1) Benzenesulfonamide, 4-amino-N-(3,4-dimethyl-5-isoxazolyl)- (sulfisoxazole); $C_{11}H_{13}N_3O_3S$; [127-69-5]

(2) 1,2,3-Propanetricarboxylic acid, 2-hydroxy- (citric acid); $C_6H_8O_7$; [77-92-9]

(3) Water; H_2O; [7732-18-5]

ORIGINAL MEASUREMENTS:

Takubo, T.; Matsumaru, H.; Tsuchiya, S.; Hiura, M. *Chem. Pharm. Bull.* 1973, *21(7)*, 1440-5.

VARIABLES:

One temperature: 37°C; one pH: 2.1

PREPARED BY:

R. Piekos

EXPERIMENTAL VALUES:

Solubility of sulfisoxazole in a citric acid solution (2.100 g citric acid per 100 ml water) of pH 2.1 at 37°C is 0.31 mg/ml solution[a] (1.16 x 10^{-3} mol dm^{-3} solution, compiler).

[a]Numerical value to the graphical one was given by one of the authors (S.T) in personal communication.

AUXILIARY INFORMATION

METHOD/APPARATUS/PROCEDURE:

Aliquots of the citric acid soln were placed in glass-stoppered flasks with excess of sulfisoxazole. The flasks were allowed to stand at 37±1°C and shaken vigorously for 4 h until equilibrium was established. One ml of the supernatant was removed by means of a filter and the sulfanilamide was assayed by the previously reported method (1).

SOURCE AND PURITY OF MATERIALS:

The sulfanilamide was of pharmaceutical grade. Source and purity of the citric acid was not specified.

Distd water was used.

ESTIMATED ERROR:

Soly and pH: not specified

Temp: ±1°C (authors)

REFERENCES:

1. Takubo, T.; Tsuchiya, S.; Hiura, M. *Yakuzaigaku* 1971, *31*, 298.

COMPONENTS:

(1) Benzenesulfonamide, 4-amino-N-(3,4-dimethyl-5-isoxazolyl)- (sulfisoxazole); $C_{11}H_{13}N_3O_3S$; [127-69-5]

(2) Phosphoric acid, disodium salt; Na_2HPO_4; [7558-94-4]

(3) 1,2,3-Propanetricarboxylic acid, 2-hydroxy- (citric acid); $C_6H_8O_7$; [77-92-9]

(4) Water; H_2O; [7732-18-5]

ORIGINAL MEASUREMENTS:

Takubo, T.; Matsumaru, H.; Tsuchiya, S.; Hiura, M. *Chem. Pharm. Bull.* 1973, *21(7)*, 1440-5.

VARIABLES:

pH

PREPARED BY:

R. Piekos

EXPERIMENTAL VALUES:

Citric acid	Na_2HPO_4	pH	Solubility at 37°C	
g/100 g water	g/100 g water		mg/ml soln[a]	10^2 mol dm^{-3} soln[b]
1.680	0.572	3.1	0.23	0.086
1.260	0.144	4.2	0.30	0.112
0.840	1.716	5.8	1.70	0.636
0.420	2.228	6.8	8.50	3.180

[a]Numerical values to the graphical ones were given by one of the authors (S. T.) in personal communication.

[b]Calculated by compiler.

AUXILIARY INFORMATION

METHOD/APPARATUS/PROCEDURE:

Aliquots of the buffer solns were placed in glass-stoppered flasks with excess of sulfisoxazole. The flasks were allowed to stand at 37±1°C and shaken vigorously for 4 h until equilibrium was established. One ml of the supernatant was removed by means of a filter pipet and sulfisoxazole was assayed by the previously reported method (1).

SOURCE AND PURITY OF MATERIALS:

The sulfisoxazole was of the pharmaceutical grade. The source and purity of Na_2HPO_4 and citric acid were not specified. Distd water was used.

ESTIMATED ERROR:

Soly and pH: not specified.

Temp: ±1°C (authors)

REFERENCES:

1. Takubo, T.; Tsuchiya, S.; Hiura, M. *Yakuzaigaku*, 1971, *31*, 298.

COMPONENTS:

(1) Benzenesulfonamide, 4-amino-N-(3,4-dimethyl-5-isoxazolyl)- (sulfisoxazole); $C_{11}H_{13}N_3O_3S$; [127-69-5]

(2) Phosphoric acid, disodium salt; Na_2HPO_4; [7558-94-4]

(3) 1,2,3-Propanetricarboxylic acid, 2-hydroxy- (citric acid); $C_6H_8O_7$; [77-92-9]

(4) Water; H_2O; [7732-18-5]

ORIGINAL MEASUREMENTS:

Biamonte, A.R.; Schneller, G.H.

J. Am. Pharm. Assoc., Sci. Ed. 1952, *41*, 341-5.

VARIABLES:

pH

PREPARED BY:

R. Piekos

EXPERIMENTAL VALUES:

Solubility of sulfafurazole* in McIlvaine's disodium phosphate - citric acid buffer solution at 37°.

Initial pH of buffer	Solubility		Final pH
	mg/100 ml	10^2mol dm^{-3} [a]	
4.5	32.3	0.121	4.5
5.0	51.6	0.193	5.0
5.5	108.7	0.407	5.5
6.0	262.0	0.980	5.9
6.5	616.0	2.304	6.3
7.0	2,135.0	7.987	6.8

[a] Calculated by compiler

*Another common trivial name is sulfisoxazole.

AUXILIARY INFORMATION

METHOD/APPARATUS/PROCEDURE:

Sulfafurazole* (500 mg) was equilibrated in a water bath with 50 ml of the buffer soln for 18 h at 37°C with agitation. The suspension was then immediately filtered through a Whatman No. 1 paper. The filtration time was approx 2 min. Sulfafurazole* in the filtrate was assayed spectrophotometrically by the Bratton and Marshall method (1) using a Beckman DU spectrophotometer, at 545 nm.

SOURCE AND PURITY OF MATERIALS:

The source of sulfafurazole* (mp 193.4 - 193.9°C) was not specified. The source and purity of the remaining materials were not specified.

ESTIMATED ERROR: pH and temp: not specified. Accuracy of the anal method was illustrated by the following values: expected 2.003, 3.004, 4.006, 5.007 mg/100 ml; found: 2.08, 3.06, 4.12, 5.10, resp.

REFERENCES:

1. Bratton, A.C.; Marshall, E.K., Jr. *J. Biol. Chem.* 1939, *128*, 537.

COMPONENTS:

(1) Benzenesulfonamide, 4-amino-N-(3,4-dimethyl-5-isoxazolyl)- (sulfisoxazole); $C_{11}H_{13}N_3O_3S$; [127-69-5]

(2) Sorbitan monolaurate, polyoxyethylene derivatives (Tween 20); [9005-64-5]

(3) Water; H_2O; [7732-18-5]

ORIGINAL MEASUREMENTS:

Khawam, M.N.; Yousef, R.T.; Czetsch-Lindenwald, H. *Sci. Pharm.* 1966, *34*, 209-13.

VARIABLES:

Concentration of Tween 20

PREPARED BY:

R. Piekos

EXPERIMENTAL VALUES:

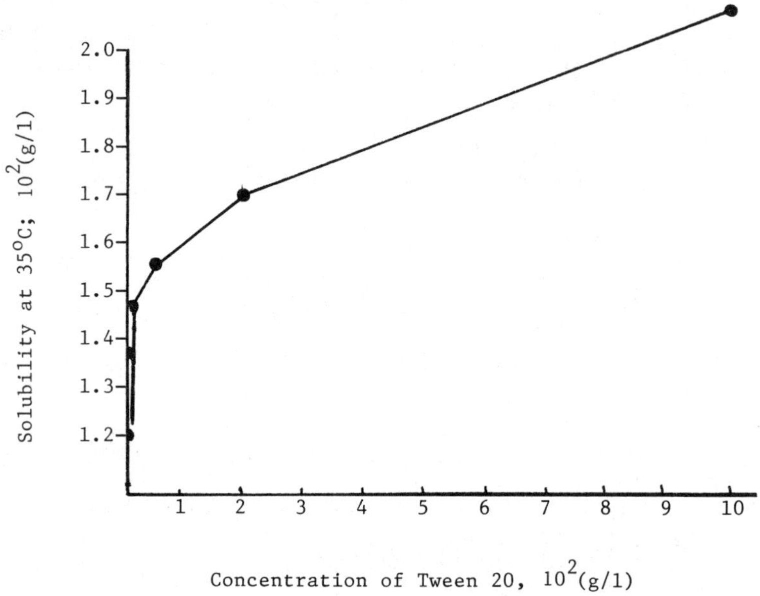

AUXILIARY INFORMATION

METHOD/APPARATUS/PROCEDURE:

An earlier described method was employed (1) whereby a 100-ml conical flask contg a Tween 20 soln was placed in a drying cabinet at 35°C and an excess of sulfisoxazole was added under stirring for 1 h. After 12 h the soln was filtered or decanted and the solute was assayed in the filtrate spectrophotometrically using a Unicam SP 500 spectrophotometer and 1-ml quartz cuvets. Results were taken from a calibration graph.

SOURCE AND PURITY OF MATERIALS:

Neither source nor purity of sulfisoxazole and water were specified. Tween 20 was supplied by Atlas-Goldschmidt A.G., Essen (purity not specified).

ESTIMATED ERROR:

Nothing specified

REFERENCES:

1. Khawam, M.N.; Tawashi, R.; Czetsch-Lindenwald, H. v. *Sci. Pharm.* 1965, *33*, 90.

COMPONENTS:

(1) Benzenesulfonamide, 4-amino-N-(3,4-dimethyl-5-isoxazolyl)- (sulfisoxazole); $C_{11}H_{13}N_3O_3S$; [127-69-5]

(2) Methanol; CH_4O; [67-56-1]

ORIGINAL MEASUREMENTS:

Mauger, J.W. ; Petersen, H. Jr.; Alexander, K. S.; Paruta, A. N.

Drug Dev. Ind. Pharm. 1977, *3(2)*, 163-83.

VARIABLES:

Temperature

PREPARED BY:

R. Piekos

EXPERIMENTAL VALUES:

t/°C	Solubility		
	mg/ml	10^3X [a]	mol dm^{-3} [b]
25	49.4	7.52	0.184
30	56.0	8.57	0.209
37	67.9	10.40	0.254

[a] X = mole fraction

[b] calculated by compiler

AUXILIARY INFORMATION

METHOD/APPARATUS/PROCEDURE:

A const temp bath contg screw-capped bottles with sulfisoxazole in excess and methanol was rotated for 24 h. Samples were withdrawn through a pledget of glass wool into a pipet, which was wiped clean and allowed to drain into a volumetric flask. Soly was detd from absorbance and previously ascertained Beer's law plots detd on a Cary model 16 spectrophotometer (1).

SOURCE AND PURITY OF MATERIALS:

Sulfisoxazole: lot 378067, Hoffman-LaRoche. Its mp agreed with the literature value. Methanol was spectrograde solvent from Mallinckrodt Chemical Works.

ESTIMATED ERROR:

Soly: av. of at least 3 detns is reported (authors).

Temp: ±0.1°C (authors).

REFERENCES:

1. Mauger, J. W.; Paruta, A. N.; Gerraughty, R. J. *J. Pharm. Sci.* 1972, *61(1)*, 94.

COMPONENTS:

(1) Benzenesulfonamide, 4-amino-N-(3,4-dimethyl-5-isoxazolyl)- (sulfisoxazole); $C_{11}H_{13}N_3O_3S$; [127-69-5]

(2) Ethanol; C_2H_6O; [64-17-5]

ORIGINAL MEASUREMENTS:

Mauger, J. W.; Petersen, H. Jr.; Alexander, K. S.; Paruta, A. N.; *Drug Dev. Ind. Pharm.* 1977, *3(2)*, 163-83.

VARIABLES:

Temperature

PREPARED BY:

R. Piekos

EXPERIMENTAL VALUES:

t/°C	Solubility		
	mg/ml	10^3X [a]	10^2 mol dm^{-3} [b]
25	19.1	4.18	7.14
30	22.6	4.99	8.45
37	26.6	5.90	9.95

[a] X = mole fraction

[b] calculated by compiler

AUXILIARY INFORMATION

METHOD/APPARATUS/PROCEDURE:

A const temp bath contg screw-capped bottles with sulfisoxazole in excess and ethanol was rotated for 24 h. Samples were withdrawn through a pledget of glass wool into a pipet, which was wiped clean and allowed to drain into a volumetric flask. Soly was detd from absorbance and previously ascertained Beer's law plots detd on a Cary model 16 spectrophotometer (1).

SOURCE AND PURITY OF MATERIALS:

Sulfisoxazole: lot 378067, Hoffman-LaRoche, Inc. Its mp agreed with the literature value. Ethanol was from the U.S. Industrial Chemical Co. Its refractive index value and density agreed with literature values.

ESTIMATED ERROR:

Soly: av of at least 3 detns is reported (authors).

Temp: ±0.1°C (authors).

REFERENCES:

1. Mauger, J. W.; Paruta, A.N.; Gerraughty, R. J. *J. Pharm. Sci.* 1972, *61(1)*, 94.

COMPONENTS:

(1) Benzenesulfonamide, 4-amino-N-(3,4-dimethyl-5-isoxazolyl)- (sulfisoxazole); $C_{11}H_{13}N_3O_3S$; [127-69-5]

(2) Ethanol; C_2H_6O; [64-17-5]

ORIGINAL MEASUREMENTS:

Sekikawa, H.; Nakano, M.; Arita, T.

Chem. Pharm. Bull. 1978, *26(1)*, 118-26.

VARIABLES:

Temperature

PREPARED BY:

R. Piekos

EXPERIMENTAL VALUES:

t/°C	Solubility[a] 10^2 mol dm^{-3} solution
10	4.43
20	5.98
30	7.86
40	11.0
50	15.2

[a]Original data are presented graphically. The numerical data are given by the authors.

AUXILIARY INFORMATION

METHOD/APPARATUS/PROCEDURE:

After attaining equilibrium, sample solns were removed by a syringe and filtered quickly through a membrane filter (pore size 0.2 μ) and sulfisoxazole was assayed spectrophotometrically at 269 nm using a Hitachi Type 200-20 spectrophotometer.

SOURCE AND PURITY OF MATERIALS:

Sulfisoxazole (Yamanouchi Pharmaceutical Co.) was of the Japanese Pharmacopeia IX grade. Abs EtOH was obtained by drying and distn of EtOH following the conventional procedures.

ESTIMATED ERROR:

Nothing specified

REFERENCES:

COMPONENTS:

(1) Benzenesulfonamide, 4-amino-N-(3,4-dimethyl-5-isoxazolyl)-(sulfisoxazole); $C_{11}H_{13}N_3O_3S$; [127-69-5]

(2) 1-Propanol; C_3H_8O; [71-23-8]

ORIGINAL MEASUREMENTS:

Mauger, J. W.; Petersen, H., Jr.; Alexander, K. S.; Paruta, A. N. *Drug Dev. Ind. Pharm.* 1977, *3(2)*, 163-83.

VARIABLES:

Temperature

PREPARED BY:

R. Piekos

EXPERIMENTAL VALUES:

$t/^{o}C$	Solubility		
	mg/ml	10^3X [a]	10^2 mol dm^{-3} [b]
25	7.95	2.23	2.97
30	9.53	2.69	3.56
37	12.2	3.44	4.56

[a] X = mole fraction

[b] calculated by compiler

AUXILIARY INFORMATION

METHOD/APPARATUS/PROCEDURE:

A const temp bath contg screw-capped bottles with sulfisoxazole in excess and 1-propanol was rotated for 24 h. Samples were withdrawn through a pledget of glass wool into a pipet, which was wiped clean and allowed to drain into a volumetric flask. Soly was detd from absorbance and previously ascertained Beer's law plots detd on a Cary model 16 spectrophotometer (1).

SOURCE AND PURITY OF MATERIALS:

Sulfisoxazole: lot 378067, Hoffman-LaRoche, Inc. Its mp agreed with the literature value. 1-Propanol was Baker Analyzed Reagent (J.T. Baker Chemical Co.). Its refractive index value and density agreed with literature values.

ESTIMATED ERROR:

Soly: av. of at least 3 detns is reported (authors).

Temp: $\pm 0.1^{o}C$

REFERENCES:

1. Mauger, J.W.; Paruta, A.N. Gerraughty, R.J. *J. Pharm. Sci.* 1972, *61(1)*, 94.

COMPONENTS:

(1) Benzenesulfonamide, 4-amino-N-(3,4-dimethyl-5-isoxazolyl)- (sulfisoxazole); $C_{11}H_{13}N_3O_3S$; [127-69-5]

(2) 1-Butanol; $C_4H_{10}O$; [71-36-3]

ORIGINAL MEASUREMENTS:

Mauger, J. W.; Petersen, H.,Jr.; Alexander, K. S.; Paruta, A. N., *Drug Dev. Ind. Pharm.* 1977 *3(2)*, 163-83.

VARIABLES:

Temperature

PREPARED BY:

R. Piekos

EXPERIMENTAL VALUES:

	SOLUBILITY		
t/°C	mg/ml	$10^3 X$[a]	10^2 mol dm^{-3} [b]
25	4.31	1.48	1.61
30	5.30	1.83	1.98
37	6.53	2.26	2.44

[a] X = mole fraction

[b] calculated by compiler

AUXILIARY INFORMATION

METHOD/APPARATUS/PROCEDURE:

Screw-capped bottles with sulfisoxazole and BuOH were rotated in a const temp bath for 24 h. Samples were withdrawn through a pledget of glass wool into a pipet, which was wiped clean and allowed to drain into a volumetric flask. Soly was detd from absorbance and previously ascertained Beer's law plots detd on a Cary Model 16 spectrophotometer (1).

SOURCE AND PURITY OF MATERIALS:

Sulfisoxazole: lot 378067, Hoffman-LaRoche, Inc. M.p. agreed with literature values. 1-Butanol was purchased from Mallinckrodt Chem Works. Refractive index value and density agreed with literature values.

ESTIMATED ERROR:

Temp: ±0.1°C (authors).

Soly: an average of at least 3 detns is reported (authors).

REFERENCES:

1. Paruta, A. N.; Mauger, J. W.; Gerraughty, R. J., *J. Pharm. Sci.* 1972, *61*, 94.

COMPONENTS:

(1) Benzenesulfonamide, 4-amino-N-(3,4-dimethyl-5-isoxazolyl)- (sulfisoxazole); $C_{11}H_{13}N_3O_3S$; [127-69-5]

(2) 1-Pentanol; $C_5H_{12}O$; [71-41-0]

ORIGINAL MEASUREMENTS:

Mauger, J. W.; Petersen, H., Jr.; Alexander, K. S.; Paruta, A. N.
Drug Dev. Ind. Pharm. 1977, *3(2)*, 163-83.

VARIABLES:

Temperature

PREPARED BY:

R. Piekos

EXPERIMENTAL VALUES:

	SOLUBILITY		
t/°C	mg/ml	10^3 X [a]	10^2 mol dm^{-3} [b]
25	2.61	1.06	0.98
30	3.20	1.30	1.20
37	3.95	1.62	1.48

[a] X = mole fraction

[b] calculated by compiler

AUXILIARY INFORMATION

METHOD/APPARATUS/PROCEDURE:

The soly was detd by the method of Paruta et al. (1): Screw-capped bottles with sulfisoxazole in excess and 1-pentanol were rotated in a const temp bath for 24 h. Samples were withdrawn through a pledget of glass wool into a pipet, which was wiped clean and allowed to drain into a volumetric flask. Soly was detd from absorbance and previously ascertained Beer's law plots detd on a Cary Model 16 spectrophotometer.

SOURCE AND PURITY OF MATERIALS:

Sulfisoxazole: lot 378067, Hoffman-LaRoche, Inc. M.p. agreed with literature values. 1-Pentanol was purchased from Fisher Scientific Co. Refractive index valu and density agreed with literature values.

ESTIMATED ERROR:

Temp: ±0.1°C

Soly: Not specified.

REFERENCES:

1. Paruta, A. N.; Mauger, J. W.; Gerraughty, R. J. *J. Pharm. Sci.* 1972, *61*, 94.

COMPONENTS:

(1) Benzenesulfonamide, 4-amino-N-(3,4-dimethyl-5-isoxazolyl)- (sulfisoxazole); $C_{11}H_{13}N_3O_3S$; [127-69-5]

(2) 1-Octanol; $C_8H_{16}O$; [111-87-5]

ORIGINAL MEASUREMENTS:

Mauger, J. W.; Petersen, H. Jr.; Alexander, K. S.; Paruta, A. N. *Drug Dev. Ind. Pharm.* 1977, *3(2)*, 163-83.

VARIABLES:

Temperature

PREPARED BY:

R. Piekos

EXPERIMENTAL VALUES:

SOLUBILITY

t/°C	mg/ml	10^3 mol dm^{-3} [b]	10^3 X [a]
25	0.94	3.52	0.55
30	1.17	4.38	0.69
37	1.40	5.24	0.83

[a] X = mole fraction

[b] calculated by compiler

AUXILIARY INFORMATION

METHOD/APPARATUS/PROCEDURE:

The soly was detd by the method of Paruta et al. (1): Screw-capped bottles with sulfisoxazole in excess and 1-Octanol were rotated in a const temp bath for 24 h. Samples were withdrawn through a pledget of glass wool into a pipet, which was wiped clean and allowed to drain into a volumetric flask. Soly was detd from absorbance and previously ascertained Beer's law plots detd on a Cary Model 16 spectrophotometer.

SOURCE AND PURITY OF MATERIALS:

Sulfisoxazole: lot 378067, Hoffman-LaRoche, Inc. M.p. agreed with the literature values. 1-Octanol was purchased from Fisher Scientific Co. Refractive index value and density agreed with literature values.

ESTIMATED ERROR:

Temp: ±0.1°C

Soly: not specified.

REFERENCES:

1. Paruta, A. N.; Mauger, J. W.; Gerraughty, R. J. *J. Pharm. Sci.* 1972, *61*, 94.

COMPONENTS:

(1) Benzenesulfonamide, 4-amino-N-(3,4-dimethyl-5-isoxazolyl)- (sulfisoxazole); $C_{11}H_{13}N_3O_3S$; [127-69-5]

(2) 1-Decanol; $C_{10}H_{22}O$; [112-30-1]

ORIGINAL MEASUREMENTS:

Mauger, J. W.; Petersen, H. Jr.; Alexander, K. S.; Paruta, A. N., *Drug, Dev. Ind. Pharm.* 1977, *3(2)*, 163-83.

VARIABLES:

Temperature

PREPARED BY:

R. Piekos

EXPERIMENTAL VALUES:

SOLUBILITY

t/°C	mg/ml	10^3 X [a]	10^3 mol dm^{-3} [b]
25	0.57	0.41	2.13
30	0.68	0.49	2.54
37	0.85	0.61	3.18

[a] X = mole fraction

b calculated by compiler

AUXILIARY INFORMATION

METHOD/APPARATUS/PROCEDURE:

The soly was detd by the method of Paruta et al. (1): Screw-capped bottles with sulfisoxazole in excess and 1-decanol were rotated in a const temp bath for 24 h. Samples were withdrawn through a pledget of glass wool into a pipet, which was wiped clean and allowed to drain into a volumetric flask. Soly was detd from absorbance and previously ascertained Beer's law plots detd on a Cary Model 16 spectrophotometer.

SOURCE AND PURITY OF MATERIALS:

Sulfisoxazole: lot 378067. Hoffman-LaRoche, Inc. M.p. agreed with that of literature. 1-decanol was purchased from Matheson, Coleman and Bell. Refractive index value and density agreed with those reported in the literature.

ESTIMATED ERROR:

Temp: ±0.1°C (authors).

Soly: not specified.

REFERENCES:

1. Paruta, A. N.; Mauger, J. W.; Gerraughty, R. J. *J. Pharm. Sci.* 1972, *61*, 94.

COMPONENTS:	ORIGINAL MEASUREMENTS:
(1) Benzenesulfonamide, 4-amino-N-(3,4-dimethyl-5-isoxazolyl)- (sulfisoxazole); $C_{11}H_{13}N_3O_3S$; [127-69-5] (2) Ethanol, 2-ethoxy-; $C_4H_{10}O_2$; [110-80-5]	Sunwoo, C.; Eisen, H. *J. Pharm. Sci.* 1971, *60*, 238-44.
VARIABLES:	PREPARED BY:
One temperature	R. Piekos

EXPERIMENTAL VALUES:

The mole fraction solubility of sulfisoxazole in 2-ethoxyethanol at 25°C is 0.0495 (13.4 g/100 g solution, compiler).

AUXILIARY INFORMATION

METHOD/APPARATUS/PROCEDURE:

Soly was detd by the method reported by Restaino and Martin. Sulfisoxazole was assayed on a Coleman-Hitachi 124 double-beam spectrophotometer at 270 nm after diln of a sample with 95% alcohol or water.

SOURCE AND PURITY OF MATERIALS:

Sulfisoxazole (Hoffman-LaRoche Inc., Nutley, N.J.) was recrystd from warm alcohol. 2-Ethoxyethanol (Cellosolve solvent, Union Carbide, New York, N.Y.) was of industrial grade.

ESTIMATED ERROR:

Temp: ±1.0°C (authors).

Soly: the mean of 3 runs was given (authors).

REFERENCES:

1. Restaino, F. A.; Martin, A. N. *J. Pharm. Sci.* 1964, *53*, 636.

COMPONENTS:

(1) Benzenesulfonamide, 4-amino-N-(3,4-dimethyl-5-isoxazolyl)- (sulfisoxazole); $C_{11}H_{13}N_3O_3S$; [127-69-5]

(2) 2-Pyrrolidinone, 1-ethenyl-,polymers (poly(vinyl pyrrolidone)); $(C_6H_9NO)_x$; [9003-39-8] K-15

(3) Ethanol; C_2H_6O; [64-17-5]

ORIGINAL MEASUREMENTS:

Sekikawa, H.; Nakano, M.; Arita, T.

Chem. Pharm. Bull. 1978, *26(1)*,118-26.

VARIABLES:

Temperature

PREPARED BY:

R. Piekos

EXPERIMENTAL VALUES:

$t/^{\circ}C$	$M \times 10^2$ sulfisoxazole solubilized by 1M vinyl-pyrrolidone equivalent
10.0	7.52
20.0	8.89
30.0	10.4
40.0	12.5
50.0	14.5

AUXILIARY INFORMATION

METHOD/APPARATUS/PROCEDURE:

After attaining equilibrium, sample solns were removed by a syringe and filtered quickly through a membrane filter (pore size 0.2 μ) and sulfisoxazole was assayed spectrophotometrically at 269 nm using a Hitachi Type 200-20 spectrophotometer. No significant absorbance was found for poly(vinyl pyrrolidone).

SOURCE AND PURITY OF MATERIALS:

Sulfisoxazole (Yamanouchi Pharmaceutical Co.) was of the Japanese Pharmacopeia IX grade. Poly(vinyl pyrrolidone) K-15 was from Daiichi Pure Chemicals Co., Tokyo. Abs EtOH was obtained by drying and distn of EtOH following the conventional procedures.

ESTIMATED ERROR:

Nothing specified

REFERENCES:

COMPONENTS:

(1) Benzenesulfonamide, 4-amino-N-(3,4-dimethyl-5-isoxazolyl)- (sulfisoxazole); $C_{11}H_{13}N_3O_3S$; [127-69-5]

(2) Acetic acid, ethyl ester (ethyl acetate); $C_4H_8O_2$; [141-78-6]

ORIGINAL MEASUREMENTS:

Grady, L.T.; Hays, S.E.; King, R.H.; Klein, H.R.; Mader, W.J.; Wyatt, D.K.; Zimmerer, R.O., Jr. *J. Pharm. Sci.* 1973, *62(3)*, 456-64.

VARIABLES:

One temperature: 25°C

PREPARED BY:

R. Piekos

EXPERIMENTAL VALUES:

Solubility of sulfisoxazole in ethyl acetate at 25°C is 15.4 mg/g (5.76×10^{-2} mol kg^{-1}, compiler).

AUXILIARY INFORMATION

METHOD/APPARATUS/PROCEDURE:

The phase solubility method was employed (1).

SOURCE AND PURITY OF MATERIALS:

Sulfisoxazole contained 0.14% impurities and produced two spots on a thin-layer chromatogram.

Purity of the ethyl acetate was not specified.

ESTIMATED ERROR:

Soly: ±0.3 mg/g (authors).

Temp: not specified.

REFERENCES:

1. *The National Formulary, 13th Ed., Mack Publishing Co., Easton, Pa.,* 1970, *pp.* 789 and 839.

COMPONENTS:	ORIGINAL MEASUREMENTS:
(1) Benzenesulfonamide, 4-amino-N-(3,4-dimethyl-5-isoxazolyl)- (sulfisoxazole); $C_{11}H_{13}N_3O_3S$; [127-69-5] (2) Methane, trichloro- (chloroform); $CHCl_3$; [67-66-3]	Riess, W. *Intern. Congr. Chemotheraphy, Proc., 3rd, Stuttgart* 1963, *1*, 627-32.
VARIABLES:	PREPARED BY:
One temperature: 20°C	R. Piekos

EXPERIMENTAL VALUES:

Solubility of sulfisoxazole in chloroform at 20°C is 80 mg% (3.0 x 10^{-3} mol dm^{-3} solution, compiler).

AUXILIARY INFORMATION

METHOD/APPARATUS/PROCEDURE:	SOURCE AND PURITY OF MATERIALS:
Nothing specified	Nothing specified
	ESTIMATED ERROR:
	Nothing specified
	REFERENCES:

COMPONENTS:	ORIGINAL MEASUREMENTS:
(1) Benzenesulfonamide, 4-amino-N-(3,4-dimethyl-5-isoxazolyl)- (sulfisoxazole); $C_{11}H_{13}N_3O_3S$; [127-69-5] (2) Methane, trichloro- (chloroform); $CHCl_3$; [67-66-3]	Yamazaki, M.; Aoki, M.; Kamada, A.; Yata, N. *Yakuzaigaku* 1967, *27(1)*, 37-40.
VARIABLES: One temperature: 30°C	PREPARED BY: R. Piekos

EXPERIMENTAL VALUES:

Solubility of sulfisoxazole in chloroform at 30°C is 2.40 mmol/L (0.641 g dm^{-3}, compiler).

AUXILIARY INFORMATION

METHOD/APPARATUS/PROCEDURE:	SOURCE AND PURITY OF MATERIALS:
Sulfisoxazole (0.5 g) was placed in an L-shaped tube together with 20 ml of chloroform. The mixt was shaken in a thermostat until equilibrium was attained. The sulfisoxazole was assayed in the supernatant spectrophotometrically at 545 nm on a Beckmann DU spectrophotometer. The results were taken from a calibration graph.	Nothing specified ESTIMATED ERROR: Soly: not specified Temp: ±1°C (authors) REFERENCES:

COMPONENTS:

(1) Acetamide,N-[(4-aminophenyl)sulfonyl]-N-(3,4-dimethyl-5-isoxazolyl)- (N^1-acetylsulfisoxazole); $C_{13}H_{15}N_3O_4S$; [80-74-0]
(2) Phosphoric acid, monopotassium salt; KH_2PO_4; [7778-77-0]
(3) Sodium hydroxide; NaOH; [1310-73-2]
(4) Water; H_2O; [7732-18-5]

ORIGINAL MEASUREMENTS:

Muzakami, S.; Nagata, K.
Ann. Rept. Shionogi Res. Lab. 1956, *6*, 58-64.

VARIABLES:

pH

PREPARED BY:

R. Piekos

EXPERIMENTAL VALUES:

pH	Solubility of N^1-acetylsulfisoxazole in Clark and Lubs buffer mixtures at 37°C	
	mg%	10^4 mol dm^{-3} [a]
6.5	3.93	1.27
7.0	3.83	1.24
7.5	4.19	1.35

[a] calculated by compiler

AUXILIARY INFORMATION

METHOD/APPARATUS/PROCEDURE:

About 100 mg of N^1-acetylsulfisoxazole was placed in a flask and 100 cm^3 of the buffer soln of a known pH was added. The mixt was vigorously agitated for 4 h in a water bath at 37.0±0.2°C and filtered through a Toyo No. 6 filter paper, keeping the temp at about 40°C. The first 10 cm^3 of the filtrate was discarded and in the following 15 cm^3, kept still at about 40°C, the sulfonamide concn was detd colorimetrically at 540 nm using a Tsuda reagent and a Beckman Model B spectrophotometer.

SOURCE AND PURITY OF MATERIALS:

N^1-acetylsulfisoxazole was synthesized by the authors and dried over $CaCl_2$.
The source and purity of the remaining materials was not specified.

ESTIMATED ERROR:

Soly: the error was below ±3% (authors).
Temp: ±0.2°C (authors); pH: not specified (a Beckman type G pH meter was used).

REFERENCES:

COMPONENTS:

(1) Acetamide, N-[(4-aminophenyl)sulfonyl]-N-(3,4-dimethyl-5-isoxazolyl)- (N^1-acetylsulfisoxazole); $C_{13}H_{15}N_3O_4S$; [80-74-0]

(2) 1,2-Benzenedicarboxylic acid, monopotassium salt; $C_8H_5KO_4$; [877-24-7]

(3) Sodium hydroxide; NaOH; [1310-73-2]

(4) Water; H_2O; [7732-18-5]

ORIGINAL MEASUREMENTS:

Muzakami, S.; Nagata, K.

Ann. Rept. Shionogi Res. Lab. 1956, *6*, 58-64.

VARIABLES:

pH

PREPARED BY:

R. Piekos

EXPERIMENTAL VALUES:

pH	Solubility of N^1-acetylsulfisoxazole in Clark and Lubs buffer mixtures at 37°C	
	mg%	10^4 mol dm^{-3} [a]
4.0	6.04	1.95
4.5	5.27	1.70
5.0	4.86	1.57
5.5	4.43	1.43

[a] calculated by compiler

AUXILIARY INFORMATION

METHOD/APPARATUS/PROCEDURE:

About 100 mg of N^1-acetylsulfisoxazole was placed in a flask and 100 cm^3 of the buffer soln of a known pH was added. The mixt was vigorously agitated for 4 h in a water bath at 37±0.2°C and filtered through a Toyo No.6 filter paper, keeping the temp at about 40°C. The first 10 cm^3 of the filtrate was discarded and in the following 15 cm^3, kept still at about 40°C, the sulfonamide concn was detd colorimetrically at 540 nm using a Tsuda reasgent and a Beckman Model B spectrophotometer.

SOURCE AND PURITY OF MATERIALS:

N^1-acetylsulfisoxazole was synthesized by the authors and dried over $CaCl_2$.
The source and purity of the remaining materials was not specified.

ESTIMATED ERROR:

Soly: the error was below ±3% (authors).
Temp: ±0.2°C (authors)
pH : not specified (a Beckman type G pH meter was used).

REFERENCES:

COMPONENTS:

(1) Acetamide, N-[4-[[(3,4-dimethyl-5-isoxazolyl)amino]sulfonyl]phenyl]- (acetyl sulfisoxazole); $C_{13}H_{15}N_3O_4S$; [4206-74-0]

(2) Phosphoric acid, disodium salt; Na_2HPO_4; [7558-94-4]

(3) Phosphoric acid, monopotassium salt; KH_2PO_4; [7778-77-0]

(4) Water; H_2O; [7732-18-5]

ORIGINAL MEASUREMENTS:

Bandelin, F. J.; Malesh, W.

J. Am. Pharm. Assoc. Sci. Ed. 1959, *48*, 177-81.

VARIABLES:

pH

PREPARED BY:

R. Piekos

EXPERIMENTAL VALUES:

Solubility of acetyl sulfisoxazole in buffers of varying mixtures of $Na_2HPO_4 \cdot 7H_2O$ (71.6 g/l distilled water; 0.27 mol dm^{-3}, compiler) and KH_2PO_4 (36.3 g/l distilled water; 0.27 mol dm^{-3}, compiler) at 37°C

Equilibrium pH	Solubility (based on sulfisoxazole) mg/100 ml	10^2 mol dm^{-3} [a]
4.5	8	0.030
5.0	12	0.045
5.5	38	0.140
6.0	105	0.393
6.4	190	0.711
6.8	375	1.400
7.2	1040	3.891

[a] calculated by compiler

AUXILIARY INFORMATION

METHOD/APPARATUS/PROCEDURE:

Solns were prepd by adding an excess of acetyl sulfisoxazole to a 10 ml of buffer soln at each pH level in 18 x 150-mm test tubes, stoppering the tubes, and placing them in water bath at 37°C with gentle agitation for 24 h. The solute was then hydrolyzed with 5% H_2SO_4 for 1 h to liberate the free sulfonamide. One-ml aliquot of the hydrolyzate was accurately pipetted into a volumetric flask for diln and analysis. The sulfonamide was assayed colorimetrically by the method of Bratton and Marshall as described in detail by Biamonte and Schneller (1). A standard curve was prepd using accurately prepd standard solutions.

SOURCE AND PURITY OF MATERIALS:

Neither source nor purity of the reagents were specified. Distilled water was used.

ESTIMATED ERROR:

Soly: av values of duplicate runs are reported (authors).
Temp and pH: not specified

REFERENCES:

1. Biamonte, A. R.; Schneller, G. E. *J. Am. Pharm. Assoc., Sci. Ed.* 1952, *41*, 341.

COMPONENTS:

(1) Acetamide, N-[4-[[(3,4-dimethyl-5-isoxazolyl)amino]sulfonyl]phenyl]- (N^4-acetylsulfafurazole)*
$C_{13}H_{15}N_3O_4S$; [4206-74-0]

(2) Phosphoric acid, disodium salt; Na_2HPO_4; [7558-94-4]

(3) Phosphoric acid, monopotassium salt; KH_2PO_4; [7778-77-0]

(4) Water; H_2O; [7732-18-5]

ORIGINAL MEASUREMENTS:

Hekster, Ch. A.; Vree, T. B.

Antibiotics Chemother. 1982, 31, 22-118.

VARIABLES:

pH

PREPARED BY:

R. Piekos

EXPERIMENTAL VALUES:

pH	Solubility at 25°C	
	mg/l	mol dm^{-3} [a]
5.5	250	8.08×10^{-4}
7.5[b]	6,893	2.228×10^{-2}

[a]Calculated by compiler

[b]Erroneous pH value of 7.0 is given in the article

*Another common trivial name is acetyl sulfisoxazole.

AUXILIARY INFORMATION

METHOD/APPARATUS/PROCEDURE:

The earlier developed method (1) was used (personal communication). Satd solns of N^4-acetylsulfafurazole* were prepd in phosphate buffers of pH 5.5 and 7.5 at 25°C. The concn of the solute was measured by means of a Spectra Physics 3500B high-performance liquid chromatograph equipped with a Model 748 column oven and a Pye-Unicam LC-UV spectrophotometric detector.

SOURCE AND PURITY OF MATERIALS:

Neither source nor the purity of the materials was specified.

ESTIMATED ERROR:

Soly: the detection limit of the solute by HPLC was 0.5 mg/l (authors). The errors in temp and pH were not specified.

REFERENCES:

1. Hekster, Y.A. Vree, T. B.; Damsma, J. E.; Friesen, W. T. *J. Antimicrob. Chemother.* 1981, 8, 133.

COMPONENTS:

(1) Acetamide, N-[4-[[(3,4-dimethyl-5-isoxazolyl)amino]sulfonyl]phenyl]- (acetyl sulfafurazole)*; $C_{13}H_{15}N_3O_4S$; [4206-74-0]

(2) Phosphoric acid, disodium salt; Na_2HPO_4; [7558-94-4]

(3) 1,2,3-Propanetricarboxylic acid, 2-hydroxy- (citric acid); $C_6H_8O_7$; [77-92-9]

(4) Water; H_2O;]7732-18-5]

ORIGINAL MEASUREMENTS:

Biamonte, A. R.; Schneller, G. H.
J. Am. Pharm. Assoc. Sci. Ed.
1952, *41*, 341-5.

VARIABLES:

pH

PREPARED BY:

R. Piekos

EXPERIMENTAL VALUES:

Solubility of acetyl sulfafurazole* in McIlvaine's disodium phosphate-citric acid buffer solution at 37°C

Initial pH of buffer	Solubility		Final pH
	mg/100 ml solution	10^3 mol dm^{-3} [a]	
4.5	6.0	0.19	4.5
5.0	17.3	0.56	5.0
6.0	126.1	4.08	6.0
7.0	757.9	24.50	6.7

[a] Calculated by compiler

*Another common trivial name is acetyl sulfisoxazole.

AUXILIARY INFORMATION

METHOD/APPARATUS/PROCEDURE:

Acetyl sulfafurazole* was equilibrated with 50 ml of the buffer soln for 18 h at 37° C with agitation. The suspension was immediately filtered through a Whatman No. 1 paper. The compd was assayed in the filtrate after boiling for 15-20 min with 30% NaOH soln as sulfafurazole* by the method of Bratton and Marshall (1) using a Beckman DU spectrophotometer, at 545 nm.

SOURCE AND PURITY OF MATERIALS:

Acetyl sulfafurazole*, mp 214.8-15.9°C, was supplied by the American Cyanamid Co, Calco Chem Div, Bound Brook, N.J. The source and purity of the remaining materials was not specified.

ESTIMATED ERROR: pH and temp: not specified. Accuracy of the anal method was illustrated by the following values: expected 2.003, 3.004, 4.006, 5.007 mg/100 ml; found 2.08; 3.06, 4.12, 5.10 resp.

REFERENCES:

1. Bratton, A. C.; Marshall, E. K. Jr. *J. Biol. Chem.* 1939, *128*, 537.

COMPONENTS:

(1) Acetamide, N-[4-[[(3,4-dimethyl-5-isoxazolyl)amino]sulfonyl]phenyl]- (acetyl sulfisoxazole); $C_{13}H_{15}N_3O_4S$; [4206-74-0]

(2) Calcium chloride; $CaCl_2$; [10043-52-4]

(3) Magnesium chloride; $MgCl_2$; [7786-30-3]

(4) Phosphoric acid, monoammonium salt; $NH_4H_2PO_4$; [7722-76-1]

(5) Potassium chloride; KCl; [7447-40-7]

(6) Sodium chloride; NaCl; [7647-15-5]

(7) Urea; CH_4N_2O; [57-13-6]

(8) Water; H_2O; [7732-18-5]

ORIGINAL MEASUREMENTS:

Bandelin, F. J.; Malesh, W.

J. Am. Pharm. Assoc., Sci. Ed. 1959, *48*, 177-81.

VARIABLES:

pH at 37° C

PREPARED BY:

R. Piekos

EXPERIMENTAL VALUES:

Solubility of acetyl sulfisoxazole in a solution containing $CaCl_2$ 0.143, $MgCl_2$ 0.121, $NH_4H_2PO_4$ 0.300, KCl 1.660, NaCl 2.950 and urea 20 g/dm^3 (synthetic urine, Mosher Vehicle) at 37°C

Equilibrium pH	Solubility mg/100 ml as sulfisoxazole	Solubility 10^2 mol dm^{-3} [a]
4.5	30	0.097
5.0	44	0.140
5.5	70	0.230
6.0	160	0.520
6.5	560	1.810
7.0	1230	3.980

[a] calculated by compiler

AUXILIARY INFORMATION

METHOD/APPARATUS/PROCEDURE:

Excess acetyl sulfisoxazole was added to aliquots of synthetic urine solns and 1% H_3PO_4 or 1% NaOH solns were used to adjust the pH to the required value. The solns were agitated for 24 h with addn of acid or base to keep them at the desired pH level until equilibrium was attained. Then the solns were filtered and in aliquots the acetyl sulfonamide was assayed spectrophotometrically by the method described by Biamonte and Schneller (1). Before detn the soln was refluxed with 5% H_2SO_4 for 1 h to liberate the free amino compound.

SOURCE AND PURITY OF MATERIALS:

Nothing specified

ESTIMATED ERROR:

Soly: average values of 2 detns were given.
Temp: not specified.
pH : not specified.

REFERENCES:

1. Biamonte, A. R.; Schneller, G. E., *J. Am. Pharm. Assoc., Sci. Ed.* 1952, *41*, 341.

COMPONENTS:

(1) Acetamide, N -[4-[[(3,4-dimethyl-5-isoxazolyl)amino[sulfonyl]phenyl]-(acetyl sulfisoxazole); $C_{13}H_{15}N_3O_4S$; [4206-74-0]

(2) Methane, trichloro- (chloroform); $CHCl_3$; [67-66-3]

ORIGINAL MEASUREMENTS:

Grady, L.T.; Hays, S.E.; King, R.H.; Klein, H. R.; Mader, W. J.; Wyatt, D. K. ; Zimmerer, R. O., Jr.

J. Pharm. Sci. 1973, *62(3)*, 456-64.

VARIABLES:

One temperature: 25°C

PREPARED BY:

R. Piekos

EXPERIMENTAL VALUES:

Solubility of acetyl sulfisoxazole in chloroform at 25°C is 15.5 mg/g

(5.01×10^{-2} mol kg^{-1}, compiler).

AUXILIARY INFORMATION

METHOD/APPARATUS/PROCEDURE:

The phase solubility method was employed (1).

SOURCE AND PURITY OF MATERIALS:

Acetyl sulfisoxazole contained 0.02% impurities and produced one spot in a thin-layer chromatogram.

Source and purity of the chloroform was not specified.

ESTIMATED ERROR:

Soly: ±0.3 mg/g (authors)

Temp: not specified.

REFERENCES:

1. *The National Formulary, 13th Ed., Mack Publishing Co., Easton, Pa.,* 1970, *pp.* 789 *and* 839.

COMPONENTS:

(1) Acetamide, N-[[4-(acetylamino)phenyl]-sulfonyl]-N-(3,4-dimethyl-5-isoxazolyl)-(N^1, N^4-diacetylsulfafurazole)*; $C_{15}H_{17}N_3O_5S$; [35943-12-5]

(2) Phosphoric acid, disodium salt; Na_2HPO_4; [7558-94-4]

(3) Phosphoric acid, monopotassium salt; KH_2PO_4; [7778-77-0]

(4) Water; H_2O; [7732-18-5]

VARIABLES:

pH

ORIGINAL MEASUREMENTS:

Hekster, Ch. A.; Vree, T. B.

Antibiotics Chemother. 1982, *31*, 22-118.

PREPARED BY:

R. Piekos

EXPERIMENTAL VALUES:

pH	Solubility at 25°C mg/l	10^5 mol dm^{-3} [a]
5.5	12.3	3.50
7.5[b]	5.2	1.50

[a] Calculated by compiler

[b] Erroneous pH value of 7.0 is given in the article

* Another common trivial name is N^1,N^4-diacetylsulfisoxazole

AUXILIARY INFORMATION

METHOD/APPARATUS/PROCEDURE:

The earlier developed method (1) was used (personal communication). Satd solns of N^1,N^4-diacetylsulfafurazole* were prepd in phosphate buffers of pH 5.5 and 7.5 at 25°C. The concn of the solute was measured by means of a Spectra Physics 3500B high-performance liquid chromatograph equipped with a Model 748 column oven and a Pye-Unicam LC-UV spectrophotometric detector.

SOURCE AND PURITY OF MATERIALS:

Neither source nor the purity of the materials was specified.

ESTIMATED ERROR:

Soly: the detection limit of the solute by HPLC was 0.5 mg/l (authors). The errors in temp and pH were not specified.

REFERENCES:

1. Hekster, Y.A.; Vree, T. B.; Damsma, J.E.; Friesen, W. T.; *J. Antimicrob. Chemother.* 1981, *8*, 133.

COMPONENTS:	ORIGINAL MEASUREMENTS:
(1) Benzenesulfonamide, 4-amino-N-(2,5-dimethyl-2,3-dihydroisoxazolyl)- $C_{11}H_{13}N_3O_3S$; [51543-32-9] (2) Water; H_2O; [7732-18-5]	Kitao, K.; Kubo, K.; Morishita, T.; Yata, N. ; Kamada, A. *Chem. Pharm. Bull.* 1973, *21*, 2417-26.
VARIABLES: One temperature: 37°C	PREPARED BY: R. Piekos

EXPERIMENTAL VALUES:

Solubility of 4-amino-N-(2,5-dimethyl-2,3-dihydroisoxazolyl)benzenesulfonamide in water at 37°C is 7.59 mmol dm^{-3} solution.

AUXILIARY INFORMATION

METHOD/APPARATUS/PROCEDURE:	SOURCE AND PURITY OF MATERIALS:
The sulfonamide was detd in the aq soln (pH 6) by diazotization. No details were given.	The sulfonamide was synthesized by the authors. its purity was not specified. Deionized water was used.
	ESTIMATED ERROR: Soly: not specified. Temp: ±1°C (authors).
	REFERENCES:

COMPONENTS:	ORIGINAL MEASUREMENTS:
(1) Benzenesulfonamide, 4-amino-N-(2,5-dimethyl-2,3-dihydroisoxazolyl)-; $C_{11}H_{13}N_3O_3S$; [51543-32-9] (2) Methane, trichloro-; $CHCl_3$; [67-66-3]	Kitao, K.; Kubo, K.; Morishita, T.; Yata, N.; Kamada, A. *Chem. Pharm. Bull.* 1973, *21*, 2417-26.
VARIABLES:	**PREPARED BY:**
One temperture: 37°C	R. Piekos

EXPERIMENTAL VALUES:

Solubility of 4-amino-N-(2,5-dimethyl-2,3-dihydroisoxazolyl)benzenesulfonamide in $CHCl_3$ at 37°C is 29.7 mmol dm^{-3} solution.

AUXILIARY INFORMATION

METHOD/APPARATUS/PROCEDURE:	SOURCE AND PURITY OF MATERIALS:
One ml of the $CHCl_3$ soln of the sulfonamide at equilibrium was taken into a test tube. After evapn of the solvent, the residue was dissolved in 1N HCl, the soln was properly dild with deionized water, and the concn of the sulfonamide was detd by diazotization.	The sulfonamide was synthesized by the authors. Its purity was not specified. Neither source nor purity of the $CHCl_3$ was specified.
	ESTIMATED ERROR: Soly: not specified. Temp: ±1°C (authors).
	REFERENCES:

COMPONENTS:	ORIGINAL MEASUREMENTS:
(1) Benzenesulfonamide, 4-amino-N-2-oxazolyl-; $C_9H_9N_3O_3S$; [17103-51-4] (2) Water; H_2O; [7732-18-5]	Anderson, G. W.; Faith, H.E.; Marson, H.W.; Winnek, P. S.; Roblin, R. O. Jr. *J. Am. Chem. Soc.* 1942, *64*, 2902-5.
VARIABLES:	PREPARED BY:
One temperature: 37°C	R. Piekos

EXPERIMENTAL VALUES:

Solubility of 4-amino-N-2-oxazolylbenzenesulfonamide in water at 37°C is 282 mg/100 cm^3 solution (1.18×10^{-2} mol dm^{-3}, compiler).

AUXILIARY INFORMATION

METHOD/APPARATUS/PROCEDURE:

Excess sulfonamide in water was heated and stirred on a steam bath for 30 min. The suspension was then agitated for 24 h in a thermostat. A sample of the satd soln was withdrawn through a glass filter, dild, and analyzed by the Marshall method (1) using a General Electric recording spectrophotometer for comparing the colors developed with those of the standards.

SOURCE AND PURITY OF MATERIALS:

The sulfonamide, mp 175-6°C (cor), was prepd by the authors. Anal. %C 45.0 (calcd 45.2); %H 3.9 (3.8); %N 17.6 (17.6).

Purity of the water was not specified.

ESTIMATED ERROR:

Nothing specified

REFERENCES:

1. Bratton, A.C.; Marshall, E. K.,Jr. *J. Pharmacol.* 1939, *66*, 4.

COMPONENTS:

(1) Benzenesulfonamide, 4-amino-N-(4,5-dimethyl-2-oxazolyl)-; $C_{11}H_{13}N_3O_3S$; [729-99-7]
(2) Hydrochloric acid: HCl; [7647-01-0]
(3) Sodium chloride; NaCl; [7647-14-5]
(4) Water; H_2O; [7732-18-5]

ORIGINAL MEASUREMENTS:

Hamlin, W.E.; Northam, J.J.; Wagner, J.G.

J. Pharm. Sci. 1965, *54*, 1651-3.

VARIABLES:

One temperature: 37°C

PREPARED BY:

R. Piekos

EXPERIMENTAL VALUES:

Solubility of 4-amino-N-(4,5-dimethyl-2-oxazolyl)benzenesulfonamide in a 0.05 N HCl (ionic strength 0.1 with NaCl; pH 1.3) solution at 37°C is 6.10 mg/ml solution (2.28 x 10^{-2} mol dm^{-3}, compiler).

AUXILIARY INFORMATION

METHOD/APPARATUS/PROCEDURE:

Excess powd compd was equilibrated in a thermostat by rotating a vial contg the suspension for at least 48 h. The soln was filtered from excess solids at 37°C. The filtrate, after appropriate diln, was assayed spectrophotometrically.

SOURCE AND PURITY OF MATERIALS:

The sulfonamide was a brand of Normark-Werke GmbH, Hamburg, Germany. Its purity was not specified.
Purity of the remaining materials was not specified.

ESTIMATED ERROR:

Soly: the average of 2 or more detns is given (authors).
Temp: not specified.

REFERENCES:

COMPONENTS:	ORIGINAL MEASUREMENTS:
(1) Benzenesulfonamide, 4-amino-N-(4,5-dimethyl-2-oxazolyl)- (sulfuno); $C_{11}H_{13}N_3O_3S$; [729-99-7] (2) Phosphoric acid, disodium salt; Na_2HPO_4; [7558-94-4] (3) Phosphoric acid, monopotassium salt; KH_2PO_4; [7778-77-0] (4) Water; H_2O; [7732-18-5]	Kuhnert-Brandstätter, M.; Martinek, A. *Microchim. Ichnoanal. Acta.* <u>1956</u>, 909-19.
VARIABLES: pH	PREPARED BY: R. Piekos

EXPERIMENTAL VALUES:

pH	Solubility of sulfuno in a 0.066 M phosphate buffer (according to Sørensen) at 20°C			
	Crystalline form I		Crystalline form II	
	mg%	10^3 mol dm^{-3} [a]	mg%	10^3 mol dm^{-3} [a]
6.0	96.1	3.595	87.6	3.277
7.3	167.7	6.274	145.6	5.447

[a] Calculated by compiler

AUXILIARY INFORMATION

METHOD/APPARATUS/PROCEDURE:

Sulfuno and the buffer soln were placed in a polyethylene vessel, agitated for 3.5 h under exclusion of oxygen, filtered, and the sulfonamide was assayed in the filtrate by uv spectrophometry. The solid phase was examd for identity of the cryst form.

SOURCE AND PURITY OF MATERIALS:

A comm available form II of sulfuno was used. Form I was obtained by recrystn of form II from 2-propanol. Distilled water was used. The source and purity of the remaining materials was not specified.

ESTIMATED ERROR:

Soly: not specified
pH : not specified
Temp: ±0.5°C (authors)

REFERENCES:

COMPONENTS:

(1) Benzenesulfonamide, 4-amino-N-(4,5-dimethyl-2-oxazolyl)- (sulfuno); $C_{11}H_{13}N_3O_3S$; [729-99-7]

(2) Hydrochloric acid; HCl; [7647-01-0]

(3) 1,2,3-Propanetricarboxylic acid, 2-hydroxy-, disodium salt; $C_6H_6Na_2O_7$; [144-33-2]

(4) Water; H_2O; [7732-18-5]

ORIGINAL MEASUREMENTS:

Kuhnert-Brandstätter, M.; Martinek, A.

Microchim. Ichnoanal. Acta 1956, 909-19.

VARIABLES:

One temperature: 20°C; one pH: 3.8

PREPARED BY:

R. Piekos

EXPERIMENTAL VALUES:

Solubility of crystalline forms I and II of sulfuno in a 0.066 M citrate buffer (according to Sørensen) of pH 3.8 at 20°C is 91.7 mg% (3.40 x 10^{-3} mol dm^{-3}, compiler) and 84.6 mg% (3.16 x 10^{-3} mol dm^{-3}, compiler), respectiviely.

AUXILIARY INFORMATION

METHOD/APPARATUS/PROCEDURE:

Sulfuno and the buffer soln were placed in a polyethylene vessel, agitated for 3.5 h under exclusion of oxygen, filtered, and the sulfonamide was assayed in the filtrate by uv spectrophotometry. The solid phase was examd thermomicroscopically for identity of the cryst form.

SOURCE AND PURITY OF MATERIALS:

A comm available form II of sulfuno was used. Form I was obtained by recrystn of form II form 2-propanol. Distilled water was used. The source and purity of the remaining materials was not specified.

ESTIMATED ERROR:

Soly: not specified
pH : not specified
Temp: ±0.5°C (authors)

REFERENCES:

COMPONENTS:

(1) Benzenesulfonamide, 4-amino-N-(5-methyl-1,2,4-oxadiazol-3-yl)-; $C_9H_{10}N_4O_3S$; [723-47-7]

(2) Water; H_2O; [7732-18-5]

ORIGINAL MEASUREMENTS:

Anderson, G. W.; Faith, H.E.; Marson, H.W.; Winnek, P. S.; Roblin, R. O., Jr.
J. Am. Chem. Soc. 1942, *64*, 2902-5.

VARIABLES:

One temperature: 37°C

PREPARED BY:

R. Piekos

EXPERIMENTAL VALUES:

Solubility of 4-amino-N-(5-methyl-1,2,4-oxadiazol-3-yl)benzenesulfonamide in water at 37°C is 113 mg/100 cm^3 solution (4.44×10^{-3} mol dm^{-3}, compiler).

AUXILIARY INFORMATION

METHOD/APPARATUS/PROCEDURE:

Excess sulfonamide in water was heated and stirred on a steam bath for 30 min. The suspension was then agitated for 24 h in a thermostat. A sample of the satd soln was withdrawn through a glass filter, dild, and analyzed by the Marshall method (1) using a General Electric recording spectrophotometer for comparing the colors developed with those of the standards.

SOURCE AND PURITY OF MATERIALS:

The sulfonamide, mp 211-13°C, was prepd by the authors. Anal: %C 42.7 (calcd 42.5); %H 3.8 (3.9); %N 22.2 (22.0). Purity of the water was not specified.

ESTIMATED ERROR:

Nothing specified

REFERENCES:

1. Bratton, A. C.; Marshall, E. K. Jr. *J. Pharmacol.* 1939, *66*, 4.

COMPONENTS:	ORIGINAL MEASUREMENTS:
(1) Benzenesulfonamide, 4-amino-N-(4-methyl-1,2,5-oxadiazol-3-yl)-; $C_9H_{10}N_4O_3S$; [17103-53-6] (2) Water; H_2O; [7732-18-5]	Anderson, G. W.; Faith, H.E.; Marson, H.W.; Winnek, P. S.; Roblin, R. O., Jr. *J. Am. Chem. Soc.* 1942, *64*, 2902-5.
VARIABLES:	PREPARED BY:
One temperature: 37°C	R. Piekos

EXPERIMENTAL VALUES:

Solubility of 4-amino-N-(4-methyl-1,2,5-oxadiazol-3-yl)benzenesulfonamide in water at 37°C is 180 mg/100 cm^3 solution (7.08×10^{-3} mol dm^{-3}, compiler).

AUXILIARY INFORMATION

METHOD/APPARATUS/PROCEDURE:

Excess sulfonamide in water was heated and stirred on a steam bath for 30 min. The suspension was then agitated for 24 h in a thermostat. A sample of the satd soln was withdrawn through a glass filter, dild, and analyzed by the Marshall method (1) using a General Electric recording spectrophotometer for comparing the colors developed with those of the standards.

SOURCE AND PURITY OF MATERIALS:

The sulfonamide, mp 148-50°C (cor), was prepd by the authors. Anal: %C 42.3 (calcd 42.5); %H 4.4 (3.9); %N 22.0 (22.0). Purity of the water was not specified.

ESTIMATED ERROR:

Nothing specified.

REFERENCES:

1. Bratton, A. C.; Marshall, E. K., Jr. *J. Pharmacol.* 1939, *66*, 4.

COMPONENTS:	EVALUATOR:
(1) Benzenesulfonamide, 4-amino-N-2-thiazolyl- (sulfathiazole) $C_9H_9N_3O_2S_2$; [72-14-0] (2) Water (3) Aqueous phosphate buffer	Anthony N. Paruta Department of Pharmaceutics University of Rhode Island Kingston, Rhode Island, USA and Ryszard Piekos Faculty of Pharmacy, University of Gdansk Gdansk, Poland 1986

CRITICAL EVALUATION:

The solubility data available for sulfathiazole cover a 44 year span and are summarized in Table I.

Table I: Solubility of Sulfathiazole in water at various temperature

Reference	10^3 mol dm^{-3} (*indicates mol kg^{-1})			
	293K	298K	303K	310K
1	-	2.35(299K)	-	-
2	-	-	-	3.7
3	-	2.35(298.9K)	-	3.56
4	-	-	-	23.5
5	-	1.96*(299K)	-	3.76*
6	1.45*	-	-	3.45*
7	2.7*	-	-	-
8	2.0(291-292K)	-	-	-
9	-	-	-	3.72
10	-	-	-	1.9
11	-	-	2.43	-
12	-	-	2.27	-
13	1.76*(Form II)	-	-	-
14	-	-	2.350(α)	3.055(α,308K)
15	1.68	-	-	-
16	-	1.837(α)	-	3.122(α,308K)
17	1.7	-	-	-
18	1.394(pm)	1.837(pm)	2.400(pm)	3.122(pm,308K)
19	-	-	2.34	-
20	1.5*	-	-	-
21	-	-	2.5	-
22	-	1.821(α)	2.326(α)	3.094(α,308K)
22	-	3.29 (β)	4.308(β)	5.354(β,308K)
23	-	-	-	2.56 (pH = 4)
24	-	-	-	3.13
25	-	-	-	3.44
26	-	3.05	-	-
27	-	-	-	3.44
28	1.3	-	-	-

pm = polymorphs

The solubility of the most stable form (mp 473K) at various temperatures is dealt with initially followed by a discussion of the polymorphic forms and their effect on solubility. In the column of values at 293K, those of Weinstein and McDonald (7) and Becher and Leya (8) should be disregarded as being substantially higher than the rest and not considered further. The solubility reported by Kuhnert-Brandstätter and Martinek (13) of a form II (probably the same as "normal" sulfathiazole) is quite similar to those given for "normal" sulfathiazole. Since there is some question as to the identity of this form (13), the value was not considered further. The values of Ito and Sekiguchi (18) at the four temperatures, are not considered for the "normal" sulfathiazole, but for the alpha (α) form solubility discussion. The remaining values (6,15,17,20,28) were considered as potentially acceptable in deriving a pool of "good" values.

Sapozhnikova and Postovskii (6) used an equilibrium time of one hour was not considered further despite their reasonable values. The values of Miseta, Kedvessy and Selmeczi (28), while no doubt at equilibrium (2 days), gave only an approximate solubility of one part in 3000, thus not considered as accurate as the other values given. The remaining values (15,17,20,28) are the pool of acceptable values. The recommended value for sulfathiazole at 293K is the average of the four values (15,17,20,28) and is given as 1.6 x 10^{-3} mol dm^{-3}. Values given at 298K by two workers (22,26) were quite high. The solubility of the beta (β) form (22) is also quite high both at 303K and 308K. The values reported at 298K-299K (1,3,5) are not sufficiently similar to allow a recommended value since the value given in (3) is a repeat of the value given in (1).

Sulfathiazole has two crystalline forms, one melts at 493K, the other at 445K, which are expected to exhibit different solubilities (14,16,18,22). Sanchez (14) and Kanke and Sekiguchi (22) specifically annotate the lower melting-point alpha form of this compound, and the values at 298K, 303K and 308K are very close to one another. Although Sekiguchi and Ito (16,18) do not specifically identify the solubility values for the alpha form, the similarity of the results leads to the probable conclusion that they refer to the lower melting point form. Thus, recommended values for the alpha form are 1.832 x 10^{-3} mol dm^{-3} at 298K, 2.338 x 10^{-3} mol dm^{-3} at 303K, and 3.098 x 10^{-3} mol dm^{-3} at 308K. At 303K, the values (11,12,19,21) were close to lead to an average value of 2.4 x 10^{-3} mol dm^{-3}. This value should be compared with the recommended value of 2.338 x 10^{-3} mol dm^{-3} for the alpha form are quite similar. While there may be some doubt as to the crystalline form in the work of Bhattacharyya and Basu (11), there is no doubt that the Higuchi and Lach (12) used the higher melting point beta (β) form, to which recrystallization usually leads. Yamazaki et al. (19) did not specify the form of the sulfathiazole. Since there is some doubt about the forms used in these reports, the average value given above can only be considered tentative.

At body temperature, 310K, there were eight values reported (2,3,5,6,9,10,24,25,27). That given by Tréfouël (4) is obviously too high, and that of Kitao et al. (23) refers to pH = 4. Neish's (10) value is too low, as is that of Dubois and Tawashi (24) and were not considered further. The remaining values (2,3,5,6,9,25,27) were averaged, and a recommended value of 3.6 x 10^{-3} mol dm^{-3} for sulfathiazole in water at 310K can be given.

The solubility of sulfathiazole in aqueous buffers have been studied at two temperatures (29,30,31) as shown in Table II. There are slight differences in the pH values reported and it is assumed that pH 5.9 and pH 6.9 can be considered as pH 6 and pH 7 for purposes of this evaluation.

Table II: Solubility of Sulfathiazole in aqueous buffer systems at various pH values at two temperatures

Reference	pH	10^3 mol dm^{-3} 293K	310K
29	6*	2.115	3.76
30	6	2.00	-
31	6**	-	3.68
29	7	2.820	6.306
30	7	2.54	-
31	7***	-	7.99
29	8	5.640	-
30	8	4.90	-

* pH = 5.906 ** pH = 5.9 *** pH = 6.9

At a pH 6,7,8 Krüger-Thiemer (29) and Pulver and Suter (30) give values at 293K, showing good agreement at pH 6 and 7, but a 15% variance at a pH = 8. At pH 6 and 7, the recommended values at 293K are 2.06 x 10^{-3} mol dm^{-3} and 2.68 x 10^{-3} mol dm^{-3} respectively. At 310K, Krüger-Thiemer (29) and Langecker (31) provided a set of values at pH 6, and the recommended value is 3.72 x 10^{-3} mol dm^{-3}.

It is instructive to compare the values in buffer and water: the value in water at 293K is 1.62 x 10^{-3} mol dm^{-3}, which is about 78% of that in buffer at pH 6. At neutrality, the water solubility is about 60% of the buffer value. At 310K, however, the aqeuous value is 96% of the buffer value. There are several problems associated with the above data at different pH (29-31). There is no recognition of the change in the pkw with temperature which would somewhat affect the pH values in the Table. Pulver and Suter (31) do not give any specific information for methods, purity and error. Krüger-Thiemer (29) use a two hour equilibrium time at 293K which may not be sufficient. In Langecker's work (31) there is an inconsistency in the tabular data which show a higher solubility at pH 4.9 compared to pH 5.9. However, Krüger-Thiemer(29) show an increasing solubility with increasing pH. Therefore, these results while interesting are magnitude directing and considered approximate.

REFERENCES:

(1) Lott, W.A., Bergeim, F.H. *J. Am. Chem. Soc.* 1939, *61*, 3593-4.
(2) Roblin, R.O., Jr.; Williams, J.H.; Winnek, P.S.; English, J.P. *J. Am. Chem. Soc.* 1940, *62*, 2002-5.
(3) Durel, M.P.; Allinne, M. *Bull. Soc. Med. Hop. Paris III* 1941, 251-9.
(4) Tréfouël, M. *Bull. Acad. Med. Paris* 1941, *124*, 546-54.
(5) Clark, W.G.; Strakosch, E.A.; Levitan, N.I. *J. Lab. Clin. Med.* 1942, *28*, 188-9.
(6) Sapozhnikova, N.V.; Postovskii, I.Ya. *Zh. Prikl. Khim.* 1944, *17*, 427-34.
(7) Weinstein, L.; McDonald, A. *Science* 1945, *101*, 44-5.
(8) Becher, R.; Leya, S. *Experientia* 1946, *2*, 459-60.
(9) Langecker, H. *Arch. Exptl. Path. Pharmakol.* 1948, *205*, 291-301.
(10) Neish, W.J.P.; *Rec. trav. chim.* 1948, *67*, 361-71.
(11) Bhattacharyya, R.; Basu, U.P. *Indian Pharmacist* 1950, *6(3)*, 77-8, 86.

REFERENCES:

(12) Higuchi, T.; Lach, J.L. *J. Amer. Pharm. Assoc., Sci. Ed.* 1954, *43*, 349-54.
(13) Kuhnert-Brandstätter, M.; Martinek, A. *Microchim. Ichnoanal. Acta* 1956, 909-19.
(14) Sánchez, F.M.E. *Rev. Fac. Farm. Univ. Central Venezuela* 1962, *3(7)*, 31-45.
(15) Likhol'ot, N.M. *FArm. Zh. (Kiev)* 1965, *20(5)*, 44-6.
(16) Sekiguchi, K. Ito, K. *Chem. Pharm. Bull.* 1965, *13(4)*, 405-13.
(17) Gusyakov, V.P.; Likhol'ot, N.M.; Kutna, I.M. *Farm. Zh. (Kiev)* 1967, *22(3)*, 34-9.
(18) Ito, K.; Sekiguchi, K. *Chem. Pharm. Bull.* 1967, *15(4)*, 420-6.
(19) Yamazaki, M. Aoki, M.; Kamada, A.; Yata, N. *Yakuzaigaku* 1967, *27(1)*, 37-40.
(20) Shkadova, A.I. *Farm. Zh. (Kiev)* 1969, *24(3)*, 39-41.
(21) Mehta, S.C.; Bernardo, P.D.; Higuchi, W.I.; Simonelli, A.P. *J. Pharm. Sci.* 1970, *59(5)*, 638-44.
(22) Kanke, M.; Sekiguchi, K. *Chem. Pharm. Bull.* 1973, *21(4)*, 878-84.
(23) Kitao, K.; Kubo, K,; Morishita, T.; Yata, N.; Kamada, A. *Chem. Pharm. Bull.* 1973, *21*, 2417-26.
(24) Dubois, S.; Tawashi, R. *Pharm. Acta Helv.* 1975, *50*, 184-7.
(25) Kaneniwa, N.; Watari, N. *Chem. Pharm. Bull.* 1978, *26(3)*, 813-26.
(26) Badawi, A.A.; El-Sayad, A.A. *J. Pharm. Sci.* 1980, *69(5)*, 492-7.
(27) Watari, N.; Kaneniwa, N.; Hanano, M. *Int. J. Pharm.* 1980, *6(2)*, 155-66.
(28) Miseta, M.; Kedvessy, G.; Selmeczi, B. *Pharmazie* 1983, *38(5)*, 326-7.
(29) Krüger-Thiemer, E. *Arch. Dermatol. Syphilis* 1942, *183*, 90-116.
(30) Pulver, R.; Suter, R. *Schweiz. Med. Wochenschr.* 1943, *73(13)*, 403-8.
(31) Langecker, H. *Arch. Exptl. Path. Pharmakol.* 1948, *205*, 291-301.

COMPONENTS:	ORIGINAL MEASUREMENTS:
(1) Benzenesulfonamide, 4-amino-N-2-thiazolyl- (sulfathiazole); $C_9H_9N_3O_2S_2$; [72-14-0] (2) Water; H_2O; [7732-18-5]	Lott, W. A.; Bergeim, F. H. *J. Am. Chem. Soc.* 1939, *61*, 3593-4.
VARIABLES: One temperature: 26°C	PREPARED BY: R. Piekos

EXPERIMENTAL VALUES:

Solubility of sulfathiazole in water at 26°C is about 60 mg/100 cm^3 (2.35 x 10^{-3} mol dm^{-3}, compiler).

AUXILIARY INFORMATION

METHOD/APPARATUS/PROCEDURE:	SOURCE AND PURITY OF MATERIALS:
Nothing specified	Sulfathiazole, mp 197-7.5°C (uncor) and 202.0-2.5°C (cor) was prepd by the authors. Purity of the water was not specified.
	ESTIMATED ERROR: Nothing specified
	REFERENCES:

COMPONENTS:	ORIGINAL MEASUREMENTS:
(1) Benzenesulfonamide, 4-amino-N-2-thiazolyl- (sulfathiazole); $C_9H_9N_3O_2S_2$; [72-14-0] (2) Water; H_2O; [7732-18-5]	Roblin, R. O.,Jr.; Williams, J. H.; Winnek, P. S.; English, J. P. *J. Am. Chem. Soc.* 1940, *62*, 2002-5.
VARIABLES:	PREPARED BY:
One temperature: 37°C	R. Piekos

EXPERIMENTAL VALUES:

Solubility of sulfathiazole in water at 37°C is 94 mg/100 cm^3 solution (3.7 x 10^{-3} mol dm^{-3}, compiler).

AUXILIARY INFORMATION

METHOD/APPARATUS/PROCEDURE:

Excess sulfathiazole in water was heated and stirred on a steam bath for 30 min. The suspension was then agitated for 24 h in a thermostat at 37°C. A sample of the satd soln was withdrawn through a glass filter, dild, and analyzed by the Marshall method (1) using a General Electric recording spectrophotometer for comparing the colors developed with those of the standards.

SOURCE AND PURITY OF MATERIALS:

Sulfathiazole had mp of 201-2°C (cor), consistent with the literature data. Purity of the water was not specified.

ESTIMATED ERROR:

Nothing specified

REFERENCES:

1. Bratton, A.C.; Marshall, E. K.,Jr. *J. Pharmacol.* 1939, *66*, 4.

COMPONENTS:	ORIGINAL MEASUREMENTS:
(1) Benzenesulfonamide, 4-amino-N-2-thiazolyl- (sulfathiazole); $C_9H_9N_3O_2S_2$; [72-14-0] (2) Water; H_2O; [7732-18-5]	Durel, M. P.; Allinne, M. *Bull, Soc. Med. Hop. Paris III* <u>1941</u>, 251-9.
VARIABLES: One temperature: 37°C	PREPARED BY: R. Piekos

EXPERIMENTAL VALUES:

Solubility of sulfathiazole in water at 37°C is 0.91 g/liter (3.56×10^{-3} mol dm^{-3}, compiler).

AUXILIARY INFORMATION

METHOD/APPARATUS/PROCEDURE:	SOURCE AND PURITY OF MATERIALS:
A mixt of sulfathiazole and water was agitated for 24 hours at 37°C.	Source and purity of sulfathiazole was not specified. Distilled water was used.
	ESTIMATED ERROR: Nothing specified
	REFERENCES:

COMPONENTS:	ORIGINAL MEASUREMENTS:
(1) Benzenesulfonamide, 4-amino-N-2-thiazolyl- (sulfathiazole); $C_9H_9N_3O_2S_2$; [72-14-0] (2) Water; H_2O; [7732-18-5]	Trefouël, M. *Bull. Acad. Med. Paris* 1941, *124*, 546-54.
VARIABLES:	PREPARED BY:
One temperature: 37°C	R. Piekos

EXPERIMENTAL VALUES:

Solubility of sulfathiazole in water at 37°C is 0.6 part per 100 parts water (2.35×10^{-2} mol kg^{-1}, water, compiler).

AUXILIARY INFORMATION

METHOD/APPARATUS/PROCEDURE:	SOURCE AND PURITY OF MATERIALS:
Sulfathiazole was diazotized, coupled with N-naphthyl-1-N-diethyl-3-propylenediamine and assayed colorimetrically.	Nothing specified
	ESTIMATED ERROR:
	Nothing specified
	REFERENCES:

COMPONENTS:

(1) Benzenesulfonamide, 4-amino-N-2-thiazolyl- (sulfathiazole); $C_9H_9N_3O_2S_2$; [72-14-0]

(2) Water; H_2O; [7732-18-5]

ORIGINAL MEASUREMENTS:

Clark, W. G.; Strakosch, E. A.; Levitan, N. I. *J. Lab. Clin. Med.* 1942, *28*, 188-9.

VARIABLES:

Temperature

PREPARED BY:

R. Piekos

EXPERIMENTAL VALUES:

t/°C	Solubility	
	g/100 g water	10^3 mol kg^{-1} water[a]
26	0.0502	1.96
37	0.0960	3.76

[a] Calculated by compiler

AUXILIARY INFORMATION

METHOD/APPARATUS/PROCEDURE:

A small tinted glass container contg excess sulfathiazole in water was shaken in a water bath thermostat for 24 h. The satd soln was then filtered by aspiration through a washed and dried asbestos filter stick into a weighed weighing bottle. The entire app was kept at the temp at which the compd was dissolved. The amt dissolved was then detd by the method of Bratton and Marshall (1), using a photoelectric colorimeter.

SOURCE AND PURITY OF MATERIALS:

Neither source nor purity of sulfathiazole was specified.

CO_2-free distd water was used.

ESTIMATED ERROR:

Soly: not specified

Temp: ±0.1°C (authors)

REFERENCES:

1. Bratton, A. C.; Marshall, E. K. Jr. *J. Biol. Chem.* 1939, *128*, 537.

COMPONENTS:

(1) Benzenesulfonamide, 4-amino-N-2 thiazolyl- (sulfathiazole); $C_9H_9N_3O_2S_2$; [72-14-0]

(2) Water; H_2O; [7732-18-5]

ORIGINAL MEASUREMENTS:

Sapozhnikova, N. V.; Postovskii, I. Ya.
Zh. Prikl. Khim. 1944, *17*, 427-34.

VARIABLES:

Temperature

PREPARED BY:

R. Piekos

EXPERIMENTAL VALUES:

t/°C	Solubility	
	Weight%	10^3 mol kg^{-1} water[a]
20	0.0370	1.45
37	0.0880	3.45
50	0.1680[b]	6.59
75	0.530	20.87
99	1.20; 1.32	47.57 ; 52.39

[a] calculated by compiler

[b] calculated from the heat of dissolution

AUXILIARY INFORMATION

METHOD/APPARATUS/PROCEDURE:

Sulfathiazole was dissolved in water to form a satd soln which was occasionally agitated in a glass vessel immersed in a thermostat. The equilibrium was usually attained after 1 h. Five to 100-cm^3 samples of the satd soln were placed in Pt crucibles or dishes and evapd to dryness at temps lower than 110-115°C. The residue was dried to const wt at 105-110°C and weighed.

SOURCE AND PURITY OF MATERIALS:

Pure, recrystd sulfathiazole was used. Its mp conformed to that reported in the literature.

Purity of the water was not specified.

ESTIMATED ERROR: Soly: quite reliable results were obtained over the temp range 20-75°C. At higher temps the accuracy was poor due to evapn of water during sampling (authors). Temp: ±0.05°C (authors).

REFERENCES:

COMPONENTS:	ORIGINAL MEASUREMENTS:
(1) Benzenesulfonamide, 4-amino-N-2-thiazolyl- (sulfathiazole); [72-14-0] (2) Water; H_2O; [7732-18-5]	Weinstein, L.; McDonald, A. *Science*, 1945, *101*, 44-5.
VARIABLES: One temperature: 20°C	PREPARED BY: R. Piekos

EXPERIMENTAL VALUES:

Solubility of sulfathiazole in water at 20°C is 69 mg/100 cm^3 water (2.7×10^{-3} mol kg^{-1}, compiler).

AUXILIARY INFORMATION

METHOD/APPARATUS/PROCEDURE:	SOURCE AND PURITY OF MATERIALS:
Nothing specified	Nothing specified
	ESTIMATED ERROR: Nothing specified
	REFERENCES:

COMPONENTS:	ORIGINAL MEASUREMENTS:
(1) Benzenesulfonamide, 4-amino-N-2-thiazolyl- (sulfathiazole); $C_9H_9N_3O_2S_2$; [72-14-0] (2) Water; H_2O; [7732-18-5]	Becher, R.; Leya, S. *Experientia*, 1946, *2*, 459-60.
VARIABLES:	PREPARED BY:
One temperature: 18-19°C	R. Piekos

EXPERIMENTAL VALUES:

Solubility of sulfathiazole in water at room temperature (18-19°C) is 50 mg% (2.0×10^{-3} mol dm^{-3}, compiler).

AUXILIARY INFORMATION

METHOD/APPARATUS/PROCEDURE:	SOURCE AND PURITY OF MATERIALS:
After standing for more than two days the soln of sulfathiazole in water was filtered and sulfathiazole was assayed in the filtrate colorimetrically by the method of Druey and Oesterheld (1).	Nothing specified
	ESTIMATED ERROR:
	Nothing specified
	REFERENCES:
	1. Druey, J.; Oesterheld, G. *Helv. Chim. Acta* 1942, *25*, 753.

COMPONENTS:

(1) Benzenesulfonamide, 4-amino-N-2-thiazolyl- (sulfathiazole); $C_9H_9N_3O_2S_2$; [72-14-0]

(2) Water; H_2O; [7732-18-5]

ORIGINAL MEASUREMENTS:

Langecker, H.

Arch. Exptl. Path. Pharmakol. 1948, *205*, 291-301.

VARIABLES:

One temperature: 37°C

PREPARED BY:

R. Piekos

EXPERIMENTAL VALUES:

Solubility of sulfathiazole in water at 37°C is 95 mg% (3.721×10^{-3} mol dm^{-3}, compiler).

AUXILIARY INFORMATION

METHOD/APPARATUS/PROCEDURE:

An excess of sulfathiazole was boiled with water and left for 24 h in a vessel protected from access of CO_2. The concn of sulfanilamide was detd by the method of Bratton and Marshall (1) using a Havemann colorimeter (2), as well as by microanal detn of the solid residue.

SOURCE AND PURITY OF MATERIALS:

Nothing specified

ESTIMATED ERROR:

Nothing specified

REFERENCES:

1. Bratton, A. G.; Marshall, E. K. *J. Biol. Chem.* 1939, *128*, 537.

1. Havemann, R. *Klin. Wochenschr.* 1940, p. 503.

COMPONENTS:

(1) Benzenesulfonamide, 4-amino-N-2-thiazolyl- (sulfathiazole); $C_9H_9N_3O_3S_2$; [72-14-0]

(2) Water; H_2O; [7732-18-5]

ORIGINAL MEASUREMENTS:

Neish, W. J. P. *Rec. trav. chim.* 1948, *67*, 361-71.

VARIABLES:

One temperature: 37°C

PREPARED BY:

R. Piekos

EXPERIMENTAL VALUES:

Solubility of sulfathiazole in water at 37°C is 490 γ/ml (1.9 x 10^{-3} mol dm^{-3}, compiler).

AUXILIARY INFORMATION

METHOD/APPARATUS/PROCEDURE:

A suspension of sulfathiazole in water was kept for 5 h at 37°C and 1 h at room temp before filtration. Soly was detd by the Westfall's method (1) based on diazotization of the sulfonamide, coupling with Na 2-napthol-3,6-disulfonate and comparing the color with that of a satd soln in a Klett colorimeter.

SOURCE AND PURITY OF MATERIALS:

Sulfathiazole : not specified.

The distd water was used.

ESTIMATED ERROR:

Nothing specified.

REFERENCES:

1. Westfall, B. B. *J. Nat. Cancer Inst.* 1945, *6*, 23.

COMPONENTS:	ORIGINAL MEASUREMENTS:
(1) Benzenesulfonamide, 4-amino-N-2-thiazolyl- (sulfathiazole); $C_9H_9N_3O_2S_2$; [72-14-0] (2) Water; H_2O; [7732-18-5]	Bhattacharyya, R.; Basu, U. P. *Indian Pharmacist* 1950, *6(3)*, 77-8, 86.
VARIABLES:	PREPARED BY:
One temperature: 30°C	R. Piekos

EXPERIMENTAL VALUES:

Solubility of sulfathiazole in water at 30°C is 62 mg per 100 ml (2.43×10^{-3} mol dm^{-3}, compiler).

AUXILIARY INFORMATION

METHOD/APPARATUS/PROCEDURE:	SOURCE AND PURITY OF MATERIALS:
A weighed sample of sulfathiazole was placed in a clean reagent bottle and a known vol of water was added. The mixt was shaken in a mech shaker at 80-100 strokes/min. After at least 24 h the mixt was filtered through a clean, dried and weighed sintered-glass crucible. At the end of the filtration the crucible was washed with about 1 ml of water, dried at 105°C for 2-3 h, cooled, and weighed to const wt.	Neither source nor purity of the sulfathiazole was not specified. Doubly distd water was used. ESTIMATED ERROR: Soly: not specified Temp: ±0.2°C (authors) REFERENCES:

COMPONENTS:

(1) Benzenesulfonamide, 4-amino-N-2-thiazolyl- (sulfathiazole); $C_9H_9N_3O_2S_2$; [72-14-0]

(2) Water; H_2O; [7732-18-5]

ORIGINAL MEASUREMENTS:

Higuchi, T.; Lach, J. L.
J. Amer. Pharm. Assoc., Sci. Ed. 1945, *43*, 349-54.

VARIABLES:

One temperature: 30°C

PREPARED BY:

R. Piekos

EXPERIMENTAL VALUES:

Solubility of sulfathiazole in water at 30°C is 2.27 x 10^{-3} mol dm^{-3} solution (0.58 g dm^{-3}, compiler).

AUXILIARY INFORMATION

METHOD/APPARATUS/PROCEDURE:

Excess sulfathiazole (75 mg) was placed in a 125-ml glass-stoppered bottle together with 50 ml of water. The bottle was placed in a mech shaker in a const temp bath and equilibrated for 8 h at 30°C. Aliquot of the supernatant liquid was analyzed for the sulfonamide by the method of Bratton and Marshall (1).

SOURCE AND PURITY OF MATERIALS:

Recrystd sulfathiazole (U.S.P.), mp 201-2°C and distilled water were used.

ESTIMATED ERROR:

Nothing specified.

REFERENCES:

1. Bratton, A. C.; Marshall, E. K. Jr. *J. Biol. Chem.* 1939, *128*, 537.

COMPONENTS:

(1) Benzenesulfonamide, 4-amino-N-2-thiazolyl- (sulfathiazole); $C_9H_9N_3O_2S_2$; [72-14-0]

(2) Water; H_2O; [7732-18-5]

ORIGINAL MEASUREMENTS:

Kuhnert-Brandstätter, M.; Martinek, A. *Michrochim. Ichnoanal. Acta* 1956, 909-19.

VARIABLES:

Temperature

PREPARED BY:

R. Piekos

EXPERIMENTAL VALUES:

Saturation solubility[a]

t/°C	Crystalline form I		Crystalline form II	
	g/100 g solution	10^2 mol kg^{-1} solution[b]	g/100 g solution	10^2 mol kg^{-1} solution[b]
20.0	0.090	0.352	0.045	0.176
30.0	0.130	0.509	--	--
30.5	--	--	0.070	0.274
40.0	0.180	0.705	0.100	0.392
50.0	0.265	1.038	0.180	0.705
59.5	0.410	1.606	0.290	1.136
69.5	0.610	2.389	--	--
70.0	--	--	0.515	2.017

[a]Numerical data received from the authors

[b]Calculated by compiler

AUXILIARY INFORMATION

METHOD/APPARATUS/PROCEDURE:

Sulfathiazole and water were placed in a polyethylene vessel, agitated for 3 h, filtered, and the sulfonamide was assayed in the filtrate gravimetrically. The solid phase was examd thermomicroscopically for identity of the cryst form.

SOURCE AND PURITY OF MATERIALS:

A comm available form II of sulfathiazole was used. Form I was obtained by keeping the comm reagent at 170°C for 2 h. Distilled water was used.

ESTIMATED ERROR:

Soly: not specified.

Temp: ±0.5°C (authors).

REFERENCES:

COMPONENTS:

(1) Benzenesulfonamide, 4-amino-N-2-thiazolyl- (sulfathiazole); $C_9H_9N_3O_2S_2$; [72-14-0]

(2) Water; H_2O; [7732-18-5]

ORIGINAL MEASUREMENTS:

Sanchez, F.M.E.

Rev. Fac. Farm. Univ. Central Venezuela 1962, *3(7)*, 31-45.

VARIABLES:

Temperature

PREPARED BY:

R. Piekos

EXPERIMENTAL VALUES:

$t/^{\circ}C$	Solubility of crystalline form A of sulfathiazole	
	mg/1000 cm^3 solution	10^3 mol dm^{-3} [a]
30	600.00	2.350
35	780.00	3.055
40	1025.00	4.015
45	1310.00	5.131
50	1750.00	6.854

[a] Calculated by compiler

AUXILIARY INFORMATION

METHOD/APPARATUS/PROCEDURE:

The soly was detd by the std Hill method (1): two 0.50-mg samples of sulfathiazole were placed in a 100-ml conical test tubes together with 35 ml of water and stoppered. One of the solns was heated to 55°C and the other kept at a given temp. Both solns were then kept in a const temp bath. Five-ml samples were withdrawn through a filter into 500-ml flasks, dild to the mark with water and sulfathiazole was assayed at 283 nm using a Beckman DU spectrophotometer.

SOURCE AND PURITY OF MATERIALS:

Cryst form A of sulfathiazole was prepd by moistening a sample of a FNAR grade sulfathiazole with abs EtOH followed by drying the sample for 4 h at 60°C under vacuum. Microscopic detn of the mp showed the specimen to contain 100% of form A, mp 173-5°C. Purity of the water was not specified.

ESTIMATED ERROR:

Soly: not specified.

Temp: ±0.01°C (author).

REFERENCES:

1. Weissberger, *Pysical methods, Pt. I, third edition,* p. 677.

COMPONENTS:	ORIGINAL MEASUREMENTS:
(1) Benzenesulfonamide, 4-amino-N-2-thiazolyl- (sulfathiazole); $C_9H_9N_3O_2S_2$; [72-14-0] (2) Water; H_2O; [7732-18-5]	Likhol'ot, N.M. *Farm. Zh. (Kiev)* 1965, *20(5)*, 44-6.
VARIABLES:	PREPARED BY:
One temperature: 20°C	R. Piekos

EXPERIMENTAL VALUES:

Solubility of sulfathiazole in water at 20°C is 0.043 g/100 ml (1.68×10^{-3} mol dm^{-3}, compiler).

AUXILIARY INFORMATION

METHOD/APPARATUS/PROCEDURE:

An earlier described method was employed (1) whereby a small excess of sulfathiazole was equilibrated with 20 ml of water for 8 h in a 50-ml test tube. Aliquots were withdrawn through a filter and sulfathiazole was assayed bromatometrically.

SOURCE AND PURITY OF MATERIALS:

Nothing specified

ESTIMATED ERROR:

Soly: not specified

Temp: ±0.1°C (authors).

REFERENCES:

1. Gusyakov, V. P.; Likhol'ot, N. M. *Farm. Zh. (Kiev)* 1960, *15(8)*, 21.

COMPONENTS:

(1) Benzenesulfonamide, 4-amino-N-2-thiazolyl- (sulfathiazole); $C_9H_9N_3O_2S_2$; [72-14-0]

(2) Water; H_2O; [7732-18-5]

ORIGINAL MEASUREMENTS:

Sekiguchi, K.; Ito, K.

Chem. Pharm. Bull. 1965, *13(4)*, 405-13.

VARIABLES:

Temperature

PREPARED BY:

R. Piekos

EXPERIMENTAL VALUES:

t/°C	Solubility	
	10^3 mol dm^{-3} solution	g dm^{-3} [a]
15	1.047	0.2673
25	1.837	0.4690
35	3.122	0.7971

[a]Calculated by compiler

AUXILIARY INFORMATION

METHOD/APPARATUS/PROCEDURE:

In a 200-ml egg-plant type flask, immersed in a thermostat, an excess of sulfathiazole was placed with 100 ml of redistd water (pH 5.7 ~ 5.9) which was previously kept at appropriate temp. Immediately after addn of water, the mixt was vigorously agitated with an elec stirrer. Aliquots were withdrawn at certain time intervals with a pipet equipped with a filter, and the concn of solute was detd spectrophotometrically at 283 mμ.

SOURCE AND PURITY OF MATERIALS:

Sulfathiazole was a comm product of J.P. grade. The most stable polymorphic modification was used.

Redistd water was used.

ESTIMATED ERROR:

Soly: not specified

Temp: ±0.05°C (authors)

REFERENCES:

COMPONENTS:	ORIGINAL MEASUREMENTS:
(1) Benzenesulfonamide, 4-amino-N-2-thiazolyl- (sulfathiazole); $C_9H_9N_3O_2S_2$; [72-14-0] (2) Water; H_2O; [7732-18-5]	Gusyakov, V.P.; Likhol'ot, N.M.; Kutna, I.M. *Farm. Zh. (Kiev)* 1967, *22(3)*, 34-9.
VARIABLES:	PREPARED BY:
One temperature: 20°C	R. Piekos

EXPERIMENTAL VALUES:

Solubility of sulfathiazole in water at 20°C is 0.043 g/100 ml

(1.7 x 10^{-3} mol dm^{-3}, compiler).

AUXILIARY INFORMATION

METHOD/APPARATUS/PROCEDURE:	SOURCE AND PURITY OF MATERIALS:
An excess of sulfathiazole in water was equilibrated for 24 h in an ampul immersed in a water thermostat. Aliquots of the satd soln were withdrawn through a filter and the sulfathiazole content was assayed in the filtrate photometrically.	Sulfathiazole conformed to the requirements of the State Pharmacopeia IX. Purity of the water was not specified.
	ESTIMATED ERROR:
	Soly: not specified. Temp: ±0.1°C (authors).
	REFERENCES:

COMPONENTS:

(1) Benzenesulfonamide, 4-amino-N-2-thiazolyl- (sulfathiazole); $C_9H_9N_3O_2S_2$; [72-14-0]

(2) Water; H_2O; [7732-18-5]

ORIGINAL MEASUREMENTS:

Ito, K.; Sekiguchi, K.

Chem. Pharm. Bull. 1967, *15(4)*, 420-6.

VARIABLES:

Temperature

PREPARED BY:

R. Piekos

EXPERIMENTAL VALUES:

t/°C	Solubility	
	10^3 mol dm^{-3} soln	g dm^{-3} [a]
20	1.394	0.3559
25	1.837	0.4690
30	2.400	0.6127
35	3.122	0.7971

[a] Calculated by compiler

AUXILIARY INFORMATION

METHOD/APPARATUS/PROCEDURE:

The earlier described method (1) was used: in a 200-ml egg-plant type flask immersed in a thermostat, an excess of sulfathiazole was placed with 100 ml of distd water which was previously kept at appropriate temp. Immediately after addn of water the mixt was vigorously agitated by an elec stirrer. Aliquots were withdrawn at certain time intervals with a pipet equipped with a filter and the concn of solute was detd spectrophotometrically at 283 mμ.

SOURCE AND PURITY OF MATERIALS:

Polymorphic modifications of sulfathiazole (source not specified) were prepd by the method of Grove (2).

Distd water was used.

ESTIMATED ERROR:

Nothing specified.

REFERENCES:

1. Sekiguchi, K.; Ito, K. *Chem. Pharm. Bull,* 1965, *13(4)*, 405.
2. Grove, D. C.; Keenan, G. L. *J. Am. Chem. Soc.* 1941, *63*, 97.

COMPONENTS:	ORIGINAL MEASUREMENTS:
(1) Benzenesulfonamide, 4-amino-N-2-thiazolyl- (sulfathiazole); $C_9H_9N_3O_2S_2$; [72-14-0] (2) Water; H_2O; [7732-18-5]	Yamazaki, M.; Aoki, M.; Kamada, A. ; Yata, N. *Yakuzaigaku* 1967, *27(1)*, 37-40.
VARIABLES:	**PREPARED BY:**
One temperature: 30°C	R. Piekos

EXPERIMENTAL VALUES:

Solubility of sulfathiazole in water at 30°C is 2.34 mmol/L (0.597 g dm^{-3}, compiler).

AUXILIARY INFORMATION

METHOD/APPARATUS/PROCEDURE:

Sulfathiazole (0.5 g) was placed in an L-shaped tube together with 20 ml of water. The mixt was then shaken in a thermostat until equilibrium was attained. The sulfathiazole was then assayed in the supernatant spectrophotometrically at 545 nm on a Beckman DU spectrophotometer. The results were taken from a calibration graph.

SOURCE AND PURITY OF MATERIALS:

Nothing specified

ESTIMATED ERROR:

Soly: not specified

Temp: ±1°C (authors)

REFERENCES:

COMPONENTS:	ORIGINAL MEASUREMENTS:
(1) Benzenesulfonamide, 4-amino-N-2-thiazolyl- (sulfathiazole); $C_9H_9N_3O_2S_2$; [72-14-0] (2) Water; H_2O; [7732-18-5]	Shkadova, A. I. *Farm. Zh. (Kiev)* 1969, *24(3)*, 39-41.
VARIABLES:	PREPARED BY:
One temperature: 20°C	R. Piekos

EXPERIMENTAL VALUES:

Solubility of sulfathiazole in water at 20°C is 0.15×10^{-2} mol/kg (3.8×10^{-2} g/100 g, compiler).

AUXILIARY INFORMATION

METHOD/APPARATUS/PROCEDURE:	SOURCE AND PURITY OF MATERIALS:
A satd aqueous soln of sulfathiazole was equilibrated in a water thermostat at 20±0.1°C. The concn of sulfathiazole was detd by alkalimetric titration.	Purity of sulfathiazole conformed to the requirements of the State Pharmacopeia IX. Distd water was used.
	ESTIMATED ERROR:
	Soly: not specified. Temp: ±0.1°C (author).
	REFERENCES:

COMPONENTS:	ORIGINAL MEASUREMENTS:
(1) Benzenesulfonamide, 4-amino-N-2-thiazolyl- (sulfathiazole); $C_9H_9N_3O_2S_2$; [72-14-0] (2) Water; H_2O; [7732-18-5]	Mehta, S. C.; Bernardo, P. D.; Higuchi, W. I.; Simonelli, A. P. *J. Pharm. Sci.* 1970, *59(5)*, 638-44.
VARIABLES:	PREPARED BY:
One temperature: 30°C	R. Piekos

EXPERIMENTAL VALUES:

Solubility of sulfathiazole in water at 30°C is 0.065 g/100 g (2.5×10^{-3} mol kg^{-1}, compiler).

AUXILIARY INFORMATION

METHOD/APPARATUS/PROCEDURE:

An excess of the amount of the recrystd sulfathiazole needed to produce a satd soln was placed in a volumetric flask with water and agitated in a water bath at 30°C. Duplicate samples were withdrawn at 12-24 h intervals, filtered through a 0.45-μ Millipore filter, and analyzed spectrophotometrically.

SOURCE AND PURITY OF MATERIALS:

Sulfathiazole (source not specified) was purified by crystallization.
Purity of the water was not specified.

ESTIMATED ERROR:

Nothing specified.

REFERENCES:

COMPONENTS:

(1) Benzenesulfonamide, 4-amino-N-2-thiazolyl- (sulfathiazole); $C_9H_9N_3O_2S_2$; [72-14-0]

(2) Water; H_2O; [7732-18-5]

ORIGINAL MEASUREMENTS:

Kanke, M.; Sekiguchi, K.

Chem. Pharm. Bull. 1973, *21(4)*, 878-84.

VARIABLES:

Temperature

PREPARED BY:

R. Piekos

EXPERIMENTAL VALUES:

$t/^{\circ}C$	Solubility			
	α - form		β - form	
	g/liter	10^3 mol dm^{-3} [a]	g/liter	10^3 mol dm^{-3} [a]
25	0.465	1.821	0.840	3.290
30	0.594	2.326	1.100	4.308
35	0.790	3.094	1.367	5.354
40	1.040	4.073	1.690	6.619
45	1.350	5.288	2.115	8.284
49	1.683	6.592	2.544	9.964

[a] Calculated by compiler

AUXILIARY INFORMATION

METHOD/APPARATUS/PROCEDURE:

About 1.5 g of sulfathiazole was placed in 100 ml of water and agitated at 600 rpm. Aliquots of the soln were taken with a glass syringe at short time intervals at the beginning of each expt and then at about hourly intervals until equilibrium was attained. The sample soln was then immediately filtered through a 0.45-μ membrane filter and a carefully measured aliquot was dild for spectrophotometric assay on a Hitachi Perkin-Elmer 139 spectrophotometer at 283 nm.

SOURCE AND PURITY OF MATERIALS:

α-Sulfathiazole: comm product of the JP VII grade was recrystd from distd water. β-Sulfathiazole, mp 200-2°C, was recrystd from PrOH. Both forms were characterized by instrumental method.

Purity of the water was not specified.

ESTIMATED ERROR:

Nothing specified.

REFERENCES:

COMPONENTS:	ORIGINAL MEASUREMENTS:
(1) Benzenesulfonamide, 4-amino-N-2-thiazolyl- (sulfathiazole); $C_9H_9N_3O_2S_2$; [72-14-0] (2) Water; H_2O; [7732-18-5]	Kitao, K.; Kubo, K.; Morishita, T.; Yata, N.; Kamada, A. *Chem. Pharm. Bull.* 1973, *21*, 2417-26.
VARIABLES: One temperature: 37°C	PREPARED BY: R. Piekos

EXPERIMENTAL VALUES:

Solubility of sulfathiazole in water at 37°C is 2.56 mmol dm^{-3} solution.

AUXILIARY INFORMATION

METHOD/APPARATUS/PROCEDURE:

Soly was detd by continuously adjusting the pH of the aq soln to 4 with 0.05 N NaOH. The concn of sulfathiazole was detd by diazotization after proper diln.

SOURCE AND PURITY OF MATERIALS:

Comm available sulfathiazole (source not specified) was used as supplied. Deionized water was used.

ESTIMATED ERROR:

Soly: not specified.

Temp: ±1°C (authors).

REFERENCES:

COMPONENTS:	ORIGINAL MEASUREMENTS:
(1) Benzenesulfonamide, 4-amino-N-2-thiazolyl- (sulfathiazole); $C_9H_9N_3O_2S_2$; [72-14-0] (2) Water; H_2O; [7732-18-5]	Dubois, S.; Tawashi, R. *Pharm. Acta Helv.* 1975, *50*, 184-7.
VARIABLES:	PREPARED BY:
One temperature: 37°C	R. Piekos

EXPERIMENTAL VALUES:

Solubility of sulfathiazole in water at 37°C is 8.00×10^{-4} g/ml (3.13×10^{-3} mol dm^{-3}, compiler).

AUXILIARY INFORMATION

METHOD/APPARATUS/PROCEDURE:

Samples were filtered through a Millipore filter 0.45 μm and the amount of dissolved sulfathiazole was detd spectrophotometrically at 280 nm.

SOURCE AND PURITY OF MATERIALS:

USP grade sulfathiazcle without further treatment was used and distilled water.

ESTIMATED ERROR:

Nothing specified

REFERENCES:

COMPONENTS:	ORIGINAL MEASUREMENTS:
(1) Benzenesulfonamide, 4-amino-N-2-thiazolyl- (sulfathiazole); $C_9H_9N_3O_2S_2$; [72-14-0] (2) Water; H_2O; [7732-18-5]	Kaneniwa, N.; Watari, N. *Chem. Pharm. Bull.* 1978, *26(3)*, 813-26.

VARIABLES:	PREPARED BY:
One temperature: 37°C	R. Piekos

EXPERIMENTAL VALUES:

Solubility of sulfathiazole in water at 37°C is 0.879 mg/ml solution (3.44×10^{-3} mol dm^{-3}, compiler).

AUXILIARY INFORMATION

METHOD/APPARATUS/PROCEDURE:

An excess of sulfathiazole was placed in a flask contg 25 ml of water. The flask was shaken (2 strokes/s at the amplitude of 3 cm) in a thermostatically controlled water bath at 37°C. One-ml sample was withdrawn every 6 h (total equilibration period was 3-5 days) using a warmed Millipore filter syringe with a filter pore size of 0.45 μ (Millipore HAWP 01300) and the filtrate was dild with water and assayed spectrophotometrically (1).

SOURCE AND PURITY OF MATERIALS:

Commercial sulfathiazole of the Japanese Pharmacopeia grade and distd water were used.

ESTIMATED ERROR:

Soly: not specified.

Temp: ±0.05°C (authors).

REFERENCES:

1. Kaneniwa, N.; Watari, N.
Chem. Pharm. Bull. 1974, *22*, 1699.

COMPONENTS:

(1) Benzenesulfonamide, 4-amino-N-2-thiazolyl- (sulfathiazole); $C_9H_9N_3O_2S_2$; [72-14-0]

(2) Water; H_2O; [7732-18-5]

ORIGINAL MEASUREMENTS:

Badawi, A. A.; El-Sayed, A. A.

J. Pharm. Sci. 1980, *69(5)*, 492-7.

VARIABLES:

One temperature: 25°

PREPARED BY:

R. Piekos

EXPERIMENTAL VALUES:

Solubility of sulfathiazole in water at 25°C is 0.78 mg/ml water

(3.05 x 10^{-3} mol dm^{-3} water, compiler).

AUXILIARY INFORMATION

METHOD/APPARATUS/PROCEDURE:

A weighed excess of sulfathiazole was placed in a 25-ml ampul contg 10 ml of water. The ampul was sealed and placed on a rotating shaft (42 rpm) immersed in a water bath at 25±1°C. Duplicate samples were withdrawn, filtered, and assayed spectrophotometrically at 283 nm.

SOURCE AND PURITY OF MATERIALS:

Sulfathiazole was of the BP 1963 purity. Purity of the water was not specified.

ESTIMATED ERROR:

Soly: not specified.

Temp: ±1°C (authors)

REFERENCES:

COMPONENTS:	ORIGINAL MEASUREMENTS:
(1) Benzenesulfonamide, 4-amino-N-2-thiazolyl- (sulfathiazole); $C_9H_9N_3O_2S_2$; [72-14-0] (2) Water; H_2O; [7732-18-5]	Watari, N.; Kaneniwa, N.; Hanano, M. *Int. J. Pharm.* 1980, *6(2)*, 155-66.
VARIABLES:	PREPARED BY:
One temperature: 37°C	R. Piekos

EXPERIMENTAL VALUES:

Solubility of sulfathiazole in water at 37°C is 87.9 mg/100 ml (3.44×10^{-3} mol dm^{-3}, compiler).

AUXILIARY INFORMATION

METHOD/APPARATUS/PROCEDURE:

The earlier developed method was employed (1), whereby an excess of sulfathiazole, required to saturate medium, was placed in a flask contg 25 ml of water. The flask was shaken (2 strokes/s) at an amplitude of 3 cm, in a thermostatically controlled bath. One-ml sample was removed every 6 h (total equilibration time was 3-5 days) using a warmed Millipore filter syringe with a filter pore size of 0.45 μ (Millipore HAWP 01300) and the fitrate was dild with water and assayed spectrophotometrically.

SOURCE AND PURITY OF MATERIALS:

Sulfathiazole was of the Japanese Pharmacopeia grade.

Distilled water was used.

ESTIMATED ERROR:

Soly: not specified

Temp: ±0.05°C (authors)

REFERENCES:

1. Kaneniwa, N.; Watari, N. *Chem. Pharm. Bull.* 1974, *22*, 1699.

COMPONENTS:

(1) Benzenesulfonamide, 4-amino-N-2-thiazolyl- (sulfathiazole); $C_9H_9N_3O_2S_2$; [72-14-0]

(2) Water; H_2O; [7732-18-5]

ORIGINAL MEASUREMENTS:

Rupprecht, H.; Ziller, K. H. *Pharmazie*, 1981, *36(4)*, 298.

VARIABLES:

One temperature: 20°C

PREPARED BY:

R. Piekos

EXPERIMENTAL VALUES:

Solubility of sulfathiazole in water at 20°C is 40.9 mg/100 ml (1.60×10^{-3} mol dm^{-3}, compiler).

AUXILIARY INFORMATION

METHOD/APPARATUS/PROCEDURE:

Fifty ml of a suspension contg 2.0 g of sulfathiazole was placed in a thermostat and stirred with a magnetic stirrer. The concn of the solute was monitored continuously after filtration through a G3 or G4 fritted-glass filters by means of a Knauer differential refractometer or a Shimadzu 100-02 UV spectrophotometer. The cuvets of the refractometer were thermostated. The variations of the refractive index or light absorption were recorded as a function of time by means of a Servogor 220 two-line recorder.

SOURCE AND PURITY OF MATERIALS:

Sulfathiazole: neither source nor purity was specified.

Distilled water was used.

ESTIMATED ERROR:

Nothing specified

REFERENCES:

COMPONENTS:	ORIGINAL MEASUREMENTS:
(1) Benzenesulfonamide, 4-amino-N-2-thiazolyl- (sulfathiazole); $C_9H_9N_3O_2S_2$; [72-14-0] (2) Water; H_2O; [7732-18-5]	Miseta, M.; Kedvessy, G.; Selmeczi, B. *Pharmazie* 1983, *38(5)*, 326-7.
VARIABLES:	**PREPARED BY:**
One temperature: 20°C	R. Piekos

EXPERIMENTAL VALUES:

Solubility of sulfathiazole in water at 20°C is 1 part in 3000 parts of water (1.3×10^{-3} mol kg^{-1} water - compiler).

AUXILIARY INFORMATION

METHOD/APPARATUS/PROCEDURE:

Soly was detd by the Pharmacopeia Hungarica V method. The equilibration time was 2 days with occasional shaking (personal communication). The concn of the solute in the satd soln was detd spectrophotometrically at 282 nm using a Spektromom 195 spectrophotometer.

SOURCE AND PURITY OF MATERIALS:

The source and purity of sulfathiazole was not specified.

Distilled water was used.

ESTIMATED ERROR:

Soly: not specified

Temp: ±2°C (personal communication).

REFERENCES:

COMPONENTS:	ORIGINAL MEASUREMENTS:
(1) Benzenesulfonamide, 4-amino-N-2-thiazolyl; (sulfathiazole); $C_9H_9N_3O_2S_2$; [72-14-0] (2) Sodium hydroxide; NaOH; [1310-73-2] (3) Water; H_2O; [7732-18-5]	Rose, F.L.; Martin, A. R.; Bevan, H.G.L. *J. Pharm. Exp. Therap.* 1943, *77*, 127-42.
VARIABLES:	PREPARED BY:
pH	R. Piekos

EXPERIMENTAL VALUES:

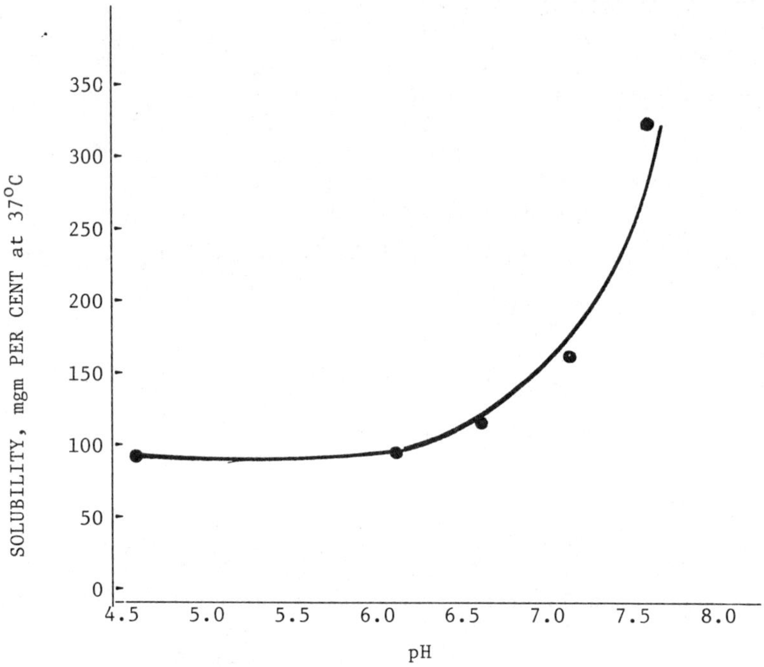

AUXILIARY INFORMATION

METHOD/APPARATUS/PROCEDURE:

An excess of sulfathiazole was stirred in boiling water, the soln was cooled to 37°C, the temp being maintained thermostatically, and 0.1 N NaOH was added to increase the pH. The pH was measured by means of a glass electrode-calomel half-cell system, and was permitted to reach equilibrium before a reading was taken. The concn of sulfathiazole in soln was detd colorimetrically by withdrawing a sample through a filter-tip into a preheated micropipet.

SOURCE AND PURITY OF MATERIALS:

The source and purity of sulfathiazole was not specified.

Water was doubly distilled.

ESTIMATED ERROR:

Nothing specified

REFERENCES:

COMPONENTS:

(1) Benzenesulfonamide, 4-amino-N-2-thiazolyl- (sulfathiazole); $C_9H_9N_3O_2S_2$; [72-14-0]

(2) Sodium hydroxide; NaOH; [1310-73-2]

(3) Water; H_2O; [7732-18-5]

ORIGINAL MEASUREMENTS:

Holz, E.; Garcia Onandia, A.; Holz, S. *Acta Cient. Venezolana* 1955, *6(2)*, 68-73.

VARIABLES:

Concentration of NaOH

PREPARED BY:

R. Piekos

EXPERIMENTAL VALUES:

Concentration of NaOH soln	Volume of the NaOH soln required to dissolve 1 g of sulfathiazole at 26°C	Solubility of sulfathiazole at 26°C
N	cm^3	mol dm^{-3} NaOH soln[a]
1/10	41.650	0.0940
1/4	16.650	0.2352
1/2	8.325	0.4705
1.0	4.175	0.9382
1.5	2.750	1.4243
1.75	2.425	1.6152
1.8	5.200	0.7530
1.9	7.200	0.5440
2.0	7.833	0.5000
2.5	46.830	0.0836

[a] calculated by compiler

AUXILIARY INFORMATION

METHOD/APPARATUS/PROCEDURE:

Nothing specified

SOURCE AND PURITY OF MATERIALS:

Nothing specified. Distd water was used.

ESTIMATED ERROR:

Nothing specified

REFERENCES:

COMPONENTS:

(1) Benzenesulfonamide, 4-amino-N-2-thiazolyl- (sulfathiazole); $C_9H_9N_3O_2S_2$; [72-14-0]

(2) Sodium chloride; NaCl; [7647-14-5]

(3) Water; H_2O; [7732-18-5]

ORIGINAL MEASUREMENTS:

Becher, R.; Leya, S. *Experientia* 1946, *2*, 459-60.

VARIABLES:

One temperature: 18-19°C

PREPARED BY:

R. Piekos

EXPERIMENTAL VALUES:

Solubility of sulfathiazole in a 5% NaCl solution at room temperature (18-19°C) is 45 mg% (1.8 x 10^{-3} mol dm^{-3}, compiler).

AUXILIARY INFORMATION

METHOD/APPARATUS/PROCEDURE:

After standing for more than two days the soln of sulfathiazole was filtered and sulfathiazole was assayed in the filtrate colorimetrically by the method of Druey and Oesterheld (1).

SOURCE AND PURITY OF MATERIALS:

Nothing specified

ESTIMATED ERROR:

Nothing specified

REFERENCES:

1. Druey, J.; Oesterheld, G. *Helv. Chim. Acta* 1942, *25*, 753.

COMPONENTS:	ORIGINAL MEASUREMENTS:
(1) Benzenesulfonamide, 4-amino-N-2-thiazolyl; (sulfathiazole); $C_9H_9N_3O_2S_2$; [72-14-0] (2) Sodium chloride, NaCl; [7647-14-5] (3) Water; H_2O; [7732-18-5]	Avico, U.; Cavazutii, G.; di Francesco, R.; Signoretti Ciranni, E.; Zuccaro, P. *Farmaco, Ed. Pratica* 1975, *30(1)*, 40-6.
VARIABLES: Temperature	PREPARED BY: R. Piekos

EXPERIMENTAL VALUES:

t/°C	Solubility of amorphous sulfathiazole in equimolal NaCl solutions	
	g/100 g water	10^3 mol kg^{-1} water[a]
25	0.627	2.456
35	1.010	3.956
40	1.214	4.755

[a] Calculated by compiler

AUXILIARY INFORMATION

METHOD/APPARATUS/PROCEDURE:

A soln of Na salt of sulfathiazole was added to a HCl soln contg stoichiometric quantity of the acid to neutralize the salt. The neutralization was carried out in a thermostat and the pH of the mixt was maintained close to that of a satd sulfathiazole soln. The procedure was repeated using various initial concns of the reagents to find the max concn of sulfathiazole at which no pptn occurred.

SOURCE AND PURITY OF MATERIALS:

Source and purity of sulfathiazole was not purified. The mp of crystalline sulfathiazole was 200-4°C.

Purity of the water was not specified.

ESTIMATED ERROR:

Nothing specified

REFERENCES:

COMPONENTS:	ORIGINAL MEASUREMENTS:
(1) Benzenesulfonamide, 4-amino-N-2-thiazolyl- (sulfathiazole); $C_9H_9N_3O_2S_2$; [72-14-0] (2) Potassium chloride; KCl; [7447-40-7] (3) Water; H_2O; [7732-18-5]	Higuchi, T.; Gupta, M.; Busse, L. W. *J. Am. Pharm. Assoc., Sci. Ed.* 1953, *52*, 157-61.
VARIABLES:	PREPARED BY:
One temperature: 25°C	R. Piekos

EXPERIMENTAL VALUES:

Solubility of sulfathiazole in a KCl solution (ionic strength 0.15 M) at 25°C is 0.0373 g/100 cm^3 saturated solution (1.46 x 10^{-3} mol dm^{-3}, compiler).

AUXILIARY INFORMATION

METHOD/APPARATUS/PROCEDURE:

A 10-30% excess of sulfathiazole was equilibrated in a sealed vial for 1-8 days. An aliquot of the supernatant was withdrawn with a hypodermic syringe, the liquid was weighed, and the sulfathiazole was detd in it spectrophotometrically at 283 nm using 0.1 M citrate buffer of pH 5 as the solvent.

SOURCE AND PURITY OF MATERIALS:

Sulfathiazole (source not specified) was recrystd from hot water. The source and purity of KCl was not specified. Distilled water was used.

ESTIMATED ERROR:

Soly: the average of the following soly values was given: 0.0372, 0.0373, 0.0384, and 0.0363 g/100 cm^3 satd soln.
Temp: not specified.

REFERENCES:

COMPONENTS:

(1) Benzenesulfonamide, 4-amino-N-2-thiazolyl- (sulfathiazole); $C_9H_9N_3O_2S_2$; [72-14-0]

(2) Phosphoric acid, disodium salt; Na_2HPO_4; [7558-94-4]

(3) Water; H_2O; [7732-18-5]

ORIGINAL MEASUREMENTS:

Krüger-Thiemer, E.

Arch. Dermatol. Syphilis 1942, *183*, 90-116.

VARIABLES:

One temperature: ca 20°C; one pH: 8.74

PREPARED BY:

R. Piekos

EXPERIMENTAL VALUES:

Solubility of sulfathiazole in a 0.705 M (10%) Na_2HPO_4 solution of pH 8.74, at room temperature (about 20°C), is 0.228 g% (8.93 x 10^{-3} mol dm^{-3} solution, compiler).

AUXILIARY INFORMATION

METHOD/APPARATUS/PROCEDURE:

Sulfathiazole (0.5 g) was dissolved in 10 cm^3 of the 0.705 M (10%) Na_2HPO_4 soln, shaken for 2 h at room temp (about 20°C), and filtered. A 1-cm^3 aliquot of the filtrate was withdrawn, cooled, acidified with 1 cm^3 of 2 N HCl, and the sulfathiazole content was detd colorimetrically by the method of Marshall modified by Kimmig (1) using an Autenrieth colorimeter. The pH was detd on an ultraionograph using a glass electrode.

SOURCE AND PURITY OF MATERIALS:

Sulfathiazole was the product manufd by Ciba under the name Cibazol.
The source and purity of the remaining materials was not specified.

ESTIMATED ERROR:

Soly: precision ±5% (author)
Temp: not specified
pH ; ±0.05 pH unit (author)

REFERENCES:

1. Kimmig, J. *Arch. Dermatol.* 1938, *176*, 722; *Erg. Hyg.* 1941, *24*, 398.

COMPONENTS:	ORIGINAL MEASUREMENTS:
(1) Benzenesulfonamide, 4-amino-N-2-thiazolyl- (sulfathiazole); $C_9H_9N_3O_2S_2$; [72-14-0] (2) Phosphoric acid, monopotassium salt; KH_2PO_4; [7778-77-0] (3) Water; H_2O; [7732-18-5]	Krüger-Thiemer, E. *Arch. Dermatol. Syphilis* 1942, *183*, 90-116.
VARIABLES:	**PREPARED BY:**
One temperature: ca 20°C; one pH: 4.37	R. Piekos

EXPERIMENTAL VALUES:

Solubility of sulfathiazole in a 0.735 M (10%) KH_2PO_4 solution of pH 4.37, at room temperature (about 20°C), is 0.029 g% (1.13 x 10^{-3} mol dm^{-3} solution, compiler).

AUXILIARY INFORMATION

METHOD/APPARATUS/PROCEDURE:

Sulfathiazole (0.5 g) was dissolved in 10 cm^3 of the 0.735 M (10%) KH_2PO_4 soln, shaken for 2 h at room temp (about 20°C), and filtered. A 1-cm^3 aliquot of the filtrate was withdrawn, cooled, acidified with 1 cm^3 of 2 N HCl, and the sulfathiazole content was detd colorimetrically by the method of Marshall modified by Kimmig (1) using an Autenrieth colorimeter. The pH was detd on an ultra-ionograph using a glass electrode.

SOURCE AND PURITY OF MATERIALS:

Sulfathiazole was the product manufd by Ciba under the name of Cibazol. The source and purity of the remaining materials was not specified.

ESTIMATED ERROR:

Soly: precision ±5% (author)
Temp: not specified
pH : ±0.05 pH unit (author)

REFERENCES:

1. Kimmig, J. *Arch. Dermatol.* *176*, 722; *Erg. Hyg.* 1941, *24*, 398.

COMPONENTS:

(1) Benzenesulfonamide, 4-amino-N-2-thiazolyl- (sulfathiazole); $C_9H_9N_3O_2S_2$; [72-14-0]
(2) Calcium chloride; $CaCl_2$; [10043-52-4]
(3) Potassium chloride; KCl; [7447-40-7]
(4) Sodium chloride; NaCl; [7647-14-5]
(5) Water; H_2O; [7732-18-5]

ORIGINAL MEASUREMENTS:

Hawking, F.

Lancet, 1941, *240*, 786-8.

VARIABLES:

Temperature

PREPARED BY:

R. Piekos

EXPERIMENTAL VALUES:

t/°C	Solubility in bicarbonate-free Locke's solution[a]	
	mg/100 ml	10^3 mol dm^{-3} [b]
17	36	1.4
36	91	3.6

[a] The solution contained NaCl 9 g, KCl 0.2 g, $CaCl_2$ 0.2 g, water 1 liter, and had a pH of 6.8.

[b] Calculated by compiler

AUXILIARY INFORMATION

METHOD/APPARATUS/PROCEDURE:

Sulfathiazole was shaken up with the bicarbonate-free Locke's soln for many hours in a tube which was corked to prevent loss of CO_2. The supernatant was filtered through a paper, dild in a hot room to prevent pptn, and sulfathiazole was detd by the method of Marshall and Litchfield (1).

SOURCE AND PURITY OF MATERIALS:

Nothing specified

ESTIMATED ERROR:

Soly: average of 3 detns has been given (authors).

Temp: not specified.

REFERENCES:

1. Marshall, E. K., Jr.; Litchfield, J. T., Jr. *Science*, 1938, *88*, 85.

COMPONENTS:	ORIGINAL MEASUREMENTS:
(1) Benzenesulfonamide, 4-amino-N-2-thiazolyl- (sulfathiazole); $C_9H_9N_3O_2S_2$; [72-14-0] (2) Phosphoric acid, disodium salt; Na_2HPO_4; [7558-94-4] (3) Phosphoric acid, monopotassium salt; KH_2PO_4; [7778-77-0] (4) Water; H_2O; [7732-18-5]	Krüger-Thiemer, E. *Arch. Dermatol. Syphilis* 1942, *183*, 90-116.
VARIABLES: Temperature, pH	PREPARED BY: R. Piekos

EXPERIMENTAL VALUES:

Composition of 1/15 M phosphate buffer solutions			pH	Solubility			
				Room temp (ca 20°C)		37°C	
Na_2HPO_4	KH_2PO_4	% content		g%	10^3 mol dm^{-3} solution[a]	g%	10^3 mol dm^{-3} solution[a]
1.0	99.0	0.91	4.944	0.053	2.076	-	-
10.0	90.0	0.91	5.906	0.054	2.115	0.096	3.760
61.1	38.9	0.93	7.005	0.072	2.820	0.161	6.306
9.5	0.5	0.733[b]	7.51	0.089	3.486	-	-
94.7	5.3	0.95	8.018	0.144	5.640	-	-

[a] Calculated by compiler

[b] Molar content; 10% buffer solution

AUXILIARY INFORMATION

METHOD/APPARATUS/PROCEDURE:

Sulfathiazole (0.5 g) was dissolved in 10 cm^3 of a buffer soln, shaken for 2 h at 20°C (or left for 48 h at 37°C), and filtered at respective temp. A 1 cm^3 aliquot of the filtrate was then withdrawn, cooled (dild for expts at 37°C), acidified with 1 cm^3 of 2 N HCl, and the sulfathiazole content was detd colorimetrically by the method of Marshall modified by Kimmig (1) using an Autenrieth colorimeter. The pH was detd on an ultra-ionograph using a glass electrode.

SOURCE AND PURITY OF MATERIALS:

Sulfathiazole was the product manufd by Ciba under the name of Cibazol.
The source and purity of the remaining reagents was not specified.

ESTIMATED ERROR:

Soly: precision ±5% (author)
Temp: not specified
pH : ±0.05 pH unit (author)

REFERENCES:

1. Kimmig, J. *Arch. Dermatol.* *176*, 722; *Erg. Hyg.* 1941, *24*, 398.

COMPONENTS:

(1) Benzenesulfonamide, 4-amino-N-2-thiazolyl- (sulfathiazole); $C_9H_9N_3O_2S_2$; [72-14-0]

(2) Phosphoric acid, disodium salt; Na_2HPO_4; [7558-94-4]

(3) Phosphoric acid, monopotassium salt KH_2PO_4; [7778-77-0]

(4) Water; H_2O; [7732-18-5]

ORIGINAL MEASUREMENTS:

Pulver R.; Suter, R.

Schweiz. Med. Wochenschr. 1943, *73(13)*, 403-8.

VARIABLES:

pH

PREPARED BY:

R. Piekos

EXPERIMENTAL VALUES:

pH	Solubility of sulfathiazole in M/15 phosphate buffers (according to Sørensen) at 20°C	
	mg%	10^3 mol dm^{-3} [a]
6.0	51	2.00
7.0	65	2.54
8.0	125	4.90

[a]calculated by compiler

AUXILIARY INFORMATION

METHOD/APPARATUS/PROCEDURE:

Nothing specified

SOURCE AND PURITY OF MATERIALS:

Nothing specified

ESTIMATED ERROR:

Nothing specified

REFERENCES:

COMPONENTS:	ORIGINAL MEASUREMENTS:
(1) Benzenesulfonamide, 4-amino-N-2-thiazolyl- (sulfathiazole); $C_9H_9N_3O_2S_2$; [72-14-0] (2) Phosphoric acid, disodium salt; Na_2HPO_4; [7558-94-4] (3) Phosphoric acid, monopotassium salt; KH_2PO_4; [7778-77-0] (4) Water; H_2O; [7732-18-5]	Sunderman, F. W.; Pepper, D. S.; Benditt, E. *J. Med. Sci.* 1940, *200*, 790-5.
VARIABLES:	PREPARED BY:
pH	R. Piekos

EXPERIMENTAL VALUES:

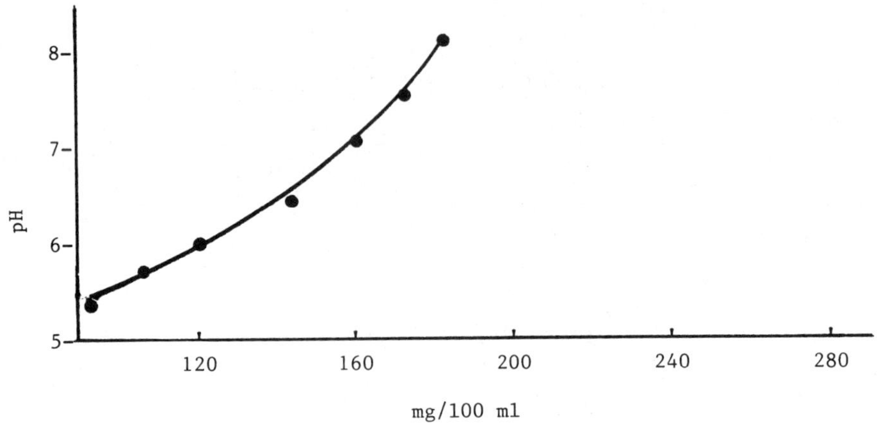

Solubility of sulfathiazole in phosphate buffer solution at 38°C.

AUXILIARY INFORMATION

METHOD/APPARATUS/PROCEDURE:

An excess of sulfathiazole was suspended in buffer solns (prepd by dilg appropriate mixts of Na_2HPO_4 and KH_2PO_4, 1 part to 10 parts of distd water), agitated and kept in a water bath at 38°C for about 2 h. The solns were then filtered and analysis for total sulfathiazole were made on the filtrates. Sulfathiazole was assayed colorimetrically after coupling with di-Me-1-napthylamine using a Bausch and Lomb colorimeter fitted with a No. 74 Wratten filter. Standards were made from a stock soln of sulfathiazole contg 200 mg/liter.

SOURCE AND PURITY OF MATERIALS:

Distd water was used.

The source and purity of the remaining reagents were not specified.

ESTIMATED ERROR:

Soly: the curve represents a composite of 3 sets of detns (authors).

Temp and pH: not specified.

REFERENCES:

COMPONENTS:	ORIGINAL MEASUREMENTS:
(1) Benzenesulfonamide, 4-amino-N-2-thiazolyl- (sulfathiazole); $C_9H_9N_3O_2S_2$; [72-14-0] (2) Phosphoric acid, disodium salt; Na_2HPO_4; [7558-94-4] (3) Phosphoric acid, monopotassium salt; KH_2PO_4; [7778-77-0] (4) Water; H_2O; [7732-18-5]	Frisk, A. R.; Hagerman, G.; Helander, S.; Sjögren, B. *Hygiea* 1946, *108(12)*, 639-51.
VARIABLES:	**PREPARED BY:**
pH	R. Piekos

EXPERIMENTAL VALUES:

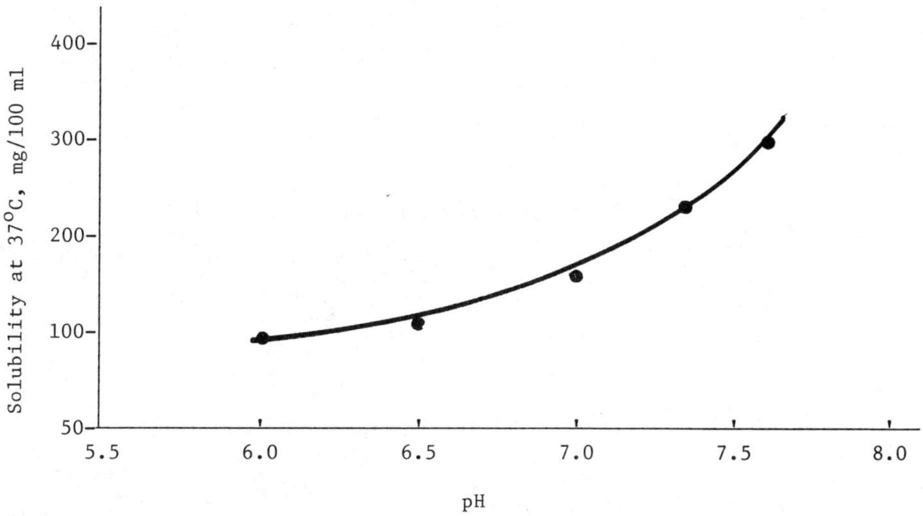

The solubility at pH 6.1 is 104 mg/100 ml solvent (4.073×10^{-3} mol dm^{-3}, compiler).

AUXILIARY INFORMATION

METHOD/APPARATUS/PROCEDURE:

An excess of sulfathiazole in the phosphate buffer was shaken at 37°C for 24 h. The concn of the solute was detd by the Bratton and Marshall method (1) using a photoelec colorimeter.

SOURCE AND PURITY OF MATERIALS:

Neither source nor purity of the materials was specified.

ESTIMATED ERROR:

Soly: precision ±8 mg/100 ml (authors).

Temp and pH: not specified.

REFERENCES:

1. Bratton, A. C.; Marshall, E. K. Jr. *J. Biol. Chem.* 1939, *128*, 537.

COMPONENTS:

(1) Benzenesulfonamide, 4-amino-N-2-thiazolyl- (sulfathiazole); $C_9H_9N_3O_2S_2$; [72-14-0]

(2) Phosphoric acid, disodium salt; Na_2HPO_4; [7558-94-4]

(3) Phosphoric acid, monopotassium salt; KH_2PO_4; [7778-77-0]

(4) Water; H_2O; [7732-18-5]

ORIGINAL MEASUREMENTS:

Frisk, A. R.; Hagerman, G.; Helander, S.; Sjögren, B. *Hygiea* 1946, *108(12)*, 639-51.

VARIABLES:

pH: 6.1

PREPARED BY:

R. Piekos

EXPERIMENTAL VALUES:

The solubility at pH 6.1 is 104 mg/100 ml solvent (4.07 x 10^{-3} mol dm^{-3}, compiler). This is the solubility value of sulfathiazole at 37°C.

AUXILIARY INFORMATION

METHOD/APPARATUS/PROCEDURE:

An excess of sulfathiazole in the phosphate buffer was shaken at 37°C for 24 h. The concn of the solute was detd by the Bratton and Marshall method (1) using a photoelec colorimeter.

SOURCE AND PURITY OF MATERIALS:

Neither source nor purity of the materials was specified.

ESTIMATED ERROR:

Soly: precision ±8 mg/100 ml (authors).

Temp and pH: not specified

REFERENCES:

1. Bratton, A. C.; Marshall, E. K. Jr. *J. Biol. Chem.* 1939, *128*, 537.

COMPONENTS:

(1) Benzenesulfonamide, 4-amino-N-2-thiazolyl- (sulfathiazole); $C_9H_9N_3O_2S_2$; [72-14-0]

(2) Phosphoric acid, disodium salt; Na_2HPO_4; [7558-94-4]

(3) Phosphoric acid, monopotassium salt; KH_2PO_4; [7778-77-0]

(4) Water; H_2O; [7732-18-5]

VARIABLES:

pH

ORIGINAL MEASUREMENTS:

Langecker, H.

Arch. Exptl. Path. Pharmakol. 1948, *205*, 291-301.

PREPARED BY:

R. Piekos

EXPERIMENTAL VALUES:

pH of the 1/15 M phosphate buffer	Solubility at 37°C	
	mg%	10^3 mol dm^{-3} [a]
4.9	108	4.23
5.9	94	3.68
6.9	204	7.99
7.5	356	13.94

[a] Calculated by compiler

AUXILIARY INFORMATION

METHOD/APPARATUS/PROCEDURE:

An excess of sulfathiazole was added to the buffer soln and boiled for 1 h in a sealed ampul followed by keeping the ampul at 37°C. The concn of sulfathiazole was detd colorimetrically by the method of Bratton and Marshall (1) using a Havemann colorimeter (2), as well as by microanal detn of the solid residue.

SOURCE AND PURITY OF MATERIALS:

Source and purity of the materials was not specified.

ESTIMATED ERROR:

Nothing specified

REFERENCES:

1. Bratton, A.G.; Marshall, E.K. Jr. *J. Biol. Chem.* 1939, *128*, 537.
2. Havemann, R. *Klin. Wochenschr.* 1940, p. 503.

COMPONENTS:

(1) Benzenesulfonamide, 4-amino-N-2-thiazolyl- (sulfathiazole); $C_9H_9N_3O_2S_2$; [72-14-0]

(2) Phosphoric acid, disodium salt; Na_2HPO_4; [7558-94-4]

(3) Phosphoric acid; monopotassium salt; KH_2PO_4; [7778-77-0]

(4) Water; H_2O; [7732-18-5]

ORIGINAL MEASUREMENTS:

Kuhnert-Brandstätter, M.; Martinek, A.
Microchim. Ichnoanal. Acta 1956, 909-19.

VARIABLES:

One temperature: 20°C; one pH: 7.3

PREPARED BY:

R. Piekos

EXPERIMENTAL VALUES:

Solubility of crystalline forms I and II of sulfathiazole in a 0.066 M phosphate buffer (according to Sørensen) of pH 7.3 at 20°C is 113.0 mg% (4.43 x 10^{-3} mol dm^{-3}, compiler) and 62.8 mg% (2.46 x 10^{-3} mol dm^{-3}, compiler), respectively.

AUXILIARY INFORMATION

METHOD/APPARATUS/PROCEDURE:

Sulfathiazole and the buffer soln were placed in a polyethylene vessel, agitated for 3 h, filtered, and the sulfonamide was assayed in the filtrate by uv spectrophotometry using water as a reference and diluent. The solid phase was examd thermomicroscopically for identity of the cryst form.

SOURCE AND PURITY OF MATERIALS:

A comm available form II of sulfathiazole was used. Form I was obtained by keeping the comm reagent at 170°C for 2 h. The source and purity of the remaining materials was not specified. Distilled water was used.

ESTIMATED ERROR:

Soly: not specified
pH : not specified
Temp: ±0.5°C (authors).

REFERENCES:

COMPONENTS:

(1) Benzenesulfonamide, 4-amino-N-2-thiazolyl- (sulfathiazole); $C_9H_9N_3O_2S_2$; [72-14-0]

(2) Phosphoric acid, disodium salt; Na_2HPO_4; [7558-94-4]

(3) Phosphoric acid, monopotassium salt; KH_2PO_4; [7778-77-0]

(4) Water; H_2O; [7732-18-5]

ORIGINAL MEASUREMENTS:

Riess, W.

Intern. Congr. Chemotherapy, Proc., 3rd, Stuttgart 1963, *1*, 627-32.

VARIABLES:

One temperature: 20°C; one pH: 7.4

PREPARED BY:

R. Piekos

EXPERIMENTAL VALUES:

Solubility of sulfathiazole in a M/15 Sörensen buffer solution (pH 7.4) at 20°C is 75 mg% (2.9 x 10^{-3} mol dm^{-3} solution, compiler).

AUXILIARY INFORMATION

METHOD/APPARATUS/PROCEDURE:

Sörensen buffer solns of pH varying between 7 and 8 were prepd, satd with sulfathiazole at 20°C, their pH was measured at equilibrium, and the sulfathiazole was assayed colorimetrically. The measured pH values were then plotted against concn, and the soly at pH 7.4 was detd by interpolation (personal communication).

SOURCE AND PURITY OF MATERIALS:

Nothing specified

ESTIMATED ERROR:

Nothing specified

REFERENCES:

COMPONENTS:	ORIGINAL MEASUREMENTS:
(1) Benzenesulfonamide, 4-amino-N-2-thiazolyl- (sulfathiazole); $C_9H_9N_3O_2S_2$; [72-14-0] (2) Phosphoric acid, disodium salt; Na_2HPO_4; [7558-94-4] (3) Phosphoric acid, monopotassium salt; KH_2PO_4; [7778-77-0] (4) Water; H_2O; [7732-18-5]	Yamazaki, M.; Aoki, M.; Kamada, A.; Yata, N. *Yakuzaigaku* 1967, *27(1)*, 37-40.
VARIABLES: One temperature: 30°C; one pH: 7.4	PREPARED BY: R. Piekos

EXPERIMENTAL VALUES:

Solubility of sulfathiazole in a phosphate buffer solution of pH 7.4 (μ = 0.17) at 30°C is 4.38 mmol/L (1.12 g dm^{-3}, compiler).

AUXILIARY INFORMATION

METHOD/APPARATUS/PROCEDURE:

Sulfathiazole (0.5 g) was placed in an L-shaped tube together with 20 ml of the buffer soln. The mixt was shaken in a thermostat until equilibrium was attained. The sulfathiazole content was assayed in the supernatant spectrophotometrically at 545 nm on a Beckman DU spectrophotometer. The results were taken from a calibration graph.

SOURCE AND PURITY OF MATERIALS:

Nothing specified

ESTIMATED ERROR:

Soly and pH: not specified

Temp: ±1°C (authors)

REFERENCES:

COMPONENTS:

(1) Benzenesulfonamide, 4-amino-N-2-thiazolyl- (sulfathiazole); $C_9H_9N_3O_2S_2$; [72-14-0]

(2) Phosphoric acid, disodium salt; Na_2HPO_4; [7558-94-4]

(3) Phosphoric acid, monopotassium salt; KH_2PO_4; [7778-77-0]

(4) Water; H_2O; [7732-18-5]

ORIGINAL MEASUREMENTS:

Hekster, Y.A.; Vree, T. B.; Damsma, J. E.; Friesen, W. T.

J. Antimicrob. Chemother. 1981, *8*, 133-44.

VARIABLES:

pH

PREPARED BY:

R. Piekos

EXPERIMENTAL VALUES:

pH	Solubility at 25°C	
	mg/l	mol dm^{-3} [a]
5.5	4565	0.01788
7.5	130868	0.51258

[a] Calculated by compiler

AUXILIARY INFORMATION

METHOD/APPARATUS/PROCEDURE:

Satd solns of sulfathiazole were prepd in phosphate buffers of pH 5.5 and 7.5 at room temp (25°C). The concn of the solute was measured by means of a Spectra Physics 3500B high-performance liquid chromatograph equipped with a column oven (Model 748) and a Pye-Unicam LC-UV spectrophotometric detector. The detector was connected to a 1-mV recorder. A stainless steel column (10 cm x 4.6 mm id.) was packed with Lichrosorb RPS, 5 µm, obtained from Chrompack. An injection loop of 100 µl was used. The oven temp was 40°C. Detection of sulfathiazole was performed at 260 nm.

SOURCE AND PURITY OF MATERIALS:

The source and purity of the materials was not specified.

ESTIMATED ERROR:

The detection limit of the solute by HPLC was 0.5 mg/l (authors).
The error in temperature and pH was not specified.

REFERENCES:

COMPONENTS:

(1) Benzensulfonamide, 4-amino-N-2-thiazolyl- (sulfathiazole); $C_9H_9N_3O_2S_2$; [72-14-0]

(2) Hydrochloric acid; HCl; [7647-01-0]

(3) 1,2,3-Propanetricarboxylic acid, 2-hydroxy-, disodium salt; $C_6H_6Na_2O_7$; [144-33-2]

(4) Water; H_2O; [7732-18-5]

ORIGINAL MEASUREMENTS:

Kuhnert- Brandstätter, M.; Martinek, A. *Microchim. Ichnoanal. Acta* 1956, 909-19.

VARIABLES:

One temperature: 20°C; one pH: 3.8

PREPARED BY:

R. Piekos

EXPERIMENTAL VALUES:

Solubility of crystalline forms I and II of sulfathiazole in a 0.066 M citrate buffer (according to Sørensen) of pH 3.8 at 20°C is 63.4 mg% (2.48 x 10^{-3} mol dm^{-3}, compiler) and 36.1 mg% (1.41 x 10^{-3} mol dm^{-3}, compiler), respectively.

AUXILIARY INFORMATION

METHOD/APPARATUS/PROCEDURE:

Sulfathiazole and the buffer soln were placed in a polyethylene vessel, agitated for 3 h, filtered, and the sulfonamide was assayed in the filtrate by uv spectrophotometry using water as a reference and diluent. The solid phase was examd thermomicroscopically for identity of the cryst form.

SOURCE AND PURITY OF MATERIALS:

A comm available form II of sulfathiazole was used. Form I was obtained by keeping the comm reagent at 170°C for 2 h. The source and purity of the remaining materials was not specified.
Distilled water was used.

ESTIMATED ERROR:

Soly: not specified
pH : not specified
Temp: ±0.5°C (authors)

REFERENCES:

COMPONENTS:

(1) Benzenesulfonamide, 4-amino-N-2-thiazolyl- (sulfathiazole); $C_9H_9N_3O_2S_2$; [72-14-0]

(2) Phosphoric acid, disodium salt; Na_2HPO_4; [7558-94-4]

(3) 1,2,3-Propanetricarboxylic acid, 2-hydroxy- (citric acid); $C_6H_8O_7$; [77-92-9]

(4) Water; H_2O; [7732-18-5]

ORIGINAL MEASUREMENTS:

Likhol'ot, N. M. *Farm. Zh. (Kiev)* 1965, *20(5)*, 44-6.

VARIABLES:

pH

PREPARED BY:

R. Piekos

EXPERIMENTAL VALUES:

pH of McIlvaine's buffer solution	Solubility at 20°C	
	g/100 ml	10^3 mol dm^{-3} [a]
4.1	0.043	1.68
5.1	0.045	1.76
5.9	0.049	1.92
6.5	0.059	2.31
6.9	0.081	3.17
7.5	0.153	5.99

[a] Calculated by compiler

AUXILIARY INFORMATION

METHOD/APPARATUS/PROCEDURE:

An earlier described method was employed (1) whereby a small excess of sulfathiazole was equilibrated with 20 ml of the McIlvaine's buffer soln for 8 h in a 50-ml test tube. Aliquots were withdrawn through a filter and sulfathiazole was assayed bromatometrically.

SOURCE AND PURITY OF MATERIALS:

Sulfathiazole: not specified.
McIlvaine's buffer solns were prepd from a 0.1 M citric acid solns. Source and purity of the buffer components were not specified.

ESTIMATED ERROR:

Soly: not specified
Temp: ±0.1°C (authors)
pH : not specified

REFERENCES:

1. Gusyakov, V. P.; Likhol'ot, N. M. *Farm. Zh. (Kiev)* 1960, *15(8)*, 21.

COMPONENTS:

(1) Benzenesulfonamide, 4-amino-N-2-thiazolyl- (sulfathiazole); $C_9H_9N_3O_2S_2$; [72-14-0]

(2) Ethanesulfonic acid, 2-[[3α,5β,7α,12α)-3,7,12-trihydroxy-24-oxocholan-24-yl]amino]-, monosodium salt (Na taurocholate); $C_{26}H_{45}NO_7S \cdot Na$; [145-42-6]

(3) Phosphoric acid, disodium salt; Na_2HPO_4; [7558-94-4]

(4) Phosphoric acid, monosodium salt; NaH_2PO_4; [7558-80-7]

(5) Water; H_2O; [7732-18-5]

ORIGINAL MEASUREMENTS:

Gasco, M. R.; Aimonetto, S.

Atti Accad. Sci. Torino, Cl, Sci. Fis., Mat. Nat. 1979, *113(1-2)*, 119-22.

VARIABLES:

Concentration of Na taurocholate; pH

PREPARED BY:

R. Piekos

EXPERIMENTAL VALUES:

Concentration of Na taurocholate mM/l solution[a]	Solubility of sulfathiazole at 25°C μm/ml solution[a]	
	pH 6.3	pH 7.2
2.0	3.65	6.41
4.0	3.28	6.27
6.0	3.42	5.77
8.0	3.60	6.05
12.0	3.70	6.18
16.0	3.89	6.73
20.0	4.21	7.02

[a] Numerical values given by the first author in personal communication.

AUXILIARY INFORMATION

METHOD/APPARATUS/PROCEDURE:

The soly of sulfathiazole was detd by the method of Hofmann (1). In a series of 15-ml glass cylinders with ground-in stoppers, 75 mg of sulfathiazole was suspended in 15 ml of phosphate buffer solns of increasing Na taurocholate concn. The suspensions were agitated for 20 h at 25°C and filtered. The quantity of sulfathiazole dissolved was detd by measuring surface tension by means of a Dognon-Abribat (Prolabo) tensiometer and spectrophotometrically by using a Perkin Elmer EPS-35 spectrophotometer.

SOURCE AND PURITY OF MATERIALS:

Neither source nor purity of the materials was specified.

The phosphate buffer was 0.3 M in respect of the Na^+ ion concn.

ESTIMATED ERROR:

Soly: precision ±2% (authors)
pH : precision ±0.02 pH unit (authors)
Temp: ±0.5°C (authors)

REFERENCES:

1. Hofmann, A. F.; *Biochem. J.* 1963, *89*, 57.

COMPONENTS:	ORIGINAL MEASUREMENTS:
(1) Benzenesulfonamide, 4-amino-N-2-thiazolyl- (sulfathiazole); $C_9H_9N_3O_2S_2$; [72-14-0] (2) Ethanesulfonic acid, 2-[[[[(3α, 5β, 7α, 12α)-3,7,12-trihydroxy-24-oxocholan-24-yl]amino]acetyl]amino]-, sodium salt (Na tauroglycocholate); $C_{28}H_{48}N_2O_8S{\cdot}Na$ [11006-55-6] (3) Phosphoric acid, disodium salt; Na_2HPO_4; [7558-94-4] (4) Phosphoric acid, monosodium salt; NaH_2PO_4; [7558-80-7] (5) Water; H_2O; [7732-18-5]	Gasco, M. R.; Aimonetto, S. *Atti Accad. Sci. Torino, Cl. Sci. Fis., Mat. Nat.* 1979, *113(1-2)*, 119-22. PREPARED BY: R. Piekos

VARIABLES:
Concentration of Na tauroglycocholate; pH

EXPERIMENTAL VALUES:

Concentration of Na tauroglycocholate mM/l solution[a]	Solubility of sulfathiazole at 25°C μM/ml solution[a]	
	pH 6.3	pH 7.2
2.0	3.80	6.69
4.0	3.52	6.50
6.0	3.62	6.32
8.0	3.70	6.41
12.0	3.90	6.43
16.0	4.08	6.50
20.0	4.17	6.52

[a] Numerical values given by the first author in personal communication.

AUXILIARY INFORMATION

METHOD/APPARATUS/PROCEDURE:

The soly of sulfathiazole was detd by the Hofmann method (1). In a series of 15-ml glass cylinders with ground-in stoppers, 75 mg of sulfathiazole was placed in 15 ml of phosphate buffer solns of increasing Na tauroglycocholate concn. The suspensions were agitated for 20 h at 25°C and filtered. The quantity of sulfathiazole dissolved was detd by measuring surface tension by means of a Dognon-Abribat (Prolabo) tensiometer and spectrophotometrically on a Perkin Elmer EPS-35 spectrophotometer.

SOURCE AND PURITY OF MATERIALS:

Neither source nor purity of the materials was specified.
The phosphate buffer was 0.3 M in respect of the Na^+ ion concn.

ESTIMATED ERROR:

Soly: precision ±2% (authors)
pH : precision ±0.02 pH unit (authors)
Temp: ±0.5°C (authors)

REFERENCES:

1. Hofmann, A. F., *Biochem. J.* 1963, *89*, 57.

COMPONENTS:

(1) Benzenesulfonamide, 4-amino-N-2-thiazolyl- (sulfathiazole); $C_9H_9N_3O_2S_2$; [72-14-0]

(2) Pectinic acid, sodium salt; $(C_{13}H_{17}NaO_{12})_n$; [9049-37-0]

(3) Water; H_2O; [7732-18-5]

ORIGINAL MEASUREMENTS:

Becher, R.; Leya, S. *Experientia* 1946, *2*, 459-60.

VARIABLES:

One temperature : 18-19°C

PREPARED BY:

R. Piekos

EXPERIMENTAL VALUES:

Solubility of sulfathiazole in a 2.6% neutral sodium pectinate solution ([sodium pectinate] = 6.7×10^{-2} mol kg^{-1} (n = 1), compiler) at room temperature (18 - 19°C) is 75 mg% (2.9×10^{-3} mol dm^{-3}, compiler).

AUXILIARY INFORMATION

METHOD/APPARATUS/PROCEDURE:

The soln was allowed to stand for more than 2 days at room temp. The soln was then filtered, and sulfathiazole assayed in the filtrate colorimetrically by the method of Druey and Oesterheld (1).

SOURCE AND PURITY OF MATERIALS:

Nothing specified.

ESTIMATED ERROR:

Nothing specified.

REFERENCES:

1. Druey, J.; Oesterheld, G.; *Helv. Chim. Acta* 1942, *25*, 753.

COMPONENTS:	ORIGINAL MEASUREMENTS:
(1) Benzenesulfonamide, 4-amino-N-2-thiazolyl- (sulfathiazole); $C_9H_9N_3O_2S_2$; [72-14-0] (2) Cholan-24-oic acid, 3,7,12-trihydroxy-, (3 α, 5 β, 7 α, 12 α)-, monosodium salt (Na cholate); $C_{24}H_{39}NaO_5$; [361-09-1] (3) Water; H_2O; [7732-18-5]	Dubois, S.; Tawashi, R. *Pharm. Acta Helv.* 1975, *50*, 184-7.
VARIABLES:	PREPARED BY:
One temperature: 37°C	R. Piekos

EXPERIMENTAL VALUES:

Solubility of sulfathiazole in a 0.04 mol dm^{-3} Na cholate solution at 37°C is 15.60 x 10^{-4} g/ml (6.110 x 10^{-3} mol dm^{-3}, compiler).

AUXILIARY INFORMATION

METHOD/APPARATUS/PROCEDURE:	SOURCE AND PURITY OF MATERIALS:
Samples were filtered through a Millipore filter 0.45 μm and the amount of dissolved sulfathiazole was detd spectrophotometrically at 280 nm.	USP grade sulfathiazole was used without further treatment. Na cholate was reagent grade. Distilled water was used.
	ESTIMATED ERROR:
	Nothing specified
	REFERENCES:

COMPONENTS:

(1) Benzenesulfonamide, 4-amino-N-2-thiozolyl- (sulfathiazole); $C_9H_9N_3O_2S_2$; [72-14-0]

(2) Cholan-24-oic acid, 3,7,12-trihydroxy-, (3α, 5β, 7α, 12α)-, monosodium salt (Na cholate); $C_{24}H_{39}NaO_5$; [361-09-1]

(3) Ext. D. and C. Blue No.1; $C_{16}H_{18}N_3S \cdot Cl$; [61-73-4]

(4) Water; H_2O; [7732-18-5]

ORIGINAL MEASUREMENTS:

Dubois, S.; Tawashi, R.

Pharm. Acta Helv. 1975, *50*, 184-7.

VARIABLES:

One temperature: 37°C

PREPARED BY:

R. Piekos

EXPERIMENTAL VALUES:

Solubility of sulfathiazole in a 0.04 mol dm^{-3} Na cholate solution containing 50 μg/ml of Ext. D. and C. Blue No. 1 at 37°C is 14.90×10^{-4} g/ml (5.84×10^{-3} mol dm^{-3}, compiler).

AUXILIARY INFORMATION

METHOD/APPARATUS/PROCEDURE:

Samples were filtered through a Millipore filter 0.45 μm and the amount of dissolved sulfathiazole was detd spectrophotometrically at 280 nm.

SOURCE AND PURITY OF MATERIALS:

USP grade sulfathiazole, reagent grade Na cholate, certified Ext. D. and C. Blue No. 1 and distd water were used.

ESTIMATED ERROR:

Nothing specified

REFERENCES:

COMPONENTS:

(1) Benzenesulfonamide, 4-amino-N-2-thiazolyl- (sulfathiazole); $C_9H_9N_3O_2S_2$; [72-14-0]

(2) Cholan-24-oic acid, 3,7,12-trihydroxy-, (3α, 5β, 7α, 12α)-, monosodium salt (Na cholate); $C_{24}H_{39}NaO_5$; [361-09-1]

(3) F.D. and C. Violet No.1; $C_{39}H_{41}N_3O_6S_2 \cdot Na$; [1694-09-3]

(4) Water; H_2O; [7732-18-5]

ORIGINAL MEASUREMENTS:

Dubois, S.; Tawashi, R.

Pharm. Acta Helv. 1975, *50*, 184-7.

VARIABLES:

One temperature: 37°C

PREPARED BY:

R. Piekos

EXPERIMENTAL VALUES:

Solubility of sulfathiazole in a 0.04 mol dm^{-3} Na cholate solution containing 50 μg/ml of F.D. and C. Violet No. 1 at 37°C is 14.50×10^{-4} g/ml (5.68×10^{-3} mol dm^{-3}, compiler).

AUXILIARY INFORMATION

METHOD/APPARATUS/PROCEDURE:

Samples were filtered through a Millipore filter 0.45 μm and the amount of dissolved sulfathiazole was detd spectrophotometrically at 280 nm.

SOURCE AND PURITY OF MATERIALS:

USP grade sulfathiazole, reagent grade Na cholate, certified F.D. and C. Violet No. 1 and distd water were used.

ESTIMATED ERROR:

Nothing specified

REFERENCES:

COMPONENTS:

(1) Benzenesulfonamide, 4-amino-N-2-thiazolyl- (sulfathiazole); $C_9H_9N_3O_2S_2$; [72-14-0]

(2) Ethanol; C_2H_6O; [64-17-5]

(3) Water; H_2O; [7732-18-5]

ORIGINAL MEASUREMENTS:

Milosovich , G.

J. Pharm. Sci. 1964, *53*, 484-7.

VARIABLES:

Temperature

PREPARED BY:

R. Piekos

EXPERIMENTAL VALUES:

Solubility in 95% v/v ethanol

$t/^{o}C$	Form I		Form II	
	g/1000 g solvent	mol kg^{-1} [a]	g/1000 g solvent	mol kg^{-1} [a]
59.1	31.50	0.1234	40.7	0.1594
48.8	19.80	0.0775	28.1	0.1101
39.4	14.00	0.0548	21.4	0.0838
29.6	9.93	0.0389	16.7	0.0654
24.1	8.15	0.0319	14.2	0.0556
20.4	7.10	0.0278	13.1	0.0513
14.5	5.70	0.0223	-	-

[a] Calculated by compiler

AUXILIARY INFORMATION

METHOD/APPARATUS/PROCEDURE:

A large excess of form I powder was added to about 500 ml of 95% EtOH in a beaker and held at a desired temp with stirring until equilibrium was obtained. Stirring was stopped, and samples were pipetted through a glass wool to remove suspended particles. The samples were weighed and quant dild with 95% EtOH for spectrophotometric assay at 288 nm. Solubilities of form II were calcd from the dissoln rate date.

SOURCE AND PURITY OF MATERIALS:

A U.S.P. sulfathiazole was used. Form I, mp 174-5°C, was obtained by slow recrystn from warm EtOH. Form II was obtained by heating form I to 180°C. The source and purity of the materials was not specified.

ESTIMATED ERROR:

Nothing specified

REFERENCES:

COMPONENTS:

(1) Benzenesulfonamide, 4-amino-N-2-thiazolyl- (sulfathiazole); $C_9H_9N_3O_2S_2$; [72-14-0]

(2) Ethanol; C_2H_6O; [64-17-5]

(3) Water; H_2O; [7732-18-5]

ORIGINAL MEASUREMENTS:

Shkadova, A. I.

Farm. Zh. (Kiev) 1969, *24(3)*, 39-41.

VARIABLES:

Concentration of ethanol

PREPARED BY:

R. Piekos

EXPERIMENTAL VALUES:

Concentration of ethanol		Solubility at 20°C	
mole %	weight %	10^2 mol kg^{-1}	g/100 g [a]
0	0.00	0.15	0.038
10	22.14	0.76	0.194
20	39.01	1.66	0.424
30	52.31	4.46	1.139
40	63.04	5.53	1.412
50	71.90	5.60	1.430
60	79.33	5.23	1.335
70	85.65	4.71	1.202
80	91.10	3.33	0.842
90	95.83	1.72	0.439

[a] Calculated by compiler

AUXILIARY INFORMATION

METHOD/APPARATUS/PROCEDURE:

Sulfathiazole was equilibrated with the solvent in a water thermostat at 20±0.1°C. The concn of sulfathiazole was detd by alkalimetric titration.

SOURCE AND PURITY OF MATERIALS:

Purity of sulfathiazole conformed to the requirements of the State Pharmacopeia IX. The EtOH - water mixts were prepd from abs EtOH (purity and source not specified) and distd water.

ESTIMATED ERROR:

Soly: not specified

Temp: ±0.1°C (author).

REFERENCES:

COMPONENTS:	ORIGINAL MEASUREMENTS:
(1) Benzenesulfonamide, 4-amino-N-2-thiazolyl- (sulfathiazole); $C_9H_9N_3O_2S_2$; [72-14-0] (2) Ethanol; C_2H_6O; [64-17-5] (3) Water; H_2O; [7732-18-5]	Mehta, S. C.; Bernardo, P. D. Higuchi, W. I.; Simonelli, A. P. *J. Pharm. Sci.* 1970, *59(5)*, 638-44.
VARIABLES:	PREPARED BY:
Concentration of ethanol	R. Piekos

EXPERIMENTAL VALUES:

Vol/vol % ethanol in water	Solubility at 30°C	
	g/100 g	10^2 mol kg^{-1} [a]
50	1.30	5.09
95	1.06	4.15

[a] Calculated by compiler

AUXILIARY INFORMATION

METHOD/APPARATUS/PROCEDURE:

An excess of the amt of the recrystd sulfathiazole needed to produce a satd soln was placed in a volumetric flask with the solvent and agitated in a water bath at 30°C. Duplicate samples were withdrawn at 12-24-h intervals, filtered through a 0.45-μ Millipore filter, and analyzed spectrophotometrically.

SOURCE AND PURITY OF MATERIALS:

Sulfathiazole (source not specified) was purified by crystallization.
The source and purity of the remaining materials were not specified.

ESTIMATED ERROR:

Nothing specified

REFERENCES:

COMPONENTS:	ORIGINAL MEASUREMENTS:
(1) Benzenesulfonamide, 4-amino-N-2-thiazolyl- (sulfathiazole); $C_9H_9N_3O_2S_2$; [72-14-0] (2) Ethanol; C_2H_6O; [64-17-5] (3) 1,2,3-Propanetriol; $C_3H_8O_3$; [56-81-5] (4) Water; H_2O; [7732-18-5]	Dolique, R. ; Foucault, J. *Trav. soc. pharm. Montpellier* 1952, *12*, 145-53.
VARIABLES:	PREPARED BY:
One temperature: 26-28°C	R. Piekos

EXPERIMENTAL VALUES:

Solubility of sulfathiazole in a mixture of 1,2,3-propanetriol and 95° ethanol (2:1 by wt) at 26-28°C is 2.08% (8.32×10^{-2} mol kg^{-1}, compiler).

AUXILIARY INFORMATION

METHOD/APPARATUS/PROCEDURE:	SOURCE AND PURITY OF MATERIALS:
The sulfathiazole content was detd by diazotization of the amine group in a cold acidified 0.1N KNO_2 soln. An excess of KNO_2 was detected by using iodinated starch.	Nothing specified ESTIMATED ERROR: Nothing specified REFERENCES:

COMPONENTS:	ORIGINAL MEASUREMENTS:
(1) Benzenesulfonamide, 4-amino-N-2-thiazolyl- (sulfathiazole); $C_9H_9N_3O_2S_2$; [72-14-0] (2) Ethanol; C_2H_6O; [64-17-5] (3) 1,2,3-Propanetriol; $C_3H_8O_3$; [56-81-5] (4) Urea; CH_4N_2O; [57-13-6] (5) Water; H_2O; [7732-18-5]	Dolique, R.; Foucault, J. *Trav. soc. pharm. Montpellier* 1952, **12**, 145-53.
VARIABLES:	PREPARED BY:
One temperature: 26-28°C	R. Piekos

EXPERIMENTAL VALUES:

Solubility of sulfathiazole at 26-28°C in a saturated solution of urea in a mixture of 1,2,3-propanetriol and 95° ethanol (2:1 by wt), containing 54.5 g of urea per 100 g of the mixture, is 2.82% (0.114 mol kg^{-1} solvent, compiler).

AUXILIARY INFORMATION

METHOD/APPARATUS/PROCEDURE:	SOURCE AND PURITY OF MATERIALS:
The sulfathiazole content was detd by diazotization of the amine group in a cold acidified 0.1N KNO_2 soln. An excess of KNO_2 was detected by using iodinated starch.	Nothing specified
	ESTIMATED ERROR:
	Nothing specified
	REFERENCES:

COMPONENTS:

(1) Benzenesulfonamide, 4-amino-N-2-thiazolyl- (sulfathiazole); $C_9H_9N_3O_2S_2$; [72-14-0]

(2) Urea; CH_4NO;]57-13-6]

(3) Water; H_2O; [7732-18-5]

ORIGINAL MEASUREMENTS:

Sobin, S. S.

J, Lab. Clin. Med. 1942, *27,* 1657-8.

VARIABLES:

Concentration of urea

PREPARED BY:

R. Piekos

EXPERIMENTAL VALUES:

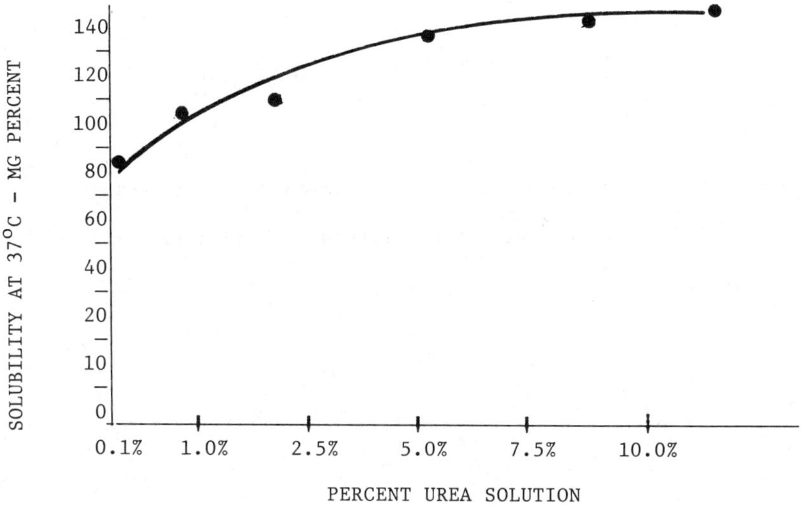

Solubility in a 10 percent urea solution at 37°C is 133.0 mg per 100 cm^3 (6.208×10^{-3} mol dm^{-3}, compiler).

AUXILIARY INFORMATION

METHOD/APPARATUS/PROCEDURE:

Urea solns of varying concns from 0.1 to 10% were incubated at 37°C with an excess of sulfathiazole, shaken at intervals, and filtered through two thicknesses of Whatman No. 42 filter paper. After appropriate diln the free sulfonamide was detd by the method of Bratton and Marshall (1) using the Evelyn colorimeter and a No. 540 filter.

SOURCE AND PURITY OF MATERIALS:

Nothing specified

ESTIMATED ERROR:

Nothing specified.

REFERENCES:

COMPONENTS:	ORIGINAL MEASUREMENTS:
(1) Benzenesulfonamide, 4-amino-N-2-thiazolyl- (sulfathiazole); $C_9H_9N_3O_2S_2$; [72-14-0] (2) Carbamic acid, ethyl ester (urethane); $C_3H_7NO_2$; [51-79-6] (3) Water; H_2O; [7732-18-5]	Weinstein, L.; McDonald, A. *Science* 1945, *101*, 44-5.
VARIABLES:	PREPARED BY:
One temperature: 20°C	R. Piekos

EXPERIMENTAL VALUES:

Solubility of sulfathiazole in a 10% aqueous urethane solution at 20°C is 200 mg/100 cm^3 urethane solution (7.83×10^{-3} mol dm^{-3}, compiler).

AUXILIARY INFORMATION

METHOD/APPARATUS/PROCEDURE:

Nothing specified

SOURCE AND PURITY OF MATERIALS:

Nothing specified

ESTIMATED ERROR:

Nothing specified

REFERENCES:

COMPONENTS:	ORIGINAL MEASUREMENTS:
(1) Benzenesulfonamide, 4-amino-N-2-thiazolyl- (sulfathiazole); $C_9H_9N_3O_2S_2$; [72-14-0] (2) 1H-Purine-2,6-dione, 3,7-dihydro-1,3,7-trimethyl- (caffeine); $C_8H_{10}N_4O_2$; [58-08-2] (3) Water; H_2O; [7732-18-5]	Higuchi, T.; Lach, J. L. *J. Amer. Pharm. Assoc., Sci. Ed.* 1954, *43*, 349-54.
VARIABLES:	PREPARED BY:
Concentration of caffeine	R. Piekos

EXPERIMENTAL VALUES:

Total solubility of sulfathiazole in water containing caffeine at 30°C

Caffeine	Sulfathiazole		Caffeine	Sulfathiazole	
10^2 mol dm^{-3}	10^3 mol dm^{-3}	g dm^{-3} [a]	10^2 mol dm^{-3}	10^{-3} mol dm^{-3}	g dm^{-3} [a]
0.000	2.27	0.58	10.485	5.01	1.28
1.419	2.66	0.68	12.250	5.35	1.37
1.514	2.63	0.67	13.342	5.45	1.39
1.674	2.74	0.70	14.069	5.54	1.41
3.457	3.21	0.82	14.074	5.46	1.39
3.922	3.35	0.85	14.908	5.62	1.43
3.944	3.27	0.83	15.089	5.69	1.45
4.573	3.47	0.89	15.907	5.58	1.42
5.468	3.68	0.94			
6.375	3.92	1.00			
7.951	4.27	1.09			
7.956	4.30	1.10			
9.017	4.59	1.17			
10.448	4.73	1.21			

[a] Calculated by compiler

AUXILIARY INFORMATION

METHOD/APPARATUS/PROCEDURE:

Sulfathiazole (75 mg) was placed in 125-ml glass-stoppered bottles together with varying but accurately weighed amts of caffeine and 50-ml portions of water. The bottles were placed in a mech shaker in a const temp bath and equilibrated for 8 h at 30°C. Aliquots of the supernatant liquid were analyzed for the sulfonamide by the method of Bratton and Marshall (1).

SOURCE AND PURITY OF MATERIALS:

Recrystd sulfathiazole (U.S.P.), mp 201-2°C and recrystd caffeine (U.S.P.), mp 235-7°C were used. The water used was distilled.

ESTIMATED ERROR:

Nothing specified

REFERENCES:

1. Bratton, A. C.; Marshall, E. K., Jr. *J. Biol. Chem.* 1939, *128*, 537.

COMPONENTS:	ORIGINAL MEASUREMENTS:
(1) Benzenesulfonamide, 4-amino-N-2-thiazolyl- (sulfathiazole); $C_9H_9N_3O_2S_2$; [72-14-0] (2) 1H-Purine-2,6-dione, 3,7-dihydro-1,3,7-trimethyl- (caffeine); $C_8H_{10}N_4O_2$; [58-08-2] (3) Water; H_2O; [7732-18-5]	Neish, W.J.P. *Rec. trav. chim.* 1948, *67*, 361-71.
VARIABLES:	PREPARED BY:
Concentration of caffeine	R. Piekos

EXPERIMENTAL VALUES:

Concentration of caffeine g/100 ml	Solubility of sulfathiazole at 37°C	
	γ /ml	10^3 mol dm^{-3} [a]
0.50	500	1.96
0.75	530	2.08
1.00	650	2.50

[a] Calculated by compiler

AUXILIARY INFORMATION

METHOD/APPARATUS/PROCEDURE:

A suspension of sulfathiazole in caffeine soln was kept for 5 h at 37°C and 1 h at room temp before filtration. Soly was detd by the Westfall's method (1) based on diazotization of the sulfonamide, coupling with Na 2-naphthol-3,6-disulfonate and comparing the color with that of a std soln in a Klett colorimeter.

SOURCE AND PURITY OF MATERIALS:

Sulfathiazole: not specified

Anhydrous caffeine was a good commercial product (source not specified).

Distilled water was used.

ESTIMATED ERROR:

Nothing specified

REFERENCES:

1. Westfall, B. B. *J. Nat. Cancer Inst.* 1945, *6*, 23.

COMPONENTS:	ORIGINAL MEASUREMENTS:
(1) Benzenesulfonamide, 4-amino-N-2-thiazolyl- (sulfathiazole); $C_9H_9N_3O_2S_2$; [72-14-0] (2) 2-Pyrrolidinone, 1-ethenyl-,polymers (PVP); $(C_6H_9NO)_x$; [9003-39-8] (3) Water; H_2O; [7732-18-5]	Rupprecht, H.; Ziller, K. H. *Pharmazie* 1981, *36(4)*, 298.
VARIABLES:	PREPARED BY:
One temperature: 20°C	R. Piekos

EXPERIMENTAL VALUES:

Solubility of sulfathiazole in a suspension containing 2.0 mg PVP/100 ml at 20°C is 37.2 mg/100 ml (1.46×10^{-3} mol dm^{-3}, compiler).

AUXILIARY INFORMATION

METHOD/APPARATUS/PROCEDURE:	SOURCE AND PURITY OF MATERIALS:
To 50 ml of a suspension of 2.0 g of sulfathiazole, 1.0 mg of PVP was added, the mixt was placed in a thermostat and stirred with a magnetic stirrer. The concn of the solute was monitored continuously after filtration through a G3 or G4 fritted-glass filters by means of a Knauer differential refractometer or a Shimadzu 100-02 UV spectrophotometer. The cuvets of the refractometer were thermostated. Variations of the refractive index or light absorption were recorded as a function of time with a Servogor 220 two-line recorder.	Sulfathiazole: neither source nor purity was specfied. PVP K30 was from BASF, Ludwigshafen. Its purity was not specified. Distilled water was used. ESTIMATED ERROR: Nothing specified REFERENCES:

COMPONENTS:	ORIGINAL MEASUREMENTS:
(1) Benzenesulfonamide, 4-amino-N-2-thiazolyl; (sulfathiazole); $C_9H_9N_3O_2S_2$; [72-14-0] (2) D-Glucose; $C_6H_{12}O_6$; [50-99-7] (3) Water; H_2O; [7732-18-5]	Becher, R.; Leya, S. *Experientia* 1946, *2*, 459-60.
VARIABLES:	**PREPARED BY:**
One temperature: 18-19°C	R. Piekos

EXPERIMENTAL VALUES:

Solubility of sulfathiazole in a 10% D-glucose solution at room temperature (18-19°C) is 57 mg% (2.2 x 10^{-3} mol dm^{-3}, compiler).

AUXILIARY INFORMATION

METHOD/APPARATUS/PROCEDURE:

After standing for more than two days the soln of sulfathiazole was filtered and the sulfonamide was assayed in the filtrate colorimetrically by the method of Druey and Oesterheld (1).

SOURCE AND PURITY OF MATERIALS:

Nothing specified

ESTIMATED ERROR:

Nothing specified

REFERENCES:

1. Druey, J.; Oesterheld, G. *Helv. Chim. Acta* 1942, *25*, 753.

COMPONENTS:	ORIGINAL MEASUREMENTS:
(1) Benzenesulfonamide, 4-amino-N-2-thiazolyl- (sulfathiazole); $C_9H_9N_3O_2S_2$; [72-14-0] (2) Pectin; $(C_{13}H_{18}O_{12})_n$; [9000-69-5] (3) Water; H_2O; [7732-18-5]	Becher, R.; Leya, S., *Experientia* 1946, *2*, 459-60.
VARIABLES:	PREPARED BY:
One temperature: 18-19°C	R. Piekos

EXPERIMENTAL VALUES:

Solubility of sulfathiazole in a 2.5% pectin solution ([pectin] = 6.8×10^{-2} mol kg^{-1}, compiler), of pH about 2.6, at room temperature (18-19°C) is 86 mg% (3.4×10^{-3} mol dm^{-3}, compiler).

AUXILIARY INFORMATION

METHOD/APPARATUS/PROCEDURE:

The soln was allowed to stand for more than 2 days at room temp. The soln was the filtered, and sulfathiazole assayed colorimetrically in the filtrate by the method of Druey and Oesterheld (1).

SOURCE AND PURITY OF MATERIALS:

A high quality apple pectin was used: the rel viscosity of a 0.5% soln was 6.2, and for neutralization of 1 g of the pectin, 1.67 cm^3 of a 1 mol dm^{-3} NaOH soln was used. The source and purity of sulfathiazole and water were not specified.

ESTIMATED ERROR:

Nothing specified

REFERENCES:

1. Druey, J.; Oesterheld, G. *Helv. Chim. Acta* 1942, *25*, 753.

COMPONENTS:

(1) Benzenesulfonamide, 4-amino-N-2-thiazolyl- (sulfathiazole); $C_9H_9N_3O_2S_2$; [72-14-0]

(2) Poly(oxy-1,2-ethanediyl), α -hydro- ω -hydroxy- (PEG 400); $(C_2H_4O)_nH_2O$; [25322-68-3] 400

(3) Water; H_2O; [7232-18-5]

ORIGINAL MEASUREMENTS:

Khawam, M. N.; Yousef, R. T.; Czetsch-Lindenwald, H. *Sci. Pharm.* 1966, *34*, 209-13.

VARIABLES:

Concentration of PEG 400

PREPARED BY:

R. Piekos

EXPERIMENTAL VALUES:

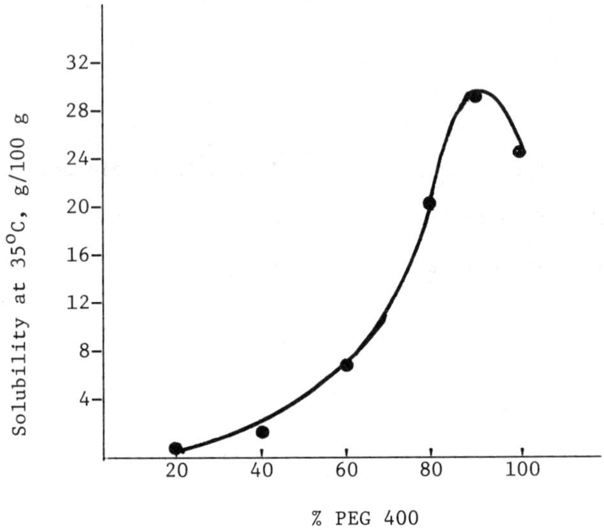

AUXILIARY INFORMATION

METHOD/APPARATUS/PROCEDURE:

An earlier described method was employed (1) whereby a 100-ml conical flask contg a PEG 400 soln was placed in a drying cabinet at 35°C and an excess of sulfathiazole was added under stirring for 1 h. After 12 h the soln was filtered or decanted and the solute was assayed in the filtrate spectrophotometrically using a Unicam SP 500 spectrophotometer and 1-ml quartz cuvets. Results were taken from a calibration graph.

SOURCE AND PURITY OF MATERIALS:

Neither source nor purity of sulfathiazole and water were specified.

PEG 400 was a product of Farbwerke Hoechst (purity not specified).

ESTIMATED ERROR:

Nothing specified

REFERENCES:

1. Khawam, M. N.; Tawashi, R.; Czetsch-Lindenwald, H.v. *Sci. Pharm.* 1965, *33*, 90.

COMPONENTS:

(1) Benzenesulfonamide, 4-amino-N-2-thiazolyl- (sulfathiazole); $C_9H_9N_3O_2S_2$; [72-14-0]

(2) Poly(oxy-1,2-ethanediyl), α-hydro- ω-hydroxy- (PEG 4000); $(C_2H_4O)_nH_2O$; [25322-68-3]

(3) Water; H_2O; [7732-18-5]

ORIGINAL MEASUREMENTS:

Khawam, M. N.; Yousef, R. T.; Czetsch-Lindenwald, H. *Sci. Pharm.* 1966, *34*, 209-13.

VARIABLES:

Concentration of PEG 4000

PREPARED BY:

R. Piekos

EXPERIMENTAL VALUES:

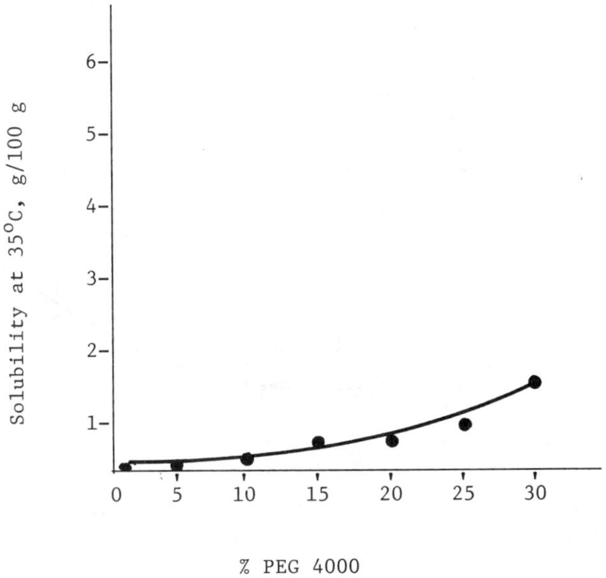

AUXILIARY INFORMATION

METHOD/APPARATUS/PROCEDURE:

An earlier described method was employed (1) whereby a 100-ml conical flask contg a PEG 4000 soln was placed in a drying cabinet at 35°C and an excess of sulfathiazole was added under stirring for 1 h. After 12 h the soln was filtered or decanted and the solute was assayed in the filtrate spectrophotometrically using a Unicam SP 500 spectrophotometer and 1-ml quartz cuvets. Results were taken from a calibration curve.

SOURCE AND PURITY OF MATERIALS:

Neither source nor purity of sulfathiazole and water were specified.

PEG 4000 was a product of Farbwerke Hoechst (purity not specified).

ESTIMATED ERROR:

Nothing specified

REFERENCES:

1. Khawam, M.N.: Tawashi, R.; Czetsch-Lindenwald, H. v. *Sci. Pharm.* 1965, *33*, 90.

COMPONENTS:

(1) Benzenesulfonamide, 4-amino-N-2-thiazolyl- (sulfathiazole); $C_9H_9N_3O_2S_2$; [72-14-0]

(2) Sorbitan monolaurate, polyoxyethylene derivatives (Tween 20); [9005-64-5]

(3) Water; H_2O; [7732-18-5]

ORIGINAL MEASUREMENTS:

Khawan, M. N.; Yousef, R. T; Czetsch-Lindenwald, H. *Sci. Pharm.* 1966, *34*, 209-13.

VARIABLES:

Concnetration of Tween 20

PREPARED BY:

R. Piekos

EXPERIMENTAL VALUES:

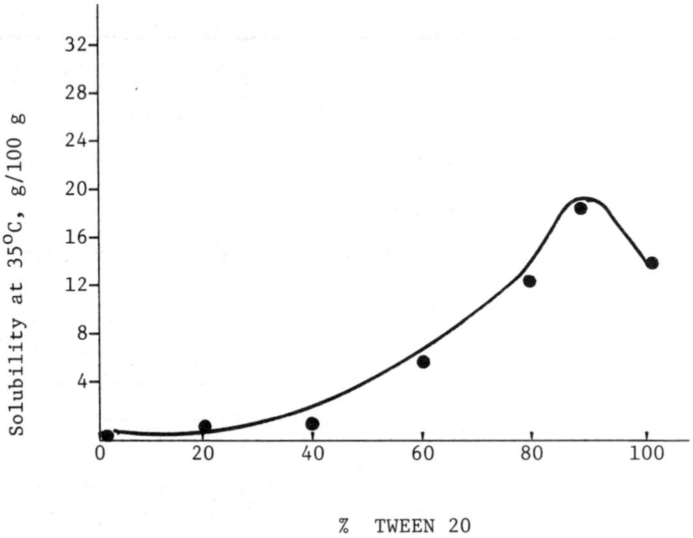

AUXILIARY INFORMATION

METHOD/APPARATUS/PROCEDURE:

An earlier described method was employed (1) whereby a 100-ml conical flask contg a Tween 20 soln was placed in a drying cabinet at 35°C and an excess of sulfathiazole was added under stirring for 1 h. After 12 h the soln was filtered or decanted and the solute was assayed in the filtrate spectrophotometrically using a Unicam SP 500 spectrophotometer and 1-ml quartz cuvets. Results were taken from a calibration graph.

SOURCE AND PURITY OF MATERIALS:

Neither source nor purity of sulfathiazole and water were specified.

Tween 20 was supplied by Atlas-Goldschmidt A.G., Essen (purity not specified).

ESTIMATED ERROR:

Nothing specified

REFERENCES:

1. Khawam, M.N.; Tawashi, R.; Czetsch-Lindenwald, H. v. *Sci. Pharm.* 1965, *33*, 90.

COMPONENTS:

(1) Benzenesulfonamide, 4-amino-N-2-thiazolyl- (sulfathiazole); $C_9H_9N_3O_2S_2$; [72-14-0]

(2) Sorbitan monolaurate, polyoxyethylene derivatives (Tween 20); [9005-64-5]

(3) Water; H_2O; [7732-18-5]

ORIGINAL MEASUREMENTS:

Gusyakov, V.P.; Likhol'ot, N. M.; Kutna, I.M.; *Farm. Zh. (Kiev)* 1967, *22(3)*, 34-9.

VARIABLES:

One temperature: 20°C

PREPARED BY:

R. Piekos

EXPERIMENTAL VALUES:

$$S/S_o = 2.0 \text{ at } 20^o$$

where S is the solubility of sulfathiazole in a 2% by weight Tween 20 solution, and

S_o is the solubility of sulfathiazole in water (0.043 g/100 ml).

Hence S = 0.086 g/100 ml (3.4×10^{-3} mol dm^{-3}), compiler.

AUXILIARY INFORMATION

METHOD/APPARATUS/PROCEDURE:

An excess of sulfathiazole in a 2% by wt aq Tween 20 soln was equilibrated for 24 h in an ampul immersed in a water thermostat. Aliquots of the satd soln were withdrawn through a filter and the sulfathiazole content was assayed in the filtrate photometrically.

SOURCE AND PURITY OF MATERIALS:

Sulfathiazole conformed to the requirements of the State Pharmacopeia IX.
Tween 20 was a product of Gee Lawson, England.
Purity of the water was not specified.

ESTIMATED ERROR:

Soly: not specified.
Temp: ±0.1°C (authors).

REFERENCES:

COMPONENTS:	ORIGINAL MEASUREMENTS:
(1) Benzenesulfonamide, 4-amino-N-2-thiazolyl- (sulfathiazole); $C_9H_9N_3O_2S_2$; [72-14-0] (2) Sorbitan monopalmitate, polyoxyethylene derivaties (Tween 40); [9005-66-7] (3) Water; H_2O; [7732-18-5]	Gusyakov, V. P.; Likhol'ot, N. M. Kutna, I.M. *Farm. Zh. (Kiev)* 1967, *22(3)*, 34-9.
VARIABLES: One temperature: 20°C	PREPARED BY: R. Piekos

EXPERIMENTAL VALUES:

$$S/S_o = 2.0 \text{ at } 20^{\circ}C$$

where S is the solubility of sulfathiazole in a 2% weight Tween 40 solution in water, and

S_o is the solubility of sulfathiazole in water (0.043 g/100 ml).

Hence S = 0.086 g/100 ml (3.4×10^{-3} mol dm^{-3}), compiler.

AUXILIARY INFORMATION

METHOD/APPARATUS/PROCEDURE:	SOURCE AND PURITY OF MATERIALS:
An excess of sulfathiazole in a 2% by wt aq Tween 40 soln was equilibrated for 24 h in an ampul immersed in a water thermostat. Aliquots of the satd soln were withdrawn through a filter and the sulfathiazole content was assayed in the filtrate photometrically.	Sulfathiazole conformed to the requirements of the State Pharmacopeia IX. Tween 40 was a product of Gee Lawson, England. Purity of the water was not specified.
	ESTIMATED ERROR: Soly: not specified. Temp: ±0.1°C (authors).
	REFERENCES:

COMPONENTS:

(1) Benzenesulfonamide, 4-amino-N-2 thaizolyl- (sulfathiazole); $C_9H_9N_3O_2S_2$; [72-14-0]

(2) Sorbitan monooleate, polyoxyethylene derivatives (Tween 80) [9005-65-6]

(3) Water; H_2O; [7732-18-5]

ORIGINAL MEASUREMENTS:

Gusyakov, V. P.; Likhol'ot, N. M.; Kutna, I.M. *Farm. Zh. (Kiev)* 1967, *22(3)*, 34-9.

VARIABLES:

One temperature: 20°C

PREPARED BY:

R. Piekos

EXPERIMENTAL VALUES:

$$S/S_o = 2.0 \text{ at } 20^oC$$

where S is the solubility of sulfathiazole in a 2% by weight aqueous Tween 80 solution, and

S_o is the solubility of sulfathiazole in water (0.043 g/100 ml).

Hence S = 0.086 g/100 ml (3.4 x 10^{-3} mol dm^{-3}), compiler.

AUXILIARY INFORMATION

METHOD/APPARATUS/PROCEDURE:

An excess of sulfathiazole in a 2% by wt aq Tween 80 soln was eqilibrated for 24 h in an ampul immersed in a water thermostat. Aliquots of the satd soln were withdrawn through a filter and the sulfathiazole content was assayed in the filtrate photometrically.

SOURCE AND PURITY OF MATERIALS:

Sulfathiazole conformed to the requirements of the State Pharmacopeia IX.

Tween 80 was a product of Gee Lawson, England.

Purity of the water was not specified.

ESTIMATED ERROR:

Soly: not specified.

Temp: ±0.1°C (authors).

REFERENCES:

COMPONENTS:

(1) Benzenesulfonamide, 4-amino-N-2-thiazolyl- (sulfathiazole); $C_9H_9N_3O_2S_2$; [72-14-0]

(2) Ethanol; C_2H_6O; [64-17-5]

ORIGINAL MEASUREMENTS:

Lott, W. A.; Bergeim, F. H.

J. Am. Chem. Soc. 1939, *61*, 3593-4.

VARIABLES:

One temperature: 26°C

PREPARED BY:

R. Piekos

EXPERIMENTAL VALUES:

Solubility of sulfathiazole in ethanol at 26°C is 525 mg/100 cm^3 (2.06×10^{-2} mol dm^{-3}, compiler).

AUXILIARY INFORMATION

METHOD/APPARATUS/PROCEDURE:

Nothing specified.

SOURCE AND PURITY OF MATERIALS:

Sulfathiazole, mp 197-7.5°C (uncor) and 202.0-2.5°C (cor) was prepd by the authors. Purity of the ethanol was not specified.

ESTIMATED ERROR:

Nothing specified.

REFERENCES:

COMPONENTS:	ORIGINAL MEASUREMENTS:
(1) Benzenesulfonamide, 4-amino-N-2-thiazolyl- (sulfathiazole); $C_9H_9N_3O_2S_2$; [72-14-0] (2) 2-Propanol; C_3H_8O; [67-63-0]	Burlage, H. M. *J. Am. Pharm. Assoc., Sci. Ed.* 1948, *37*, 345.
VARIABLES:	**PREPARED BY:**
One temperature: 25°C	R. Piekos

EXPERIMENTAL VALUES:

Solubility of sulfathiazole in 2-propanol at 25°C is 0.5750 g/100 cm^3 solution (2.252×10^{-2} mol dm^{-3}, compiler).

AUXILIARY INFORMATION

METHOD/APPARATUS/PROCEDURE:

Satd solns of sulfathiazole in 2-propanol were prepd at 25°C and definite vols of the solns were measured into tared dishes by means of standard pipets. The alcohol was allowed to evap at room temp and the residue was dried at 105°C. In the case of losses due to apparent decompn, the residue was dried in a desiccator (1).

SOURCE AND PURITY OF MATERIALS:

The sulfathiazole was manufd by Merck and was of the U.S.P. purity. The source and purity of 2-propanol was not specified.

ESTIMATED ERROR:

Nothing specified.

REFERENCES:

1. Burlage, H. M. *J. Am. Pharm. Assoc., Sci. Ed.* 1947, *36(1)*, 16.

COMPONENTS:

(1) Benzenesulfonamide, 4-amino-N-2-thiazolyl- (sulfathiazole); $C_9H_9N_3O_2S_2$; [72-14-0]

(2) 2-Propanol; C_3H_8O; [67-63-0]

ORIGINAL MEASUREMENTS:

Kuhnert-Brandstätter, M.; Martinek, A. *Microchim. Ichnoanal. Acta* 1956, 909-19.

VARIABLES:

Temperature

PREPARED BY:

R. Piekos

EXPERIMENTAL VALUES:

t/°C	Saturation solubility[a]			
	Crystalline form I		Crystalline form II	
	g/100 g solution	10^2 mol kg^{-1} solution[b]	g/100 g solution	10^2 mol kg^{-1} solution[b]
30.5	0.400	1.567	-	-
31.0	-	-	0.220	0.862
40.5	0.500	1.958	0.310	1.214
50.5	0.660	2.585	0.510	2.000
59.5	0.890	3.486	-	-
60.0	-	-	0.735	2.879
61.0	-	-	0.770	3.016
65.0	-	-	0.880	3.447
69.0	1.215	4.759	-	-
70.0	1.260	4.935	1.085	4.250

[a] Numerical data received from the authors.

[b] Calculated by compiler.

AUXILIARY INFORMATION

METHOD/APPARATUS/PROCEDURE:

Sulfathiazole and 2-propanol were placed in a polyethylene vessel, agitated for 3 h, filtered, and the sulfonamide was assayed in the filtrate gravimetrically. The solid phase was examd thermomicroscopically for identity of the cryst form.

SOURCE AND PURITY OF MATERIALS:

A comm available form II of sulfathiazole was used. Form I was obtained by keeping the comm reagent at 170°C for 2 h. The source and purity of 2-propanol was not specified.

ESTIMATED ERROR:

Soly: not specified.

Temp: ±0.5°C (authors).

REFERENCES:

COMPONENTS:	ORIGINAL MEASUREMENTS:
(1) Benzenesulfonamide, 4-amino-N-2-thiazolyl- (sulfathiazole); $C_9H_9N_3O_2S_2$; [72-14-0] (2) Ethanol, 2-ethoxy-; $C_4H_{10}O_2$; [110-80-5]	Sunwoo, C.; Eisen, H. *J. Pharm. Sci.* 1971, *60*, 238-44.
VARIABLES:	**PREPARED BY:**
One temperature: 25°C	R. Piekos

EXPERIMENTAL VALUES:

The mole fraction solubility of crystalline form II of sulfathiazole in 2-ethoxyethanol at 25°C is 0.0224 (6.09 g/100 g solution, compiler).

AUXILIARY INFORMATION

METHOD/APPARATUS/PROCEDURE:

Soly was detd by the method reported by Restaino and Martin (1). Sulfathiazole was assayed spectrophotometrically on a Coleman-Hitachi 124 double-beam spectrophotometer at 280 nm after dilg the sample with 95% EtOH or water.

SOURCE AND PURITY OF MATERIALS:

Sulfathiazole (American Cyanamid Co., Pearl River, N.Y.) was recrystd from super-satd soln of warm acetone to give cryst form II.
Industrial grade 2-ethoxyethanol (Cellosolve solvent, Union Carbide Corp., New York, N.Y.) was used.

ESTIMATED ERROR:

Temp: ±1.0°C (authors).
Soly: the mean of 3 runs was given (authors).

REFERENCES:

1. Restaino, F. A.; Martin, A. N. *J. Pharm. Sci.* 1964, *53*, 636.

COMPONENTS:	ORIGINAL MEASUREMENTS:
(1) Benzenesulfonamide, 4-amino-N-2-thiazolyl- (sulfathiazole); $C_9H_9N_3O_2S_2$; [72-14-0] (2) 2-Butanol; $C_4H_{10}O$; [78-92-2] (3) Ethanol; C_2H_6O; [64-17-5]	Mehta, S. C.; Bernardo, P. D.; Higuchi, W. I.; Simonelli, A. P. *J. Pharm. Sci.* 1970, *59(5)*, 638-44.
VARIABLES:	**PREPARED BY:**
One temperature: 30°C	R. Piekos

EXPERIMENTAL VALUES:

Solubility of sulfathiazole in a 60% v/v mixture of ethanol in 2-butanol, at 30°C, is 0.555 g/100 g (2.17×10^{-2} mol kg^{-1}, compiler).

AUXILIARY INFORMATION

METHOD/APPARATUS/PROCEDURE:	SOURCE AND PURITY OF MATERIALS:
An excess of the amt of the recrystd sulfathiazole needed to produce a satd soln was placed in a volumetric flask with the solvent and agitated in a water bath at 30°C. Duplicate samples were withdrawn at 12-24-h intervals, filtered through a 0.45-μ Millipore filter, and analyzed spectrophotometrically.	Sulfathiazole (source not specified) was purified by crystallization. The source and purity of the remaining materials was not specified. **ESTIMATED ERROR:** Nothing specified. **REFERENCES:**

COMPONENTS:

(1) Benzenesulfonamide, 4-amino-N-2-thiazolyl- (sulfathiazole); $C_9H_9N_3O_2S_2$; [72-14-0]

(2) Methane, trichloro- (chloroform); $CHCl_3$; [67-66-3]

ORIGINAL MEASUREMENTS:

Riess, W.

Intern. Congr. Chemotherapy, Proc., 3rd, Stuttgart 1963, *1*, 627-32.

VARIABLES:

One temperature: 20°C

PREPARED BY:

R. Piekos

EXPERIMENTAL VALUES:

Solubility of sulfathiazole in chloroform at 20°C is 15 mg%

(5.9 x 10^{-4} mol dm^{-3} solution, compiler).

AUXILIARY INFORMATION

METHOD/APPARATUS/PROCEDURE:

Nothing specified.

SOURCE AND PURITY OF MATERIALS:

Nothing specified.

ESTIMATED ERROR:

Nothing specified.

REFERENCES:

COMPONENTS:	ORIGINAL MEASUREMENTS:
(1) Benzenesulfonamide, 4-amino-N-2-thiazolyl- (sulfathiazole); $C_9H_9N_3O_2S_2$; [72-14-0] (2) Methane, trichloro- (chloroform); $CHCl_3$; [67 -66-3]	Yamazaki, M.; Aoki, M.; Kamada, A.; Yata, N. *Yakuzaigaku* 1967, *27(1)*, 37-40.
VARIABLES: One temperature: 30°C	PREPARED BY: R. Piekos

EXPERIMENTAL VALUES:

Solubility of sulfathiazole in chloroform at 30°C is 0.48 mmol/L.

AUXILIARY INFORMATION

METHOD/APPARATUS/PROCEDURE:	SOURCE AND PURITY OF MATERIALS:
Sulfathiazole (0.5 g) was placed in an L-shaped tube together with 20 ml of chloroform. The mixt was then shaken in a thermostat until equilibrium was attained. The sulfathiazole was then assayed in the supernatant spectrophotometrically at 545 nm on a Beckmann DU spectrophotometer. The results were taken from a calibration graph.	Nothing specified. ESTIMATED ERROR: Soly: not specified. Temp: ±1°C (authors). REFERENCES:

COMPONENTS:	ORIGINAL MEASUREMENTS:
(1) Benzenesulfonamide, 4-amino-N-2-thiazolyl- (sulfathiazole); $C_9H_9N_3O_2S_2$; [72-14-0] (2) Methane, trichloro-; $CHCl_3$; [67-66-3]	Kitao, K.; Kubo, K.; Morishita, T.; Yata, N.; Kamada, A. *Chem. Pharm. Bull.* 1973, *21*, 2417-26.
VARIABLES: One temperature: 37°C	PREPARED BY: R. Piekos

EXPERIMENTAL VALUES:

Solubility of sulfathiazole in $CHCl_3$ at 37°C is 0.843 mmol dm^{-3} solution.

AUXILIARY INFORMATION

METHOD/APPARATUS/PROCEDURE:	SOURCE AND PURITY OF MATERIALS:
One ml of the sulfathiazole soln in $CHCl_3$ at equilibrium was taken into a test tube. After evapn of the solvent, the residue was dissolved in 1N NaOH, the soln was properly dild with deionized water and the concn of sulfathiazole was detd by diazotization.	Comm available sulfathiazole (source not specified) was used as supplied. Neither source nor purity of the $CHCl_3$ was specified.
	ESTIMATED ERROR: Soly: not specified. Temp: ±1°C (authors).
	REFERENCES:

COMPONENTS:

(1) Benzenesulfonamide, 4-amino-N-2-thiazolyl- (sulfathiazole); $C_9H_9N_3O_2S_2$; [72-14-0]

(2) 2-Propanone (acetone); C_3H_6O; [67-64-1]

ORIGINAL MEASUREMENTS:

Gutierrez, F. H.

Anales fis. quim. (Madrid) 1945, *41*, 537-60.

VARIABLES:

Temperature

PREPARED BY:

R. Piekos

EXPERIMENTAL VALUES:

t/°C	G[a]	E[b]	X_g/1[c]	mol/1[d] acetone	mmol/mol acetone	1:X_g[e]	1 + X_{cc}[f]
0	0.994	0.984	8.097	31.7	2.26	100.60	123.50
5	1.247	1.212	10.086	39.5	2.84	80.19	99.15
10	1.506	1.484	12.093	47.4	3.43	64.41	82.72
15	1.728	1.699	13.774	53.9	3.93	57.87	72.60
20	2.025	1.985	16.022	62.7	4.61	49.38	62.41
25	2.349	2.295	18.484	72.4	5.34	42.58	54.10
30	2.653	2.584	20.675	80.9	6.03	37.69	48.37
35	3.000	2.913	23.199	90.8	6.82	33.33	43.11
40	3.380	3.269	25.938	101.6	7.70	29.58	38.71
45	3.704	3.571	28.200	110.4	8.43	26.99	35.46
50	4.133	3.969	31.225	122.3	9.40	24.19	32.02

[a] $G = \frac{p\ 100}{P - p}$, where p and P are the weights of solute and solution, resp.

[b] $E = \frac{G\ 100}{G + 100}$; [c] g/1 acetone; [d] should be mmol/1 acetone (compiler);

[e] g of acetone required to dissolved 1 g of solute; [f] volume (cm^3) of acetone required to dissolve 1 g of solute.

AUXILIARY INFORMATION

METHOD/APPARATUS/PROCEDURE:

A special all-glass app was constructed enabling the prepn of satd solns, agitation by bubbling a stream of acetone-satd N, filtration and distn off the solvent without contact with air. Two exchangeable dissoln vessels of 15 and 8 cm^3 working capacity were used depending on the soly of solute. The app was immersed in a thermostat. The vols of acetone used were 15 or 5 cm^3, and the equilibration time was 2-2.5 h. The satd solns were filtered, weighed, the solvent was distd off, the residues were dried at 105°C, weighed and examd for the presence of solvated acetone.

SOURCE AND PURITY OF MATERIALS:

The source of the materials was not specified. Pure, anhyd acetone was used. The absence of impurities and water was confirmed by procedures of the German Pharmacopeia VI and Spanish Pharmacopeia VIII. The purity of sulfathiazole was not specified.

ESTIMATED ERROR:

Soly: measurements were repeated until 2 values not differing in the second decimal were obtained (author).
Temp: ±0.1°C (author).

REFERENCES:

COMPONENTS:	ORIGINAL MEASUREMENTS:
(1) Benzenesulfonamide, 4-amino-N-2-thiazolyl- (sulfathiazole); $C_9H_9N_3O_2S_2$; [72-14-0] (2) Methylcyclohexanone; $C_7H_{12}O$; [1331-22-2]	Barber, H. J.; Wilkinson, J. H. *Quart. J. Pharm. Pharmacol.* 1946, *19*, 248-55.
VARIABLES:	PREPARED BY:
One temperature: 25°C	R. Piekos

EXPERIMENTAL VALUES:

Approximate solubility of sulfathiazole in methylcyclohexanone at 37°C is 8.5 percent w/v (0.33 mol dm^{-3} solution, compiler).

AUXILIARY INFORMATION

METHOD/APPARATUS/PROCEDURE:	SOURCE AND PURITY OF MATERIALS:
Nothing specified.	Nothing specified
	ESTIMATED ERROR:
	Nothing specified
	REFERENCES:

COMPONENTS:

(1) Benzenesulfonamide, 4-amino-N-2-thiazolyl- (sulfathiazole); $C_9H_9N_3O_2S_2$; [72-14-0]

(2) Methylcyclohexanone; $C_7H_{12}O$; [1331-22-2]

ORIGINAL MEASUREMENTS:

Barber, H. J.; Wilkinson, J. H.
Pharm. J. 1946, 105-6.

VARIABLES:

One temperature: 25°C

PREPARED BY:

R. Piekos

EXPERIMENTAL VALUES:

Approximate solubility of sulfathiazole in methylcyclohexanone at 25°C is 8.5 percent w/v (0.33 mol dm^{-3} solution, compiler).

AUXILIARY INFORMATION

METHOD/APPARATUS/PROCEDURE:

Nothing specified

SOURCE AND PURITY OF MATERIALS:

Nothing specified

ESTIMATED ERROR:

Nothing specified

REFERENCES:

COMPONENTS:

(1) Benzenesulfonamide, 4-amino-N-2-thiazolyl- (sulfathiazole); $C_9H_9N_3O_2S_2$; [72-14-0]

(2) Poly(oxy-1,2-ethanediyl), α-hydro-ω-hydroxy-(PEG 400); $(C_2H_4O)_nH_2O$; [25322-68-3]

ORIGINAL MEASUREMENTS:

Wahlgren, S.; *Svensk farm. tidskr.* 1962, *66*, 585-91.

VARIABLES:

Temperature

PREPARED BY:

R. Piekos

EXPERIMENTAL VALUES:

	Solubility in PEG 400	
$t/^{\circ}C$	weight %	mol kg^{-1} [a]
20	22	1.1
60	22	1.1

[a] Calculated by compiler

AUXILIARY INFORMATION

METHOD/APPARATUS/PROCEDURE:

The soly detns were made in 100-cm^3 Erlenmeyer flasks immersed in a const-temp bath. The suspension was stirred with an electrically driven propeller stirrer for a least 4 h.

SOURCE AND PURITY OF MATERIALS:

The source and purity of sulfathiazole was not specified. PEG 400: pH 4.7 (1.00 g in 20.0 g of water); ash content 0.030%; free acid: 0.30 cm^3 of 0.1N NaOH was required to neutralize free acids in 5.0 g of PEG 400 dissolved in 20 cm^3 of EtOH; average mol wt 400; water content 0.2%.

ESTIMATED ERROR:

Temp: ±0.5°C (author).
Soly: duplicate tests were made of concns on both sides of the borderline value (author).

REFERENCES:

COMPONENTS:

(1) Benzenesulfonamide, 4-amino-N-2-thiazolyl- (sulfathiazole); $C_9H_9N_3O_2S_2$; [72-14-0]

(2) Poly(oxy-1,2-ethanediyl), α-hydro-ω-hydroxy- (PEG 400); $(C_2H_4O)_nH_2O$; [25322-68-3]

ORIGINAL MEASUREMENTS:

Gusyakov, V.P.; Likhol'ot, N.M.; Kutna, I.M. *Farm. Zh. (Kiev)* 1968, *23(6)*, 56-61.

VARIABLES:

One temperature: 21-25°C

PREPARED BY:

R. Piekos

EXPERIMENTAL VALUES:

Solubility of sulfathiazole in α-hydro-ω-hydroxypoly(oxy-1,2-ethanediyl) 400 at room temperature (21-25°C) is 28% by weight (1.5 mol kg^{-1} PEG 400, compiler).

AUXILIARY INFORMATION

METHOD/APPARATUS/PROCEDURE:

Small quanitites (2-4 mg) of sulfathiazole were added to a known quantity of PEG 400 under stirring until satn was attained.

SOURCE AND PURITY OF MATERIALS:

Sulfathiazole: neither source nor purity was specified. PEG 400: source not specified; sp. gr. 1.127 g cm^{-3}; temp of solidification approx 6°C; refractive index 1.466 (temp not indicated).

ESTIMATED ERROR:

Nothing specified.

REFERENCES:

COMPONENTS:

(1) Benzenesulfonamide, 4-amino-N-2-thiazolyl- (sulfathiazole); $C_9H_9N_3O_2S_2$; [72-14-0]

(2) Poly(oxy-1,2-ethanediyl), α-hydro- ω -hydrozy- (poly(ethylene glycol) 3000)); $(C_2H_4O)_nH_2O$; [25322-68-3]

ORIGINAL MEASUREMENTS:

Wahlgren, S.; *Svensk farm. tidskr.* 1962, *66*, 585-91.

VARIABLES:

One temperature: 60°C

PREPARED BY:

R. Piekos

EXPERIMENTAL VALUES:

Solubility of sulfathiazole in poly(ethylene glycol) 3000 at 60°C is 20% by weight (0.98 mol kg^{-1}, compiler).

AUXILIARY INFORMATION

METHOD/APPARATUS/PROCEDURE:

The soly detns were made in 100-cm^3 Erlenmeyer flasks immersed in a const-temp bath. The suspension was stirred with an electrically driven propeller stirrer for at least 4 h.

SOURCE AND PURITY OF MATERIALS:

The source and purity of sulfathiazole was not specified. PEG 3000: mp 56°C; pH 6.4 (1.00 g in 20.0 g of water); ash content 0.025%; free acid: 0.05 cm^3 of 0.1N NaOH was required to neutralize free acids in 5.0 g of PEG dissolved in 20 cm^3 of EtOH against phenolphthalein; average mol wt 3000: water content 0.4%.

ESTIMATED ERROR:

Temp: ±0.5°C (author).
Soly: duplicate tests were made of concns on both sides of the borderline value (author).

REFERENCES:

COMPONENTS:	ORIGINAL MEASUREMENTS:
(1) Benzenesulfonamide, 4-amino-N-2-thiazolyl- (sulfathiazole); $C_9H_9N_3O_2S_2$; [72-14-0] (2) Cottonseed oil	Whitworth, C. W.; Becker, C. H. *J. Pharm. Sci.* 1965, *54(4)*, 569-73.
VARIABLES:	PREPARED BY:
One temperature: 37.5°C	R. Piekos

EXPERIMENTAL VALUES:

Solubility of sulfathiazole in cottonseed oil at 37.5°C is 0.863 mg% (3.38×10^{-5} mol dm^{-3} solution, compiler).

AUXILIARY INFORMATION

METHOD/APPARATUS/PROCEDURE:

A satd soln of sulfathiazole in cottonseed oil was made and filtered carefully at a const temp to remove suspended particles. A portion of the soln was shaken for 4 h with 100 ml of EtOH. The alcoholic layer was centrifuged for 30 min. Aliquot portions of the alcoholic soln were allowed to evap to dryness, a trichloroacetic acid soln added, and subsequently the Marshall reagents. From the intensity of the color developed it was possible to det the amt of the drug extd by the process utilized. A Klett-Summerson colorimeter with a No 54 filter was employed to det the color intensity, which was compared to that of standard solns.

SOURCE AND PURITY OF MATERIALS:

Sulfathiazole (N.F. grade) was from Eli Lilly and Co.
Neither source nor purity of the cottonseed oil was specified.

ESTIMATED ERROR:

Soly: not specified
Temp: ±1°C (authors)

REFERENCES:

COMPONENTS:

(1) Benzenesulfonamide, 4-amino-N-2-thiazolyl- (sulfathiazole); $C_9H_9N_3O_2S_2$; [72-14-0]

(2) Sorbitan, (Z)-9-octadecenoate (2:3) (Arlacel 83); [8007-43-0]

(3) White petrolatum (liquid petrolatum)

ORIGINAL MEASUREMENTS:

Whitworth, C.W.; Becker, C. H.

J. Pharm. Sci. 1965, *54(4)*, 569-73.

VARIABLES:

Concentration of Arlacel 83

PREPARED BY:

R. Piekos

EXPERIMENTAL VALUES:

Concentration of Arlacel 83 %	Solubilty at 37.5°C mg%	Solubilty at 37.5°C 10^5 mol dm^{-3} soln[a]
1	2.178	8.531
5	2.272	8.899
10	17.136	67.119

[a] Calculated by compiler

AUXILIARY INFORMATION

METHOD/APPARATUS/PROCEDURE:

A satd soln of sulfathiazole in the solvent was made and filtered carefully at a const temp to remove all suspended particles. A 5-ml portion of the soln was shaken for 4 h with 100 ml of EtOH. The alcoholic layer was centrifuged for 30 min. Aliquot portions of the alcoholic solns were allowed to evap to dryness, a trichloroacetic acid soln was added, and subsequently the Marshall reagents. From the intensity of the color developed it was possible to det the amt of the drug extd by the process utilized. A Klett-Summerson colorimeter with a No 54 filter was employed to det the color intensity, which was compared to that of standard solns.

SOURCE AND PURITY OF MATERIALS:

Sulfathiazole (N.F. grade) was from Eli Lilly and Co.

Arlacel 83 (Lot No 129) was from Atlas Powder Co. (purity not specified).

White petrolatum (liquid petrolatum) (U.S.P. grade) was from Fisher Scientific Co.

ESTIMATED ERROR:

Soly: not specified

Temp: ±1°C (authors)

REFERENCES:

COMPONENTS:

(1) Benzenesulfonamide, 4-amino-N-2-thiazolyl- (sulfathiazole); $C_9H_9N_3O_2S_2$; [72-14-0]

(2) Cottonseed oil

(3) Sorbitan, (Z)-9-octadecenoate (2:3) (Arlacel 83); [8007-43-0]

ORIGINAL MEASUREMENTS:

Whitworth, C. W.; Becker, C. H.

J. Pharm. Sci. 1965, *54(4)*, 569-73.

VARIABLES:

Concentration of Arlacel 83

PREPARED BY:

R. Piekos

EXPERIMENTAL VALUES:

Concentration of Arlacel 83 %	Solubility at 37.5°C mg%	10^5 mol dm^{-3} soln[a]
1	0.798	3.120
5	8.098	31.710
10	19.953	78.152

[a] Calculated by compiler

AUXILIARY INFORMATION

METHOD/APPARATUS/PROCEDURE:

A satd soln of sulfathiazole in the solvent was made and filtered carefully at a const temp to remove all suspended particles. A 5-ml portion of the soln was shaken for 4 h with 100 ml of EtOH. The alcoholic layer was centrifuged for 30 min. Aliquot portions of the alcoholic soln were allowed to evap to dryness, a trichloroacetic acid soln was added, and subsequently the Marshall reagents. From the intensity of the color developed it was possible to det the amt of the drug extd by the process utilized. A Klett-Summerson colorimeter with a No 54 filter was employed to det the color intensity , which was compared to that of standard solns.

SOURCE AND PURITY OF MATERIALS:

Sulfathiazole (N.F. grade) was from Eli Lilly and Co.

Neither source nor purity of the cottonseed oil was specified.

Arlacel 83 (Lot No 129) was from Atlas Powder Co. (purity not specified).

ESTIMATED ERROR:

Soly: not specified

Temp: ±1°C (authors)

REFERENCES:

COMPONENTS:	ORIGINAL MEASUREMENTS:
(1) Benzenesulfonamide, 4-amino-N-2-thiazolyl- (sulfathiazole); $C_9H_9N_3O_2S_2$; [72-14-0] (2) Hydrochloric acid; HCl; [7647-01-0] (3) Sodium chloride; NaCl; [7647-14-5] (4) Water; H_2O; [7732-18-5]	Miseta, M.; Kedvessy, G.; Selmeczi, B. *Pharmazie* 1983, *38(5)*, 326-7.
VARIABLES:	**PREPARED BY:**
One temperature: 20°	R. Piekos

EXPERIMENTAL VALUES:

Solubility of sulfathiazole in a simulated gastric juice (composition: 2.0 g NaCl, 25.0 g 10% HCl, and distilled water up to 1000 cm^3; pH 1.2), at 20°C, is 1 part sulfathiazole in 240 parts of the gastric juice (1.6×10^{-2} mol kg^{-1} gastric juice - compiler).

AUXILIARY INFORMATION

METHOD/APPARATUS/PROCEDURE:

Soly was detd by the Pharmacopeia Hungarica V method. The equilibration time was 2 days with occasional shaking (personal communication). The concn of the solute in the satd soln was detd spectrophotometrically at 282 nm using a Spektromom 195 spectrophotometer.

SOURCE AND PURITY OF MATERIALS:

The source and purity of sulfathiazole was not specified. The simulated gastric juice was prepd by the authors.
The source and purity of the components was not specified.
Distilled water was used.

ESTIMATED ERROR:

Soly: not specified
Temp: ±2°C (personal communication).

REFERENCES:

COMPONENTS:	ORIGINAL MEASUREMENTS:
(1) Benzenesulfonamide, 4-amino-N-2-thiazolyl- (sulfathiazole); $C_9H_9N_3O_2S_2$; [72-14-0] (2) Cellulose, ethers, 2-hydroxypropyl ether (Klucel MF) [9004-64-2] (3) Hydrochloric acid; HCl; [7647-01-0] (4) Sodium chloride; NaCl; [7647-14-5] (5) Water; H_2O; [7732-18-5]	Miseta, M.; Kedvessy, G.; Selmeczi, B. *Pharmazie* 1983, *38(5)*, 326-7.
VARIABLES:	**PREPARED BY:**
One temperature: 20°C	R. Piekos

EXPERIMENTAL VALUES:

Solubility of sulfathiazole in a simulated gastric juice (composition: 2.0 g NaCl; 25.0 g 10% HCl, and distilled water up to 1000 cm^3; pH 1.2), containing 0.5% Klucel MF, at 20°C, is 1 part sulfathiazole in 135 parts of the simulated gastric juice containing 0.5% Klucel MF (2.9×10^{-2} mol kg^{-1} simulated gastric juice containing 0.5% Klucel MF - compiler).

AUXILIARY INFORMATION

METHOD/APPARATUS/PROCEDURE:	SOURCE AND PURITY OF MATERIALS:
Soly was detd by the Pharmacopeia Hungarica V method. The equilibration time was 2 days with occasional shaking (personal communication). The concn of the solute in the satd soln was detd spectrophotometrically at 282 nm using a Spektromom 195 spectrophotometer.	The source and purity of sulfathiazole was specified. The simulated gastric juice contg 0.5% Klucel MF was prepd by the authors. The source and purity of the components was not specified. Distilled water was used. **ESTIMATED ERROR:** Soly: not specified Temp: ±2°C (personal communication). **REFERENCES:**

COMPONENTS:

(1) Benzenesulfonamide, 4-amino-N-2-thiazolyl- (sulfathiazole); $C_9H_9N_3O_2S_2$; [72-14-0]

(2) Cellulose, ethers, 2-hydroxypropyl methyl ether (Methocel 65 HG) [9004-65-3]

(3) Hydrochloric acid; HCl; [7647-01-0]

(4) Sodium chloride; NaCl; [7647-14-5]

(5) Water; H_2O; [7732-18-5]

ORIGINAL MEASUREMENTS:

Miseta, M.; Kedvessy, G.; Selmeczi, B.

Pharmazie 1983, *38(5)*, 326-7.

VARIABLES:

One temperature: 20°C

PREPARED BY:

R. Piekos

EXPERIMENTAL VALUES:

Solubility of sulfathiazole in a simulated gastric juice (composition: 2.0 g NaCl, 25.0 g 10% HCl, and distilled water up to 1000 cm^3; pH 1.2), containing 0.5% Methocel 65 HG, at 20°C, is 1 part sulfathiazole in 160 parts of the simulated gastric juice containing 0.5% Methocel 65 HG (2.4×10^{-2} mol kg^{-1} simulated gastric juice containing 0.5% Methocel 65 HG - compiler).

AUXILIARY INFORMATION

METHOD/APPARATUS/PROCEDURE:

Soly was detd by the Pharmacopeia Hungarica V method. The equilibration time was 2 days with occasional shaking (personal communication). The concn of the solute in the satd soln was detd spectrophotometrically at 282 nm using a Spektromom 195 spectrophotometer.

SOURCE AND PURITY OF MATERIALS:

The source and purity of sulfathiazole was not specified. The simulated gastric juice contg 0.5% Methocel 65 HG was prepd by the authors. The source and purity of the components was not specified.
Distilled water was used.

ESTIMATED ERROR:

Soly: not specified

Temp: ±2°C (personal communication)

REFERENCES:

COMPONENTS:

(1) Benzenesulfonamide, 4-amino-N-2-thiazolyl- (sulfathiazole); $C_9H_9N_3O_2S_2$; [72-14-0]

(2) Benzenesulfonamide, 4-amino-N-2-thiazolyl- -2-pyrrolidinone, 1-ethenyl-, homopolymer, complex; $C_9H_9N_3O_2S_2 \cdot (C_6H_9NO)_x$; [*]

(3) 2-Pyrrolidinone, 1-ethenyl-, homopolymer (povidone); $(C_6H_9NO)_x$; [9003-39-8]

(4) Water; H_2O; [7732-18-5]

VARIABLES: Concentration of povidone

ORIGINAL MEASUREMENTS:

Badawi, A. A.; El-Sayed, A. A.

J. Pharm. Sci. 1980, *69(5)*, 492-7.

PREPARED BY:

R. Piekos

EXPERIMENTAL VALUES:

Amount of povidone %	Amount of complexed sulfathiazole %	Amount of complex %	Solubility at 25°C expressed as mg sulfathiazole per ml of water	Solubility at 25°C 10^2 mol dm^{-3} water [a]
20	11.48	31.48	2.45	0.960
40	22.96	62.96	5.02	1.966
60	34.44	84.44[b]	7.60	2.977

[a] Calculated by compiler

[b] Should be 94.44 - compiler

[*] This complex is not registrable by the conventions of the CAS Registry System (information from Knox Hazelton, Exptl. Services of Ca.; Jan. 12, 1981) - compiler.

AUXILIARY INFORMATION

METHOD/APPARATUS/PROCEDURE:

Mixts contg a weighed excess of sulfathiazole and a povidone-sulfathiazole coacervated system were placed in 25-ml ampuls contg 10 ml of water. The ampuls were sealed and placed on a rotating shaft (42 rpm) immersed in a water bath at 25±1°C. Duplicate samples were withdrawn, filtered, and assayed spectrophotometrically at 283 nm.

SOURCE AND PURITY OF MATERIALS:

Sulfathiazole was of the BP 1963 purity. The povidone-sulfathiazole coacervated systems were prepd by the authors. Povidone (mol wt 25,000) was manufd by BASF (West Germany).

Purity of the water was not specified.

ESTIMATED ERROR:

Soly: not specified.

Temp: ±1°C (authors).

REFERENCES:

COMPONENTS:

(1) Benzenesulfonamide, 4-amino-N-2-thiazolyl- monohydrochloride (sulfathiazole hydrochloride); $C_9H_9N_3O_2S_2 \cdot HCl$; [23325-73-7]

(2) Water; H_2O; [7732-18-5]

ORIGINAL MEASUREMENTS:

Lott, W. A.; Bergeim, F. H.

J. Am. Chem. Soc. 1939, *61*, 3593-4.

VARIABLES:

One temperature: 26°C

PREPARED BY:

R. Piekos

EXPERIMENTAL VALUES:

Solubility of sulfathiazole hydrochloride in water at 26°C is less than 2% (7 x 10^{-2} mol kg^{-1} solution, compiler).

AUXILIARY INFORMATION

METHOD/APPARATUS/PROCEDURE:

Nothing specified

SOURCE AND PURITY OF MATERIALS:

Sulfathiazole hydrochloride, mp 193-7°C (uncor), was prepd by the authors by adding alcoholic HCl to an alcoholic soln of sulfathiazole and adding ether. Purity of the water was not specified.

ESTIMATED ERROR:

Nothing specified

REFERENCES:

COMPONENTS:

(1) Cobalt, bis(4-amino-N-2-thiazolyl-benzenesulfonamidato-N^N,O)-, hydrate; $C_{18}H_{16}CoN_6O_4S_4 \cdot nH_2O$; [86729-22-8]

(2) Hydrochloric acid; HCl; [7647-01-0]

(3) Water; H_2O; [7732-18-5]

ORIGINAL MEASUREMENTS:

Tskitishvili, M. G.; Mikadze, I. I. *Soobshch. Akad. Nauk Gruz. SSR* 1978, *89(3)*, 589-92.

VARIABLES:

pH

PREPARED BY:

R. Piekos

EXPERIMENTAL VALUES:

K_{so} over the HCl concentration range 2.5×10^{-2} - 2.5×10^{-5} mol cm^{-3}, at 25°C, is 2.46×10^{-12}.

AUXILIARY INFORMATION

METHOD/APPARATUS/PROCEDURE:

In a glass vessel, a mixt of 100 ml of HCl of appropriate concn and the solute was placed and shaken for 6 h in a water thermostat at 25°C. After attaining equilibrium, the pH of the soln was measured and the Co^{2+} and S content detd to calculate K_{so}.

SOURCE AND PURITY OF MATERIALS:

Nothing specified.

ESTIMATED ERROR:

Nothing specified.

REFERENCES:

COMPONENTS:	ORIGINAL MEASUREMENTS:
(1) Copper, bis(4-amino-N-2-thiazolylbenzenesulfonamidato-N^N,O)-,hydrate; $C_{18}H_{16}CuN_6O_4S_4 \cdot nH_2O$; [86729-21-7] (2) Hydrochloric acid; HCl; [7647-01-0] (3) Water; H_2O; [7732-18-5]	Tskitishvili, M. G.; Mikadze, I. I. *Soobshch. Akad. Nauk Gruz. SSR* 1978, *89(3)*, 589-92.
VARIABLES:	PREPARED BY:
pH	R. Piekos

EXPERIMENTAL VALUES:

K_{so} over the HCl concentration range 2.5×10^{-2} - 2.5×10^{-5} mol dm^{-3}, at 25°C, is 2.17×10^{-17}.

AUXILIARY INFORMATION

METHOD/APPARATUS/PROCEDURE:	SOURCE AND PURITY OF MATERIALS:
In a glass vessel, a mixt of 100 ml of HCl of appropriate concn and the solute was placed and shaken for 6 h in a water thermostat at 25°C. After attaining equilibrium, the pH of the soln was measured and the Cu^{2+} and S content was detd to caluculate K_{so}.	Nothing specified. ESTIMATED ERROR: Nothing specified. REFERENCES:

COMPONENTS:	ORIGINAL MEASUREMENTS:
(1) Magnesium, (T-4)-bis(4-amino-N-2-thiazolylbenzenesulfonamidato-N^{N},O)- hydrate; $C_{18}H_{16}MgN_6O_4S_4 \cdot nH_2O$; [84812-78-2] (2) Hydrochloric acid; HCl; [7647-01-0] (3) Water; H_2O; [7732-18-5]	Tskitishvili, M. G.; Shvelashvili, A. E.; Mikadze, I. I; Zhorzholiani, N. B.; Chrelashvili, M. V. *Izv. Akad. Nauk Gruz. SSR, Ser. Khim.* 1981, *7(4)*, 300-4.
VARIABLES:	PREPARED BY:
pH	R. Piekos

EXPERIMENTAL VALUES:

K_{so} over the HCl concentration range 5.0×10^{-3} to 1.5×10^{-5} mol dm^{-3} at 25°C is 4.05×10^{-4}.

AUXILIARY INFORMATION

METHOD/APPARATUS/PROCEDURE:

The earlier described apparatus and method was used (1): in a glass vessel, a mixt of 100 ml of HCl of appropriate concn and the solute were placed and shaken for 6 h in a water thermostat at 25°C. After attaining equilibrium, the pH of the soln was measured and the Mg^{2+} and S content was determined to calculate K_{so}. The pH was measured on a pH-673 pH meter.

SOURCE AND PURITY OF MATERIALS:

0.1M solns of chem pure $Mg(OAc)_2$, monosodium salt of sulfathiazole, and HCl as well as doubly distd water were used. The source of the materials was not specified.

ESTIMATED ERROR:

Nothing specified.

REFERENCES:

1. Tskitishvili, M. G.; Mikadze, I. I. *Soobshch. Akad. Nauk. Gruz. SSR* 1978, *89(3)*, 589.

COMPONENTS:

(1) Manganese, bis(4-amino-N-2-thiazolyl-benzenesulfonamidato-$\underline{N}^N$,$\underline{O}$)-hydrate; $C_{18}H_{16}MnN_6O_4S_4 \cdot nH_2O$; [84812-77-1]

(2) Hydrochloric acid; HCl; [7647-01-0]

(3) Water; H_2O; [7732-18-5]

ORIGINAL MEASUREMENTS:

Tskitishvili, M.G.; Shvelashvili, A. E.; Mikadze, I. I.; Zhorzholiani, N. B.; Chrelashvili, M. V. *Izv. Akad. Nauk Gruz. SSR, Ser. Khim.* 1981, *7(4)*, 300-4.

VARIABLES:

pH

PREPARED BY:

R. Piekos

EXPERIMENTAL VALUES:

Concentration of HCl (mol/l)	pH	10^9 K_{so} at 25°C
5.0×10^{-3}	5.54	1.50
2.5×10^{-3}	5.57	1.48
1.0×10^{-3}	5.65	1.52
5.0×10^{-4}	5.79	1.46
2.5×10^{-4}	6.08	1.49
1.0×10^{-4}	6.29	1.51
5.0×10^{-5}	6.45	1.49
1.5×10^{-5}	6.72	1.47
	Mean	1.49

AUXILIARY INFORMATION

METHOD/APPARATUS/PROCEDURE:

The earlier described apparatus and method was used (1): in a glass vessel a mixt of 100 ml of HCl of appropriate concn and the solute were placed and shaken for 6 h in a water thermostat at 25°C. After attaining equilibrium, the pH of the soln was measured and the Mn^{2+} and S content was detd to calculate K_{so}. The pH was measured on a pH-673 pH meter.

SOURCE AND PURITY OF MATERIALS:

0.1M solns of chemically pure $Mn(OAc)_2$, monosodium salt of sulfathiazole and HCl as well as doubly distd water were used. The source of the materials was not specified.

ESTIMATED ERROR:

K_{so}: std deviation 2×10^{-11} (compiler)

Temp and pH: not specified.

REFERENCES:

1. Tskitishvili, M. G.; Mikadze, I. I.; *Soobshch. Akad. Nauk Gruz. SSR* 1978, *89(3)*, 589.

COMPONENTS:	ORIGINAL MEASUREMENTS:
(1) Nickel, bis(4-amino-N-2-thiazolylbenzenesulfonamidato-$\underline{N}^N$,$\underline{O}$)- hydrate; $C_{18}H_{16}N_6NiO_4S_4 \cdot nH_2O$; [84812-76-0] (2) Hydrochloric acid; HCl; [7647-01-0] (3) Water; H_2O; [7732-18-5]	Tskitishvili, M. G.; Shvelashvili, A. E.; Mikadze, I. I.; Zhorzholiani, N. B.; Chrelashvili, M. V. *Izv. Akad. Nauk Gruz. SSR. Ser. Khim.* 1981, *7(4)*, 300-4.
VARIABLES:	PREPARED BY:
pH	R. Piekos

EXPERIMENTAL VALUES:

Concentration of HCl (mol/l)	pH	10^{12} K_{so} at 25°C
2.5×10^{-2}	6.62	3.29
1.0×10^{-2}	7.28	3.24
5.0×10^{-3}	8.01	3.23
2.5×10^{-3}	8.30	3.22
1.0×10^{-3}	8.78	3.22
5.0×10^{-4}	8.80	3.30
2.5×10^{-4}	8.89	3.28
1.0×10^{-4}	8.90	3.25
5.0×10^{-5}	8.90	3.23
2.5×10^{-5}	8.90	3.22
	Mean	3.25

AUXILIARY INFORMATION

METHOD/APPARATUS/PROCEDURE:

The earlier described apparatus and method was used (1): in a glass vessel, a mixt of 100 ml of HCl of appropriate concn and the solute were placed and shaken for 6 h in a water thermostat at 25°C. After attaining equilibrium, the pH of the soln was measured and the Ni^{2+} and S content was determined to calculate K_{so}. The pH was measured on a pH-673 pH meter.

SOURCE AND PURITY OF MATERIALS:

0.1M solns of chemically pure $Ni(OAc)_2$, monosodium salt of sulfathiazole, and HCl as well as doubly distd water were used. The source of the materials was not specified.

ESTIMATED ERROR:

K_{so}: std deviation 1×10^{-13} (compiler).
Temp and pH: not specified.

REFERENCES:

1. Tskitishvili, M. G.; Mikadze, I. I. *Soobshch. Akad. Nauk Gruz. SSR* 1978, *89(3)*, 589.

COMPONENTS:	ORIGINAL MEASUREMENTS:
(1) Benzenesulfonamide, 4-amino-N-2-thiazolyl-, monosodium salt (sodium sulfathiazole); $C_9H_8N_3NaO_2S_2$; [144-74-1] (2) Water; H_2O; [7732-18-5]	Clark, W. G.; Strakosch, E. A.; Levitan, N. I. *J. Lab. Clin. Med.* 1942, *28*, 188-9.
VARIABLES:	PREPARED BY:
Temperature	R. Piekos

EXPERIMENTAL VALUES:

t/°C	Solubility	
	g/100 g water	mol kg^{-1} water[a]
25	45.0	1.62
37	60.0	2.16

[a]Calculated by compiler

AUXILIARY INFORMATION

METHOD/APPARATUS/PROCEDURE:

A small tinted glass container contg excess Na sulfathiazole in water was shaken in a water bath thermostat for 24 h. The satd soln was then filtered by aspiration through a washed and dried asbestos filter stick into a weighed weighing bottle. The entire app was kept at the temp at which the compd was dissolved. The amt dissolved was the detd by the method of Bratton and Marshall (1), using a photoelectric colorimeter.

SOURCE AND PURITY OF MATERIALS:

Neither source nor purity of Na sulfathiazole was specified.
CO_2-free distd water was used.

ESTIMATED ERROR:

Soly: not specified.
Temp: ±0.1°C (authors).

REFERENCES:

1. Bratton, A.C.; Marshall, E. K. Jr. *J. Biol. Chem.* 1939, *128*, 537.

COMPONENTS:

(1) Benzenesulfonamide, 4-amino-N-2-thiazolyl-, monosodium salt, hexahydrate; $C_9H_8N_3NaO_2S_2 \cdot 6H_2O$; [71119-42-1]

(2) Water; H_2O; [7732-18-5]

ORIGINAL MEASUREMENTS:

Sapozhnikova, N. V.; Postovskii, I. Ya.
Zh. Prikl. Khim. 1944, *17*, 427-34.

VARIABLES:

Temperature

PREPARED BY:

R. Piekos

EXPERIMENTAL VALUES:

t/°C	Solubility	
	Weight%	mol kg^{-1} water[a]
0	9.7	0.280
5	11.0	0.321
20	23.7	0.806
37	43.8	2.022

[a] Calculated by compiler

AUXILIARY INFORMATION

METHOD/APPARATUS/PROCEDURE:

The salt was dissolved in water to form a satd soln which was occasionally agitated in a glass vessel immersed in a thermostat. The equilibrium was usually attained after 1 h. Five to 100-cm^3 samples of the satd soln were placed in Pt crucibles or dishes and evapd to dryness at temps lower than 110-115°C. The residue was dried to const wt at 105-110°C and weighed.

SOURCE AND PURITY OF MATERIALS:

Pure, recrystd salt was used.

Purity of the water was not specified.

ESTIMATED ERROR:

Soly: quite reliable results were obtained (authors).

Temp: ±0.05°C (authors).

REFERENCES:

COMPONENTS:

(1) Zinc, (T-4)-bis(4-amino-N-2-thiazolyl-benzenesulfonamidato-N^N,O)-(Zn(II) sulfathiazole); $C_{18}H_{16}N_6O_4S_4Zn$; [12286-43-0]

(2) Water; H_2O; [7732-18-5]

ORIGINAL MEASUREMENTS:

Fox, Ch. L., Jr.; Modak, S.; Stanford, J. W.; Fox, P. L.
Scand. J. Plast. Reconstr. Surg. 1979, *13(1)*, 89-94.

VARIABLES:

One temperature: 28-30°C

PREPARED BY:

R. Piekos

EXPERIMENTAL VALUES:

Solubility of Zn(II) sulfathiazole in water at room temperature (28-30°C)[a] is 50.4 mg% (8.78×10^{-4} mol dm^{-3} solution, compiler).

[a]Value given by one of the authors (S.M.) in personal communication.

AUXILIARY INFORMATION

METHOD/APPARATUS/PROCEDURE:

Satd soln of Zn(II) sulfadiazine was prepd in water and after 24 h aliquots from the clear supernatant were assayed for sulfathiazole content using the colorimetric method of Bratton and Marshall (1). The soly value was then calculated from the molecular formula.

SOURCE AND PURITY OF MATERIALS:

The Zn(II) sulfathiazole was prepd by the authors as follows: an inorg Zn salt was reacted with Na salt of sulfathiazole and the ppt was analyzed and characterized. No details were given, however.
Purity of the materials was not specified.

ESTIMATED ERROR:

Nothing specified

REFERENCES:

1. Bratton, A. C.; Marshall, E. K. Jr. *J. Biol. Chem.* 1939, *120*, 537.

COMPONENTS:

(1) Benzenesulfonamide, 4-amino-N-methyl-N-2-thiazolyl-; $C_{10}H_{11}N_3O_2S_2$; [51203-19-1]

(2) Water; H_2O; [7732-18-5]

ORIGINAL MEASUREMENTS:

Shepherd, R. G.; Bratton, A. C.; Blanchard, K. C. *J. Am. Chem. Soc.* <u>1942</u>, *64*, 2532-7.

VARIABLES:

One temperature: 37°C

PREPARED BY:

R. Piekos

EXPERIMENTAL VALUES:

Solubility of 4-amino-N-methyl-N-2-thiazolylbenzenesulfonamide in water at 37°C is 57 mg% (2.1 x 10^{-3} mol dm^{-3} solution, compiler).

AUXILIARY INFORMATION

METHOD/APPARATUS/PROCEDURE:

The sulfonamide was assayed colorimetrically (1). No details were given.

SOURCE AND PURITY OF MATERIALS:

The sulfonamide, mp 111-2°C, was synthesized by the authors. Analysis: %C 45.00 (calcd 44.60); %H 4.19 (4.12); %N 15.36 (15.60). Colorimetric factor 0.618 (0.639). Purity of the water was not specified.

ESTIMATED ERROR:

Nothing specified

REFERENCES:

1. Bratton, A. C.; Marshall, E. K., Jr. *J. Biol. Chem.* <u>1939</u>, *128*, 537.

COMPONENTS:

(1) Benzenesulfonamide, 4-amino-N-methyl-N-2-thiazolyl-; $C_{10}H_{11}N_3O_2S_2$; [51203-19-1]

(2) Water; H_2O; [7732-18-5]

ORIGINAL MEASUREMENTS:

Kitao, K.; Kubo, K.; Morishita, T.; Yata, N.; Kamada, A.

Chem. Pharm. Bull, 1973, *21,* 2417-26.

VARIABLES:

One temperature: 37°C

PREPARED BY:

R. Piekos

EXPERIMENTAL VALUES:

Solubility of 4-amino-N-methyl-N-2-thiazolylbenzenesulfonamide in water at 37°C is 1.15 mmol dm^{-3} solution.

AUXILIARY INFORMATION

METHOD/APPARATUS/PROCEDURE:

The sulfonamide was assayed by diazotization. No details were given.

SOURCE AND PURITY OF MATERIALS:

The sulfonamide was synthesized by the authors. Its purity was not specified. Deionized water was used.

ESTIMATED ERROR:

Soly: not specified.

Temp: ±1°C (authors).

REFERENCES:

COMPONENTS:	ORIGINAL MEASUREMENTS:
(1) Benzenesulfonamide, 4-amino-N-methyl-N-2-thiazolyl-; $C_{10}H_{11}N_3O_2S_2$; [51203-19-1] (2) Methane, trichloro-; $CHCl_3$; [67-66-3]	Kitao, K.; Kubo, K.; Morishita, T.; Yata, N.; Kamada, A. *Chem. Pharm. Bull.* 1973, *21*, 2417-26.
VARIABLES:	**PREPARED BY:**
One temperature: 37°C	R. Piekos

EXPERIMENTAL VALUES:

Solubility of 4-amino-N-methyl-N-2-thiazolylbenzenesulfonamide in $CHCl_3$ at 37° is 1415 mmol dm^{-3} solution.

AUXILIARY INFORMATION

METHOD/APPARATUS/PROCEDURE:	SOURCE AND PURITY OF MATERIALS:
One ml of the sulfonamide soln in $CHCl_3$ at equilibrium was taken into a test tube. After evapn of the solvent, the residue was dissolved in EtOH, the soln was properly dild with deionized water and the concn of the sulfonamide was detd by diazotization.	The sulfonamide was sysnthesized by the authors. Its purity was not specified. Neither source nor purity of $CHCl_3$ was specified.
	ESTIMATED ERROR:
	Soly: not specified. Temp: ±1°C (authors).
	REFERENCES:

COMPONENTS:

(1) Acetamide, N-[4-[(2-thiazolylamino)sulfonyl]phenyl]- (acetyl sulfathiazole); $C_{11}H_{11}N_3O_3S_2$; [127-76-4]

(2) Water; H_2O; [7732-18-5]

ORIGINAL MEASUREMENTS:

Roblin, R. O., Jr.; Williams, J. H. Winnek, P. S.; English, J. P.
J. Am. Chem. Soc. 1940, *62*, 2002-5.

VARIABLES:

One temperature: 37°C

PREPARED BY:

R. Piekos

EXPERIMENTAL VALUES:

Solubility of acetyl sulfathiazole in water at 37°C is 7.1 mg/100 cm^3 solution (2.4×10^{-4} mol dm^{-3}, compiler).

AUXILIARY INFORMATION

METHOD/APPARATUS/PROCEDURE:

Excess acetyl sulfathiazole in water was heated and stirred on a steam bath for 30 min. The suspension was then agitated for 24 h in a thermostat at 37°C. A sample of the satd soln was withdrawn through a glass filter, dild and analyzed by the Marshall method (1) using a General Electric spectrophotometer for comparing the colors developed with those of the standards.

SOURCE AND PURITY OF MATERIALS:

Acetyl sulfathiazole was prepd by the authors by condensing recrystd acetylsulfanilyl chloride with 2-aminothiazole in AcOEt or dioxane.

Purity of the water was not specified.

ESTIMATED ERROR:

Nothing specified

REFERENCES:

1. Bratton, A. C.; Marshall, E. K., Jr. *J. Pharmacol.* 1939, *66*, 4.

COMPONENTS:

(1) Acetamide, N-[4-[(2-thiazolylamino)sulfonyl]phenyl]- (acetyl sulfathiazole); $C_{11}H_{11}N_3O_3S_2$; [127-76-4]

(2) Water; H_2O; [7732-18-5]

ORIGINAL MEASUREMENTS:

Durel, M.P.; Allinne, M.

Bull. Soc. Med. Hop. Paris III 1941, 251-9.

VARIABLES:

One temperature: 37°C

PREPARED BY:

R. Piekos

EXPERIMENTAL VALUES:

Solubility of acetyl sulfathiazole in water at 37°C is 0.10 g/liter (3.4×10^{-4} mol dm^{-3}, compiler).

AUXILIARY INFORMATION

METHOD/APPARATUS/PROCEDURE:

A mixt of acetyl sulfathiazole and water was agitated for 24 hours at 37°C.

SOURCE AND PURITY OF MATERIALS:

Source and purity of acetyl sulfathiazole was not specified.

Distilled water was used.

ESTIMATED ERROR:

Nothing specified.

REFERENCES:

COMPONENTS:

(1) Acetamide, N-[4-[(2-thiazolylamino)sulfonyl]phenyl]- (acetyl sulfathiazole); $C_{11}H_{11}N_3O_3S_2$; [127-76-4]

(2) Water; H_2O ; [7732-18-5]

ORIGINAL MEASUREMENTS:

Sapozhnikova, N.V.; Postovskii, I. Ya. *Zh. Prikl. Khim.* 1944, *17*, 427-34.

VARIABLES:

Temperature

PREPARED BY:

R. Piekos

EXPERIMENTAL VALUES:

t/°C	Solublity	
	Weight%	10^3 mol kg^{-1} water[a]
50	0.013	0.44
75	0.047	1.58
99	0.126[b]	4.24

[a] Calculated by compiler

[b] Calculated from the heat of dissolution

AUXILIARY INFORMATION

METHOD/APPARATUS/PROCEDURE:

Acetyl sulfathiazole was dissolved in water to form a satd soln which was occasionally agitated in a glass vessel immersed in a thermostat. The equilibrium was usually attained after 1 h. Five- to 100-cm^3 samples of the satd soln were placed in Pt crucibles or dishes and evapd to dryness at temps lower than 110-115°C. The residue was dried to const wt at 105-110°C and weighed.

SOURCE AND PURITY OF MATERIALS:

Pure, recrystd acetyl sulfathiazole was used. Its mp conformed to that reported in the literature.

Purity of the water was not specified.

ESTIMATED ERROR:

Soly: quite reliable results were obtained at 50 and 75°C. At 99°C the accuracy was poor due to evapn of water during sampling (authors). Temp: ±0.05°C (authors).

REFERENCES:

COMPONENTS:	ORIGINAL MEASUREMENTS:
(1) Acetamide, N-[4-[(2-thiazolylamino-sulfonyl]phenyl]- (acetyl sulfathiazole); $C_{11}H_{11}N_3O_3S_2$; [127-76-4] (2) Sodium hydroxide; NaOH; [1310-73-2] (3) Water; H_2O; [7732-18-5]	Rose, F. L.; Martin, A. R.; Bevan, H.G.L. *J. Pharm. Exp. Therap.* 1943, *77*, 127-42.
VARIABLES:	PREPARED BY:
pH	R. Piekos

EXPERIMENTAL VALUES:

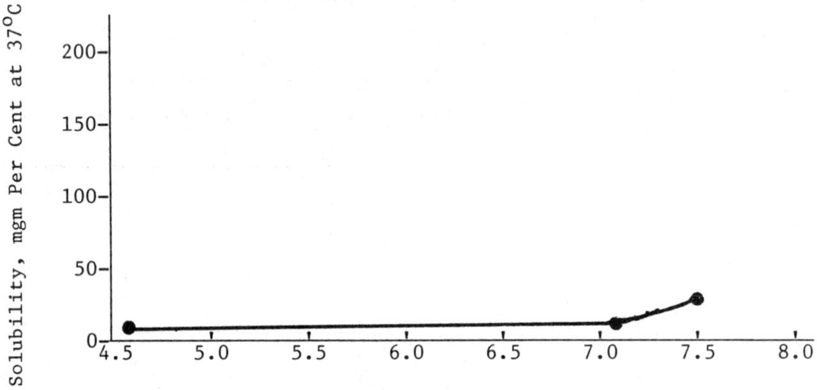

AUXILIARY INFORMATION

METHOD/APPARATUS/PROCEDURE:

An excess of acetyl sulfathiazole was stirred in boiling water, the soln was cooled to 37°C, the temp being maintained thermostatically, and 0.1N NaOH was added to increase the pH. The pH was measured by means of a glass electrode-calomel half-cell system and was permitted to reach equilibrium before a reading was taken. The drug was then de-acetylated and the concn of sulfathiazole in soln was detd colorimetrically by withdrawing a sample through a filter-tip into a preheated micropipet.

SOURCE AND PURITY OF MATERIALS:

The source and purity of acetyl sulfathiazole were not specified.

Water was doubly distilled.

ESTIMATED ERROR:

Nothing specified.

REFERENCES:

COMPONENTS:	ORIGINAL MEASUREMENTS:
(1) Acetamide, N-[4-[(2-thiazolylamino)-sulfonyl]phenyl]- (acetyl sulfathiazole); $C_{11}H_{11}N_3O_3S_2$; [127-76-4] (2) Phosphoric acid, disodium salt; Na_2HPO_4; [7558-94-4] (3) Water; H_2O; [7732-18-5]	Krüger-Thiemer, E. *Arch. Dermatol. Syphilis* 1942, *183*, 90-116.

VARIABLES:	PREPARED BY:
One temperature: ca 20°C; one pH: 8.74	R. Piekos

EXPERIMENTAL VALUES:

Solubility of acetyl sulfathiazole in a 0.705 M (10%) Na_2HPO_4 solution of pH 8.74 at room temperature (about 20°C) is 0.060 g% (2.02×10^{-3} mol dm^{-3} solution, compiler).

AUXILIARY INFORMATION

METHOD/APPARATUS/PROCEDURE:

Acetyl sulfathiazole (0.5 g) was dissolved in 10 cm^3 of the 0.705 M (10%) Na_2HPO_4 soln of pH 8.74, shaken for 2 h at room temp (about 20°C), and filtered. The filtrate was treated with equal vol of 2N HCl and refluxed for 15 min. After proper diln, a 1-cm^3 aliquot was witdrawn, acidified, cooled, and the sulfonamide content was detd colorimetrically (as sulfathiazole) by the Marshall method modified by Kimmig (1) using an Autenrieth colorimeter. The pH was detd on an ultraionograph using a glass electrode.

SOURCE AND PURITY OF MATERIALS:

Acetyl sulfathiazole (source not specified) gave no coloration upon diazotization of its satd soln, thus showing absence of sulfathiazole.
The source and purity of the remaining materials was not specified.

ESTIMATED ERROR:

Soly: precision ±5% (author)
Temp: not specified
pH : ±0.05 pH unit (author)

REFERENCES:

1. Kimmig, J. *Arch. Dermatol.* 1938, *176*, 722; *Erg. Hyg.* 1941, *24*, 398.

COMPONENTS:

(1) Acetamide, N-[4-[(2-thiazolylamino)-sulfonyl]phenyl]- (acetyl sulfathiazole); $C_{11}H_{11}N_3O_3S_2$; [127-76-4]

(2) Phosphoric acid, monopotassium salt; KH_2PO_4 [7778-77-0]

(3) Water; H_2O; [7732-18-5]

ORIGINAL MEASUREMENTS:

Krüger-Thiemer, E.

Arch. Dermatol. Syphilis 1942, *183*, 90-116.

VARIABLES:

One temperature: ca 20°C; one pH: 4.37

PREPARED BY:

R. Piekos

EXPERIMENTAL VALUES:

Solubility of acetyl sulfatiazole in a 0.735 M (10%) KH_2PO_4 solution of pH 4.37 at room temperature (about 20°C) is 0.0027 g% (9.08 x 10^{-5} mol dm^{-3} solution, compiler).

AUXILIARY INFORMATION

METHOD/APPARATUS/PROCEDURE:

Acetyl sulfathiazole (0.5 g) was dissolved in 10 cm^3 of the 0.735 M (10%) KH_2PO_4 soln of pH 4.37, shaken for 2 h at room temp (about 20°C), and filtered. The filtrate was treated with equal vol of 2N HCl, and refluxed for 15 min. After proper diln, a 1-cm^3 aliquot was withdrawn, acidified, cooled, and the sulfonamide content was detd colorimetrically (as sulfathiazole) by Marshall method modified by Kimmig (1) using an Authenrieth colorimeter. The pH was detd on an ultraionograph using a glass electrode.

SOURCE AND PURITY OF MATERIALS:

Acetyl sulfathiazole (source not specified) gave no coloration upon diazotization of its satd soln, thus showing absence of sulfathiazole.

The source and purity of the remaining materials was not specified.

ESTIMATED ERROR:

Soly: precision ±5% (author)
Temp: not specified
pH : ±0.05 pH unit (author)

REFERENCES:

1. Kimmig, J. *Arch. Dermatol.* 1938, *176*, 722; *Erg. Hyg.* 1941, *24*, 398.

COMPONENTS:	ORIGINAL MEASUREMENTS:
(1) Acetamide, N-[4-[(2-thiazolylamino-sulfonyl]phenyl]- (acetyl sulfathiazole); $C_{11}H_{11}N_3O_3S_2$; [127-76-4] (2) Phosphoric acid, disodium salt; Na_2HPO_4; [7558-94-4] (3) Phosphoric acid, monopotassium salt; KH_2PO_4; [7778-77-0] (4) Water; H_2O; [7732-18-5]	Sunderman, F. W.; Pepper, D. S.; Benditt, E. *J. Med. Sci.* 1940, *200*, 790-5.
VARIABLES:	PREPARED BY:
pH	R. Piekos

EXPERIMENTAL VALUES:

Solubility of acetyl sulfathiazole in phosphate buffer solutions at 38°C

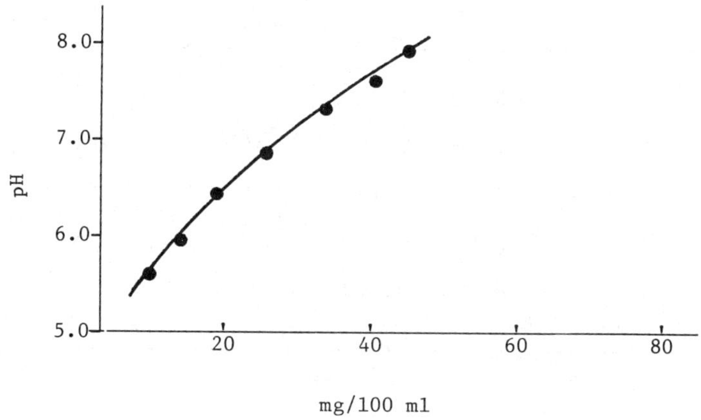

AUXILIARY INFORMATION

METHOD/APPARATUS/PROCEDURE:

An excess of acetyl sulfathiazole was suspended in buffer solns (prepd by dilg appropriate mixts of Na_2HPO_4 and KH_2PO_4, 1 part to 10 parts of distd water), agitated and kept in a water bath at 38°C for about 2 h. The solns were then filtered and analyses for total sulfathiazole were made on the filtrates. Acetyl sulfathiazole was assayed colorimetrically after coupling with di-Me-1-naphthylamine using a Bausch and Lomb colorimeter fitted with a No. 74 Wratten filter. Standards were made from a stock soln of acetyl sulfathiazole contg 200 mg/liter.

SOURCE AND PURITY OF MATERIALS:

Distd water was used.

The source and purity of the remaining reagents were not specified.

ESTIMATED ERROR:

Soly: the curve represents a composite of 3 sets of detns (authors).

Temp and pH: not specified.

REFERENCES:

COMPONENTS:	ORIGINAL MEASUREMENTS:
(1) Acetamide,N-[4-[(2-thiazolylamino)-sulfonyl]phenyl]- (acetyl sulfathiazole); $C_{11}H_{11}N_3O_3S_2$; [127-76-4] (2) Phosphoric acid, disodium salt; Na_2HPO_4; [7558-94-4] (3) Phosphoric acid, monopotassium salt; KH_2PO_4; [7778-77-0] (4) Water; H_2O; [7732-18-5]	Krüger- Thiemer, E. *Arch. Dermatol. Syphilis* 1942, *183*, 90-116.
VARIABLES:	PREPARED BY:
Temperature, pH	R. Piekos

EXPERIMENTAL VALUES:

Composition of 1/15 M phosphate buffer solutions			pH	Solubility			
				Room temp (ca 20°C)		37°C	
Na_2HPO_4	KH_2PO_4	%content		g%	10^4 mol dm^{-3} solution	g%	10^4 mol dm^{-3} solution[a]
1.0	99.0	0.91	4.944	0.0080	2.70	-	-
10.0	90.0	0.91	5.906	0.0073	2.40	0.0092	3.1
61.1	38.9	0.93	7.005	0.0101	3.40	0.0188	6.32
9.5	0.5	0.733[b]	7.51	0.0101	3.40	-	-
94.7	5.3	0.95	8.018	0.0360	12.00	-	-

[a] Calculated by compiler

[b] Molar content; 10% buffer solution

AUXILIARY INFORMATION

METHOD/APPARATUS/PROCEDURE:

Acetyl sulfathiazole (0.5 g) was dissolved in 10 cm^3 of a buffer soln, shaken for 2 h at 20°C (or left for 48 h at 37°C), and filtered at respective temp. The filtrate was treated with equal vol of 2N HCl and refluxed for 15 min. After proper diln, a 1-cm^3 aliquot was withdrawn, acidified, cooled, and the sulfonamide content was detd colorimetrically (as sulfathiazole) by the Marshall method modified by Kimmig (1) using an Authenrieth colorimeter. The pH was detd on an ultraionograph using a glass electrode.

SOURCE AND PURITY OF MATERIALS:

Acetyl sulfathiazole (source not specified) gave no coloration upon diazotization of its satd soln, thus showing absence of sulfathiazole.

The source and purity of the remaining materials was not specified.

ESTIMATED ERROR:

Soly: precision ±5% (author)
RTemp: not specified
pH : ±0.05 pH unit (author)

REFERENCES:

1. Kimmig, J. *Arch. Dermatol.* 1938, *176*, 722; *Erg. Hyg.* 1941, *24*, 398.

COMPONENTS:

(1) Acetamide, N-[4-[(2-thiazolylamino)-sulfonyl]phenyl]- (acetyl sulfathiazole); $C_{11}H_{11}N_3O_3S_2$ [127-76-4]

(2) Phosphoric acid, disodium salt; Na_2HPO_4; [7558-94-4]

(3) Phosphoric acid; monopotassium salt; KH_2PO_4; [7778-77-0]

(4) Water; H_2O; [7732-18-5]

ORIGINAL MEASUREMENTS:

Pulver, R.; Suter, R.

Schweiz. Med. Wochenschr. 1943, *73(13)*, 403-8.

VARIABLES:

pH

PREPARED BY:

R. Piekos

EXPERIMENTAL VALUES:

pH	Solubility of acetylsulfathiazole in M/15 phosphate buffers (according to Sørensen) at 20°C	
	mg%	10^3 mol dm^{-3} [a]
6.0	8	0.30
7.0	11	0.37
8.0	35	1.2

[a] Calculated by compiler

AUXILIARY INFORMATION

METHOD/APPARATUS/PROCEDURE:

Nothing specified

SOURCE AND PURITY OF MATERIALS:

Nothing specified

ESTIMATED ERROR:

Nothing specified

REFERENCES:

COMPONENTS:

(1) Acetamide, N-[4-[(2-thiazolylamino)-sulfonyl]phenyl]-(acetyl sulfathiazole); $C_{11}H_{11}N_3O_3S_2$; [127-76-4]

(2) Phosphoric acid, disodium salt; Na_2HPO_4; [7558-94-4]

(3) Phosphoric acid, monopotassium salt; KH_2PO_4; [7778-77-0]

(4) Water; H_2O; [7732-18-5]

ORIGINAL MEASUREMENTS:

Frisk, A. R.; Hagerman, G.; Helander, S.; Sjögren, B. *Hygiea* 1946, *108(12)* 639-51.

VARIABLES:

One temperature: 37°C; one pH: 6.1

PREPARED BY:

R. Piekos

EXPERIMENTAL VALUES:

Solubility of acetyl sulfathiazole in M/30 phosphate buffer of pH 6.1 at 37°C is 8.4 mg/100 ml solvent (2.8×10^{-4} mol dm^{-3}, compiler).

AUXILIARY INFORMATION

METHOD/APPARATUS/PROCEDURE:

An excess of acetyl sulfathiazole in the phosphate buffer was shaken at 37°C for 24 h. The concn of acetyl sulfathiazole was detd by the Bratton and Marshall method (1) using a photoelec colorimeter.

SOURCE AND PURITY OF MATERIALS:

Neither source nor purity of the materials was specified.

ESTIMATED ERROR:

Soly: precision ±0.7 mg/100 ml (authors).

Temp and pH: not specified.

REFERENCES:

1. Bratton, A. C.; Marshall, E. K., Jr. *J. Biol. Chem.* 1939, *128*, 537.

COMPONENTS:

(1) Acetamide, N-[4-[(2-thiazolylamino)-sulfonly]phenyl]-(N^4-acetylsulfathiazole); $C_{11}H_{11}N_3O_3S_2$; [127-76-4]
(2) Phosphoric acid, disodium salt; Na_2HPO_4; [7558-94-4]
(3) Phosphoric acid, monopotassium salt; KH_2PO_4; [7778-77-0]
(4) Water; H_2O; [7732-18-5]

ORIGINAL MEASUREMENTS:

Hekster, Y. A.; Vree, T. B.; Damsma, J. E.; Friesen, W. T.

J. Antimicrob. Chemother. 1981, *8*, 133-44.

VARIABLES:

pH

PREPARED BY:

R. Piekos

EXPERIMENTAL VALUES:

pH	Solubility at 25°C mg/l	10^4 mol dm^{-3} [a]
5.5	54	1.8
7.5	233	7.83

AUXILIARY INFORMATION

METHOD/APPARATUS/PROCEDURE:

Satd solns of N^4-acetylsulfathiazole were prepd in phosphate buffers of pH 5.5 and 7.5 at room temp (25°C). The concn of the solute was measured by means of a Spectra Physics 3500B high-performance liquid chromatograph equipped with a column oven (Model 748) and a Pye-Unicam LC-UV spectrophotometric detector. The detector was connected to a 1-mV recorder. A stainless steel column (10 cm x 4.6 mm i.d.) was packed with Lichrosorb RPS, 5 μm, obtained from Chrompack. An injection loop of 100 ul was used. The oven temp was 40°C. Detection of the solute was performed at 260 nm.

SOURCE AND PURITY OF MATERIALS:

The source and purity of the materials was not specified.

ESTIMATED ERROR:

The detection limit of the solute by HPLC was 0.5 mg/l (authors). The error in temperature and pH was not specified.

REFERENCES:

COMPONENTS:	ORIGINAL MEASUREMENTS:
(1) Butanoic acid, 4-oxo-4-[[4-[(2-thiazolylamino)sulfonyl]phenyl]amino]-(sulfasuxidine); $C_{13}H_{13}N_3O_5S_2$; [116-43-8] (2) 2-Propanol; C_3H_8O; [67-63-0]	Burlage, H.M. *J. Am. Pharm. Assoc. Sci. Ed.* 1948, *37*, 345.
VARIABLES:	PREPARED BY:
One temperature: 25°C	R. Piekos

EXPERIMENTAL VALUES:

Solubility of sulfasuxidine in 2-propanol at 25°C is 0.5690 g/100 cm^3 solution (1.601×10^{-2} mol dm^{-3}, compiler).

AUXILIARY INFORMATION

METHOD/APPARATUS/PROCEDURE:

Satd soln of sulfasuxidine in 2-propanol were prepd at 25°C and definite vols of the solns were measured into tared dishes by means of standard pipets. The alcohol was allowed to evap at room temp and the residue was dried at 105°C. In the case of losses due to apparent decompn, the residue was dried in a desiccator.

SOURCE AND PURITY OF MATERIALS:

The sulfasuxidine N.N.R. was manufd by Sharp and Dohme. The source and purity of 2-propanol was not specified.

ESTIMATED ERROR:

Nothing specified

REFERENCES:

1. Burlage, H. M. *J. Am. Pharm. Assoc. Sci. Ed.* 1947, *36(1)*, 16.

COMPONENTS:	EVALUATOR:
(1) Benzenesulfonamide, 4-amino-N-(4-methyl-2-thiazolyl)- (sulfamethylthiazole); $C_{10}H_{11}N_3O_2S_2$; [515-59-3] (2) Water	Anthony N. Paruta Department of Pharmaceutics University of Rhode Island Kingston, Rhode Island, USA and Ryszard Piekos Faculty of Pharmacy, University of Gdansk Gdansk, Poland 1986

CRITICAL EVALUATION:

Sulfamethylthiazole solubilities in water at 310K have been reported by three groups (1-3). In 1940, Roblin et al. (1), using a correct technique gave a solubility value of 1.07×10^{-3} mol dm^{-3}. Durel and Allinne's (2) is 0.965×10^{-3} mol dm^{-3}. In this case, however, no details are provided for the analytical method or error estimate, but it is considered of sufficient accuracy to use in the average result. Sapozhnikova and Postovskii (3) provide a value at 310K of 0.869×10^{-3} mol kg^{-1}. The short equilibrium time of one hour probably mitigate against being a saturation value and was not considered further. The simple average of the two acceptable values allow for a recommended value of 1.02×10^{-3} mol dm^{-3} in water at 310K.

REFERENCES:

(1) Roblin, R.O., Jr.; Williams, J.H.; Winnek, P.S.; English, J.P. *J. Am. Chem. Soc.* 1940, *62*, 2002-5.

(2) Durel, M.P.; Allinne, M. *Bull. Soc. Med. Hop. Paris III* 1941, 251-9.

(3) Sapozhnikova, N.V.; Postovskii, I.Ya. *Zh. Prikl. Khim.* 1944, *17*, 427-34.

COMPONENTS:	ORIGINAL MEASUREMENTS:
(1) Benzenesulfonamide, 4-amino-N-(4-methyl-2-thiazolyl)-(sulfamethylthiazole); $C_{10}H_{11}N_3O_2S_2$; [515-59-3] (2) Water; H_2O; [7732-18-5]	Roblin, R. O., Jr.; Williams, J. H.; Winnek, P. S.; English, J. P. *J. Am. Chem. Soc.* 1940, *62*, 2002-5.
VARIABLES:	PREPARED BY:
One temperature: 37°C	R. Piekos

EXPERIMENTAL VALUES:

Solubility of sulfamethylthiazole in water at 37°C is 28.9 mg/100 cm^3 solution (1.07 x 10^{-3} mol dm^{-3}, compiler).

AUXILIARY INFORMATION

METHOD/APPARATUS/PROCEDURE:

Excess sulfamethylthiazole in water was heated and stirred on a steam bath for 30 min. The suspension was then agitated for 24 h in a thermostat at 37°C. A sample of the satd soln was withdrawn through a glass filter, dild, and analyzed by the Marshall method (1) using a General Electric recording spectrophotometer for comparing the colors developed with those of the standards.

SOURCE AND PURITY OF MATERIALS:

Sulfamethylthiazole had mp of 237-8°C (cor) consistent with the literature data. Purity of the water was not specified.

ESTIMATED ERROR:

Nothing specified

REFERENCES:

1. Bratton, A. C.; Marshall, E. K., Jr. *J. Pharmacol.* 1939, *66*, 4.

COMPONENTS:	ORIGINAL MEASUREMENTS:
(1) Benzenesulfonamide, 4-amino-N-(4-methyl-2-thiazolyl)-(sulfamethylthiazole); $C_{10}H_{11}N_3O_2S_2$; [515-59-3] (2) Water; H_2O; [7732-18-5]	Durel, M.P. ; Allinne, M. *Bull. Soc. Med. Hop. Paris III* 1941, 251-9.
VARIABLES:	PREPARED BY:
One temperature: 37°C	R. Piekos

EXPERIMENTAL VALUES:

Solubility of sulfamethylthiazole in water at 37°C is 0.26 g/liter (9.65×10^{-4} mol dm^{-3}, compiler).

AUXILIARY INFORMATION

METHOD/APPARATUS/PROCEDURE:	SOURCE AND PURITY OF MATERIALS:
A mixt of sulfamethylthiazole and water was agitated for 24 hours at 37°C.	Source and purity of sulfamethylthiazole was not specified. Distilled water was used.
	ESTIMATED ERROR:
	Nothing specified
	REFERENCES:

COMPONENTS:	ORIGINAL MEASUREMENTS:
(1) Benzenesulfonamide, 4-amino-N-(4-methyl-2-thiazolyl)-(sulfamethylthiazole); $C_{10}H_{11}N_3O_2S_2$; [515-59-3] (2) Water; H_2O; [7732-18-5]	Sapozhnikova, N. V.; Postovskii, I. Ya. *Zh. Prikl. Khim.* 1944, *17*, 427-34.
VARIABLES:	PREPARED BY:
Temperature	R. Piekos

EXPERIMENTAL VALUES:

t/°C	Solubility	
	Weight%	10^2 mol kg^{-1} water[a]
20	0.0088	3.47
37	0.0220	8.69
50	0.0423	1.67
75	0.130	5.14
99	0.390 ; 0.333[b]	15.46; 13.19

[a] Calculated by compiler

[b] Calculated from the heat of dissolution (9,866 cal mol^{-1})

AUXILIARY INFORMATION

METHOD/APPARATUS/PROCEDURE:

Sulfamethylthiazole was dissolved in water to form a satd soln which was occasionally agitated in a glass vessel immersed in a thermostat. The equilibrium was usually attained after 1 h. Five- to 100-cm^3 samples of the satd soln were placed in Pt crucibles or dishes and evapd to dryness at temps lower than 110-115°C. The residue was dried to const wt at 105-110°C and weighed.

SOURCE AND PURITY OF MATERIALS:

Pure, recrystd sulfamethylthiazole was used. Its mp conformed to that reported in the literature.

Purity of the water was not specified.

ESTIMATED ERROR:

Soly: quite reliable results were obtained over the temp range 20-75°C. At higher temps the accuracy was pure due to evapn of water during sampling (authors). Temp: ±0.05°C (authors).

REFERENCES:

COMPONENTS:	ORIGINAL MEASUREMENTS:
(1) Benzenesulfonamide, 4-amino-N-(4-methyl-2-thiazolyl)- (sulfamethylthiazole); $C_{10}H_{11}N_3O_2S_2$; [515-59-3] (2) Phosphoric acid, disodium salt; Na_2HPO_4; [7558-94-4] (3) Water; H_2O; [7732-18-5]	Krüger-Thiemer, E. *Arch. Dermatol. Syphilis* 1942, *183*, 90-116.

VARIABLES:	PREPARED BY:
One temperature: ca 20°C; one pH: 8.74	R. Piekos

EXPERIMENTAL VALUES:

Solubility of sulfamethylthiazole in a 0.705 M (10%) Na_2HPO_4 solution of pH 8.74, at room temperature (about 20°C), is 0.058 g% (2.15 x 10^{-3} mol dm^{-3} solution, compiler).

AUXILIARY INFORMATION

METHOD/APPARATUS/PROCEDURE:

Sulfamethylthiazole (0.5 g) was dissolved in 10 cm^3 of the 0.705 M (10%) Na_2HPO_4 soln of pH 8.74, shaken for 2 h at room temp (about 20°C), and filtered. A 1-cm^3 aliquot of the filtrate was withdrawn, cooled, acidified with 1 cm^3 of 2N HCl, and the sulfonamide content was detd colorimetrically by the method of Marshall modified by Kimmig (1) using an Authenrieth colorimeter. The pH was detd on an ultraionograph using a glass electrode.

SOURCE AND PURITY OF MATERIALS:

Sulfamethylthiazole was the product manufd by Sanebo under the name Ultraseptyl. The source and purity of the remaining materials was not specified.

ESTIMATED ERROR:

Soly: precision ±5% (author)
Temp: not specified
pH : ±0.05 pH unit (author)

REFERENCES:

1. Kimmig, J. *Arch. Dermatol.* 1938, *176*, 722; *Erg. Hyg.* 1941, *24*, 398.

COMPONENTS:	ORIGINAL MEASUREMENTS:
(1) Benzenesulfonamide, 4-amino-N-4-methyl-2-thiazolyl)- (sulfamethylthiazole); $C_{10}H_{11}N_3O_2S_2$; [515-59-3] (2) Phosphoric acid, monopotassium salt; KH_2PO_4; [7778-77-0] (3) Water; H_2O; [7732-18-5]	Krüger-Thiemer, E. *Arch. Dermatol. Syphilis* 1942, *183*, 90-116.

VARIABLES:	PREPARED BY:
One temperature: ca 20°C; one pH: 4.37	R. Piekos

EXPERIMENTAL VALUES:

Solubility of sulfamethylthiazole in a 0.735 M (10%) KH_2PO_4 solution of pH 4.37, at room temperature (about 20°C), is 0.0094 g% (3.5×10^{-4} mol dm^{-3} solution, compiler).

AUXILIARY INFORMATION

METHOD/APPARATUS/PROCEDURE:

Sulfamethylthiazole (0.5 g) was dissolved in 10 cm^3 of the 0.735 M (10%) KH_2PO_4 soln of pH 4.37, shaken for 2 h at room temp (about 20°C), and filtered. A 1-cm^3 aliquot of the filtrate was withdrawn, cooled, acidified with 1 cm^3 of 2N HCl, and the sulfonamide content was detd colorimetrically by the method of Marshall modified by Kimmig (1) using an Authenrieth colorimeter. The pH was detd on an ultraionograph using a glass electrode.

SOURCE AND PURITY OF MATERIALS:

Sulfamethylthiazole was the product manufd by Sanebo under the name Ultraseptyl. The source and purity of the remaining materials was not specified.

ESTIMATED ERROR:

Soly: precision ±5% (author)
Temp: not specified
pH : ±0.05 pH unit (author)

REFERENCES:

1. Kimmig, J. *Arch. Dermatol.* 1938, *176*, 722; *Erg. Hyg.* 1941, *24*, 398.

COMPONENTS:

(1) Benzenesulfonamide, 4-amino-N-(4-methyl-2-thiazolyl)- (sulfamethylthiazole); $C_{10}H_{11}N_3O_2S_2$; [515-59-3]

(2) Phosphoric acid, disodium salt; Na_2HPO_4; [7558-94-4]

(3) Phosphoric acid, monopotassium salt; KH_2PO_4; [7778-77-0]

(4) Water; H_2O; [7732-18-5]

ORIGINAL MEASUREMENTS:

Krüger-Thiemer, E.

Arch. Dermatol. Syphilis 1942, *183*, 90-116.

VARIABLES:

Temperature, pH

PREPARED BY:

R. Piekos

EXPERIMENTAL VALUES:

Composition of 1/15 M phosphate buffer solutions			pH	Solubility			
				Room temp (ca 20°C)		37°C	
Na_2HPO_4	KH_2PO_4	%content		g%	10^3 mol dm^{-3} solution[a]	g%	10^3 mol dm^{-3} solution[a]
1.0	99.0	0.91	4.944	0.021	0.780	-	-
10.0	90.0	0.91	5.906	0.021	0.780	0.023	0.854
61.1	38.9	0.93	7.005	0.022	0.817	0.028	1.040
9.5	0.5	0.733[b]	7.510	0.0178	0.661	-	-
94.7	5.3	0.95	8.018	0.042	1.559	-	-

[a] Calculated by compiler

[b] Molar content: 10% buffer solution

AUXILIARY INFORMATION

METHOD/APPARATUS/PROCEDURE:

Sulfamethylthiazole (0.5 g) was dissolved in 10 cm^3 of a buffer soln, shaken for 2 h at 20°C (or left for 48 h at 37°C), and filtered at respective temp. A 1-cm^3 aliquot of the filtrate was withdrawn, cooled (dild for expts at 37°C), acidified with 1 cm^3 of 2N HCl, and the sulfonamide contents was detd colorimetrically by the method of Marshall modified by Kimmig (1) using an Autenrieth colorimeter. The pH was detd on an ultraionograph using a glass electrode.

SOURCE AND PURITY OF MATERIALS:

Sulfamethylthiazole was the product manufd by Sanebo under the name Ultraseptyl. The source and purity of the remaining materials was not specified.

ESTIMATED ERROR:

Soly: precision ±5% (author)
Temp: not specified
pH : ±0.05 pH unit (author)

REFERENCES:

1. Kimmig, J. *Arch. Dermatol.* 1938, *176*, 722; *Erg. Hyg.* 1941, *24*, 398.

COMPONENTS:	ORIGINAL MEASUREMENTS:
(1) Benzenesulfonamide, 4-amino-N-(4-methyl-2-thiazolyl)- (sulfamethylthiazole); $C_{10}H_{11}N_3O_2S_2$; [515-59-3] (2) 2-Propanone (acetone); C_3H_6O; [67-64-1]	Gutierrez, F. H. *Anales fis. quim. (Madrid)* 1945, *41*, 537-60.
VARIABLES:	PREPARED BY:
Temperature	R. Piekos

EXPERIMENTAL VALUES:

$t/^{\circ}C$	G^a	E^b	$X_g/1^c$	mol/1[d] acetone	mmol/mol acetone	$1:X_g^e$	$1 + X_{cc}^f$
0	1.068	1.050	8.700	32.3	2.30	93.63	114.90
5	1.125	1.112	9.099	33.7	2.42	88.88	109.90
10	1.410	1.361	11.322	42.0	3.04	70.92	88.32
15	1.504	1.482	11.988	44.5	3.24	66.49	83.35
20	1.677	1.649	13.268	49.3	3.64	59.63	75.37
25	1.813	1.783	14.236	52.9	3.93	55.16	70.24
30	2.152	2.107	16.771	62.1	4.64	46.47	60.44
35	2.505	2.445	19.371	71.9	5.40	39.92	51.62
40	2.903	2.822	22.278	82.7	6.26	34.44	44.84
45	3.513	3.394	26.798	99.5	7.57	28.46	37.32
50	4.524	4.328	34.179	126.9	9.76	23.10	29.25

[a] $G = \frac{p\,100}{P - p}$, where p and P are the weights of solute and solution, resp.

[b] $E = \frac{G\,100}{G + 100}$; [c] g/1 acetone; [d] should be mmol/1 acetone (compiler);
[e] g of acetone required to dissolved 1 g of solute; [f] volume (cm^3) of acetone required to dissolve 1 g of solute.

AUXILIARY INFORMATION

METHOD/APPARATUS/PROCEDURE:

A special all-glass app was constructed enabling the prepn of satd solns, agitation by bubbling a stream of acetone-satd N, filtration, and distn off the solvent without contact with air. Two exhangeable dissoln vessels of 15 and 8 cm^3 working capacity were used depending on the soly of solute. The app was immersed in a thermostat. The vols of acetone used were 15 or 5 cm^3, and the equilibration time was 2-2.5 h. The satd solns were filtered, weighed, the solvent was distd off, the residues were dried at $105^{\circ}C$, weighed, and examd for the presence of solvated acetone.

SOURCE AND PURITY OF MATERIALS:

The source of the materials was not specified. Pure, anhyd acetone was used. The absence of impurities and water was confirmed by procedures of the German Pharmacopeia VI and Spanish Pharmacopeia VIII.
The purity of sulfamethylthiazole was not specified.

ESTIMATED ERROR:

Soly: measurements were repeated until 2 values not differing in the second decimal were obtained (author).
Temp: $\pm 0.1^{\circ}C$ (author).

REFERENCES:

COMPONENTS:	EVALUATOR:
(1) Acetamide, N-[4-[[(4-methyl)-2-thiazolylamino]sulfonyl]phenyl]-(acetyl sulfamethylthiazole); $C_{12}H_{13}N_3O_3S_2$; [71119-13-6] (2) Water	Anthony N. Paruta Department of Pharmaceutics University of Rhode Island Kingston, Rhode Island, USA and Ryszard Piekos Faculty of Pharmacy, University of Gdansk Gdansk, Poland 1986

CRITICAL EVALUATION:

For this compound, the acetyl derivative of the previously evaluated sulfonamide, two values were available (1,2) in water at 310K. Roblin et al. (1) give a value of 1.8×10^{-4} mol dm^{-3}, and Durel and Allinne (2) 2×10^{-4} mol dm^{-3}. Both groups used quite adequate equilibrium times, though Durel and Allinne (2) do not specify the analytical technique. The similarity of the two values is considered to be evidence of accuracy and an average value of 1.9×10^{-4} mol dm^{-3} is the recommended value in water at 310K. It is interesting to note that the acetyl-derivative possesses a solubility of about one fifth (20%) of the parent compound, sulfamethylthiazole. This is usually the case, decreasing solubility for acetyl-derivative compounds.

REFERENCES:

(1) Roblin, R.O., Jr.; Williams, J.H.; Winnek, P.S.; English, J.P. *J. Am. Chem. Soc.* 1940, *62*, 2002-5.

(2) Durel, M.P.; Allinne, M. *Bull. Soc. Med. Hop. Paris III* 1941, 251-9.

COMPONENTS:

(1) Acetamide, N-[4-[[(4-methyl)-2-thiazolylamino]sulfonyl]phenyl]- (acetyl sulfamethylthiazole); $C_{12}H_{13}N_3O_3S_2$; [71119-13-6]

(2) Water; H_2O; [7732-18-5]

ORIGINAL MEASUREMENTS:

Roblin, R. O., Jr.; Williams, J. H.; Winnek, P.S.; English, J. P.

J. Am. Chem. Soc. 1940, *62*, 2002-5.

VARIABLES:

One temperature: 37°C

PREPARED BY:

R. Piekos

EXPERIMENTAL VALUES:

Solubility of acetyl sulfamethylthiazole in water at 37°C is 5.5 mg/100 cm^3 solution (1.8×10^{-4} mol dm^{-3}, compiler).

AUXILIARY INFORMATION

METHOD/APPARATUS/PROCEDURE:

Excess acetyl sulfamethylthiazole in water was heated and stirred on a steam bath for 30 min. The suspension was then agitated for 24 h in a thermostat at 37°C. A sample of the satd soln was withdrawn through a glass filter, dild, and analyzed by the Marshall method (1) using a General Electric spectrophotometer for comparing the colors developed with those of the standards.

SOURCE AND PURITY OF MATERIALS:

Acetyl sulfamethylthiazole was prepd by treating 2 moles of 2-amino-4-methylthiazole with 1 mole of acetylsulfanilyl chloride in an AcOEt or dioxane soln. Purity of the water was not specified.

ESTIMATED ERROR:

Nothing specified

REFERENCES:

1. Bratton, A. C.; Marshall, E. K., Jr. *J. Pharmacol.* 1939, *66*, 4.

COMPONENTS:	ORIGINAL MEASUREMENTS:
(1) Acetamide, N-[4-[[(4-methyl-2-thiazolyl)amino]sulfonyl]phenyl]- (acetyl sulfamethylthiazole); $C_{12}H_{13}N_3O_3S_2$; [71119-13-6] (2) Water; H_2O; [7732-18-5]	Durel, M.P.; Allinne, M. *Bull. Soc. Med. Hop. Paris III* 1941, 251-9.
VARIABLES:	PREPARED BY:
One temperature: 37°C	R. Piekos

EXPERIMENTAL VALUES:

Solubility of acetyl sulfamethylthiazole in water at 37°C is 0.07 g/liter (2×10^{-4} mol dm^{-3}, compiler).

AUXILIARY INFORMATION

METHOD/APPARATUS/PROCEDURE:

A mixt of acetyl sulfamethylthiazole and water was agitated for 24 hours at 37°C.

SOURCE AND PURITY OF MATERIALS:

Source and purity of acetyl sulfamethylthiazole was not specified.

Distilled water was used.

ESTIMATED ERROR:

Nothing specified.

REFERENCES:

COMPONENTS:

(1) Acetamide, N-[4-[[(4-methyl-2-thiazolyl-amino)sulfonyl]phenyl]- (acetyl sulfamethylthiazole); $C_{12}H_{13}N_3O_3S_2$; [71119-13-6]

(2) Water; H_2O; [7732-18-5]

ORIGINAL MEASUREMENTS:

Sapozhnikova, N. V.; Postovskii, I. Ya.
Zh. Prikl. Khim. 1944, *17*, 427-34.

VARIABLES:

Temperature

PREPARED BY:

R. Piekos

EXPERIMENTAL VALUES:

t^o/C	Solubility	
	Weight%	10^3 mol kg^{-1} water[a]
20	0.0022	0.071
50	0.0080; 0.0100[b]	0.260; 0.320
75	0.0350	1.100
99	0.0860[b]	2.800

[a] Calculated by compiler

[b] Calculated from the heat of dissolution (10,548 cal mol^{-1}).

AUXILIARY INFORMATION

METHOD/APPARATUS/PROCEDURE:

Acetyl sulfamethylthiazole was dissolved in water to form a satd soln which was occasionally agitated in a glass vessel immersed in a thermostat. The equilibrium was usually attained after 1 h. Five- to 100-cm^3 samples of the satd soln were placed in Pt crucibles or dishes and evapd to dryness at temps lower than 110-115°C. The residue was dried to const wt at 105-110°C and weighed.

SOURCE AND PURITY OF MATERIALS:

Pure, recrystd acetyl sulfamethylthiazole was used. Its mp conformed to that reported in the literature.

Purity of the water was not specified.

ESTIMATED ERROR: Soly: quite reliable results were obtained over the temp range 20-75°C. At higher temps the accuracy was poor due to evapn of water during sampling (authors).

Temp: ±0.05° C (authors).

REFERENCES:

COMPONENTS:

(1) Acetamide,N-[4-[[(4-methyl-2-thiazolyl)-amino]sulfonyl]phenyl]- (acetyl sulfamethylthiazole); $C_{12}H_{13}N_3O_3S_2$; [71119-13-6]

(2) Phosphoric acid, disodium salt; Na_2HPO_4; [7558-94-4]

(3) Water; H_2O; [7732-18-5]

ORIGINAL MEASUREMENTS:

Krüger-Thiemer, E.

Arch. Dermatol. Syphilis 1942, *183*, 90-116.

VARIABLES:

One temperature: ca 20°C; one pH: 8.74

PREPARED BY:

R. Piekos

EXPERIMENTAL VALUES:

Solubility of acetyl sulfamethylthiazole in a 0.705 M (10%) Na_2HPO_4 solution of pH 8.74 at room temperature (about 20°C) is 0.052 g% (1.67×10^{-3} mol dm^{-3} solution, compiler).

AUXILIARY INFORMATION

METHOD/APPARATUS/PROCEDURE:

Acetyl sulfamethylthiazole (0.5 g) was dissolved in 10 cm^3 of the 0.705 M (10%) Na_2HPO_4 soln, shaken for 2 h at room temp (about 20°C), and filtered. The filtrate was treated with equal vol of 2N HCl and refluxed for 15 min. After proper diln, a 1-cm^3 aliquot was withdrawn, acidified, cooled, and the sulfonamide content was detd colorimetrically (as sulfamethylthiazole) by the Marshall method modified by Kimmig (1) using an Autenrieth colorimeter. The pH was detd on an ultraionograph using a glass electrode.

SOURCE AND PURITY OF MATERIALS:

Acetyl sulfamethylthiazole (source not specified) gave no coloration upon diazotization of its satd soln, thus showing absence of sulfamethylthiazole. The source and purity of the remaining materials was not specified.

ESTIMATED ERROR:

Soly: precision ±5% (author)
Temp: not specified
pH : ±0.05 pH unit (author)

REFERENCES:

1. Kimmig, J. *Arch. Dermatol.* 1938, *176*, 722; *Erg. Hyg.* 1941, *24*, 398.

COMPONENTS:	ORIGINAL MEASUREMENTS:
(1) Acetamide, N-[4-[[(4-methyl-2-thiazolyl)-amino]sulfonyl]phenyl]- (acetyl sulfamethylthiazole); $C_{12}H_{13}N_3O_3S_2$; [71119-13-6] (2) Phosphoric acid, monopotassium salt; KH_2PO_4; [7778-77-0] (3) Water; H_2O; [7732-18-5]	Krüger-Thiemer, E. *Arch. Dermatol. Syphilis* <u>1942</u>, *183*, 90-116.
VARIABLES:	PREPARED BY:
One temperature: ca 20°C; one pH: 4.37	R. Piekos

EXPERIMENTAL VALUES:

Solubility of acetyl sulfamethylthiazole in a 0.735 M (10%) KH_2PO_4 solution of pH 4.37 at room temperature (about 20°C) is 0.0039 g% (1.25×10^{-4} mol dm^{-3}, compiler).

AUXILIARY INFORMATION

METHOD/APPARATUS/PROCEDURE:

Acetyl sulfamethylthiazole (0.5 g) was dissolved in 10 cm^3 of the 0.735 M (10%) KH_2PO_4 soln, shaken for 2 h at room temp (about 20°C), and filtered. The filtrate was treated with equal vol of 2N HCl and refluxed for 15 min. After proper diln, a 1-cm^3 aliquot was withdrawn, acidified, cooled, and the sulfonamide content was detd colorimetrically (as sulfamethylthiazole) by the Marshall method modified by Kimmig (1) using an Autenrieth colorimeter. The pH was detd on an ultraionograph using a glass electrode.

SOURCE AND PURITY OF MATERIALS:

Acetyl sulfamethylthiazole (source not specified) gave no coloration upon diazotization of its satd soln, thus showing absence of sulfamethylthiazole. The source and purity of the remaining materials was not specified.

ESTIMATED ERROR:

Soly: precision ±5% (author)
Temp: not specified
pH : ±0.05 pH unit (author)

REFERENCES:

1. Kimmig, J. *Arch. Dermatol.* <u>1938</u>, *176*, 722, *Erg. Hyg.* <u>1941</u>, *24*, 398.

COMPONENTS:	ORIGINAL MEASUREMENTS:
(1) Acetamide, N-[4-[[(4-methyl-2-thiazolyl)-amino]sulfonyl]phenyl]- (acetyl sulfamethylthiazole); $C_{12}H_{13}N_3O_3S_2$; [71119-13-6] (2) Phosphoric acid, disodium salt; Na_2HPO_4; [7558-94-4] (3) Phosphoric acid, monopotassium salt; KH_2PO_4; [7778-77-0] (4) Water; H_2O; [7732-18-5]	Krüger-Thiemer, E. *Arch. Dermatol. Syphilis* 1942, *183*, 90-116.
VARIABLES: Temperature; pH	PREPARED BY: R. Piekos

EXPERIMENTAL VALUES:

Composition of 1/15 M phosphate buffer solutions			pH	Solubility			
				Room temp (ca 20°C)		37°C	
Na_2HPO_4	KH_2PO_4	%content		g%	10^4 mol dm^{-3} solution	g%	10^4 mol dm^{-3} solution
1.0	99.0	0.91	4.944	0.0069	2.215	-	-
10.0	90.0	0.91	5.906	0.0070	2.248	0.0092	2.954
61.0	38.9	0.93	7.005	0.0078	2.505	0.0188	6.037
9.5	0.5	0.733[b]	7.510	0.0097	3.115	-	-
94.7	5.3	0.95	8.018	0.0199	6.391	-	-

[a]Calculated by compiler

[b]Molar content; 10% buffer solution

AUXILIARY INFORMATION

METHOD/APPARATUS/PROCEDURE:

Acetyl sulfamethylthiazole (0.5 g) was dissolved in 10 cm^3 of a buffer soln, shaken for 2 h at 20°C (or left for 48 h at 37°C), and filtered at respective temp. The filtrate was treated with equal vol of 2N HCl and refluxed for 15 min. After proper diln, a 1-cm^3 aliquot was withdrawn, acidified, cooled, and the sulfonamide content was detd colorimetrically (as sulfamethylthiazole) by the Marshall method modified by Kimmig (1) using an Authenrieth colorimeter. The pH was detd on an ultraionograph using a glass electrode.

SOURCE AND PURITY OF MATERIALS:

Acetyl sulfamethylthiazole (source not specified) gave no coloration upon diazotization of its satd soln, thus showing absence of sulfamethylthiazole. The source and purity of the remaining materials were not specified.

ESTIMATED ERROR:

Soly: precision ±5% (author)
Temp: not specified
pH : ±0.05 pH unit (author)

REFERENCES:

1. Kimmig, J. *Arch. Dermatol.* 1938, *176*, 722; *Erg. Hyg.* 1941, *24*, 398.

COMPONENTS:	ORIGINAL MEASUREMENTS:
(1) Benzenesulfonamide, 4-amino-N-(3-methyl-2,3-dihydro-2-thiazolyl)-; $C_{10}H_{11}N_3O_2S_2$; [51203-20-4] (2) Water; H_2O; [7732-18-5]	Shepherd, R. G.; Bratton, A. C.; Blanchard, K. C.; *J. Am. Chem. Soc.* 1942, *64*, 2532-7.
VARIABLES:	PREPARED BY:
One temperature: 37°C	R. Piekos

EXPERIMENTAL VALUES:

Solubility of 4-amino-N-(3-methyl-2,3-dihydro-2-thiazolyl) benzenesulfonamide in water at 37°C is 22 mg% (8. 2 x 10^{-4} mol dm^{-3} solution, compiler).

AUXILIARY INFORMATION

METHOD/APPARATUS/PROCEDURE:

The sulfonamide was assayed colorimetrically (1). No details were given.

SOURCE AND PURITY OF MATERIALS:

The sulfonamide, mp 250-1°C, was synthesized by the authors. Analysis: %C 44.49 (calcd 44.60); %H 4.13 (4.12); %N 15.54 (15.60). Colorimetric factor: 0.656 (calcd 0.639).
Purity of the water was not specified.

ESTIMATED ERROR:

Nothing specified

REFERENCES:

1. Bratton, A. C.; Marshall, E. K., Jr. *J. Biol. Chem.* 1939, *128*, 537.

COMPONENTS:	ORIGINAL MEASUREMENTS:
(1) Benzenesulfonamide, 4-amino-N-(3-methyl-2,3-dihydro-2-thiazolyl)-; $C_{10}H_{11}N_3O_2S_2$; [51203-20-4] (2) Water; H_2O; [7732-18-5]	Kitao, K.; Kubo, K.; Morishita, T.; Yata, N.; Kamada, A. *Chem. Pharm. Bull.* 1973, *21*, 2417-26.
VARIABLES:	PREPARED BY:
One temperature: 37°C	R. Piekos

EXPERIMENTAL VALUES:

Solubility of 4-amino-N-(3-methyl-2,3-dihydro-2-thiazolyl)benzenesulfonamide in water at 37°C is 0.569 mmol dm^{-3} solution.

AUXILIARY INFORMATION

METHOD/APPARATUS/PROCEDURE:	SOURCE AND PURITY OF MATERIALS:
The sulfonamide was assayed by diazotization. No details were given.	The sulfonamide was synthesized by the authors. Its purity was not specified. Deionized water was used.
	ESTIMATED ERROR: Soly: not specified. Temp: ±1°C (authors).
	REFERENCES:

COMPONENTS:	ORIGINAL MEASUREMENTS:
(1) Benzenesulfonamide, 4-amino-N-(3-methyl-2,3-dihydro-2-thiazolyl)- $C_{10}H_{11}N_3O_2S_2$; [51203-20-4] (2) Methane-, trichloro-; $CHCl_3$; [67-66-3]	Kitao, K.; Kubo, K.; Morishita, T.; Yata, N.; Kamada, A. *Chem. Pharm. Bull.* 1973, *21*, 2417-26.
VARIABLES:	PREPARED BY:
One temperature: 37°C	R. Piekos

EXPERIMENTAL VALUES:

Solubility of 4-amino-N-(3-methyl-2,3-dihydro-2-thiazolyl)benzenesulfonamide in $CHCl_3$ at 37°C is 3.15 mmol dm^{-3} solution.

AUXILIARY INFORMATION

METHOD/APPARATUS/PROCEDURE:

One ml of the sulfonamide soln in $CHCl_3$ at equilibrium was taken into a test tube. After evapn of the solvent, the residue was dissolved in 1N HCl, the soln was properly dild with dionized water and the concn of the sulfonamide was detd by diazotization.

SOURCE AND PURITY OF MATERIALS:

The sulfonamide was synthesized by the authors. Its purity was not specified. Neither source nor purity of the $CHCl_3$ was specified.

ESTIMATED ERROR:

Soly: not specified.

Temp: ±1°C (authors).

REFERENCES:

COMPONENTS:	ORIGINAL MEASUREMENTS:
(1) Benzenesulfonamide, 4-amino-N-2-[3-(2-hydroxyethyl)-2,3-dihydro-2-thiazolyl]-; $C_{11}H_{15}N_3O_3S_2$; [71119-27-2] (2) Water; H_2O; [7732-18-5]	Shepherd, R. G.; Bratton, A. C.; Blanchard, K.C.; *J. Am. Chem. Soc.* 1942, *64*, 2532-7.
VARIABLES:	PREPARED BY:
One temperature: 37°C	R. Piekos

EXPERIMENTAL VALUES:

Solubility of 4-amino-N-2-[3-(2-hydroxyethyl)-2,3-dihydro-2-thiazolyl]-benzenesulfonamide in water at 37°C is 169 mg% (5.96×10^{-3} mol dm^{-3} solution, compiler).

AUXILIARY INFORMATION

METHOD/APPARATUS/PROCEDURE:

The sulfonamide was assayed colorimetrically (1). No details were given.

SOURCE AND PURITY OF MATERIALS:

The sulfonamide, mp 159-60°C, was synthesized by the authors. Analysis: %C 44.42 (calcd 44.13); %H 4.35 (4.38); %N 14.36 (14.04). Colorimetric factor: 0.600 (calcd 0.575).
Purity of the water was not specified.

ESTIMATED ERROR:

Nothing specified.

REFERENCES:

1. Bratton, A. C.; Marshall, E. K. Jr. *J. Biol. Chem.* 1939, *128*, 537.

COMPONENTS:

(1) Benzenesulfonamide, 4-amino-N-2-benzothiazolyl-; $C_{13}H_{11}N_3O_2S_2$; [6138-01-8]

(2) Water; H_2O; [7732-18-5]

ORIGINAL MEASUREMENTS:

Roblin, R. O.,Jr.; Williams, J. H.; Winnek, P.S.; English, J. P.
J. Am. Chem. Soc. 1940, *62*, 2002-5.

VARIABLES:

One temperature: 37°C

PREPARED BY:

R. Piekos

EXPERIMENTAL VALUES:

Solubility of 4-amino-N-2-benzothiazolylbenzenesulfonamide in water at 37°C is 0.1 mg/100 cm^3 solution (3 x 10^{-6} mol dm^{-3}, compiler).

AUXILIARY INFORMATION

METHOD/APPARATUS/PROCEDURE:

Excess sulfonamide in water was heated and stirred on a steam bath for 30 min. The suspension was then agitated for 24 h in a thermostat at 37°C. A sample of the satd soln was withdrawn through a glass filter, dild, and analyzed by the Marshall method (1) using a General Electric recording spectrophotometer for comparing the colors developed with those of the standards.

SOURCE AND PURITY OF MATERIALS:

The sulfonamide, mp 304-5°C (cor, dec) was prepd by the authors. Anal: %C 38.2 (calcd 37.5); %H 2.9 (3.1); %N 21.0 (21.8). Purity of the water was not specified.

ESTIMATED ERROR:

Nothing specified.

REFERENCES:

1. Bratton, A. C.; Marshall, E. K.,Jr. *J. Pharmacol.* 1939, *66*, 4.

COMPONENTS:	ORIGINAL MEASUREMENTS:
(1) Benzenesulfonamide, 4-amino-N-4-[4-(4-biphenylyl)-2-thiazolyl]-; $C_{21}H_{17}N_3O_2S_2$; [71119-15-8] (2) Water; H_2O; [7732-18-5]	Roblin, R. O., Jr.; Williams, J. H. Winnek, P. S.; English, J. P. *J. Am. Chem. Soc.* 1940, *62*, 2002-5.
VARIABLES:	PREPARED BY:
One temperature: 37°C	R. Piekos

EXPERIMENTAL VALUES:

Solubility of 4-amino-N-4-[4-(4-biphenylyl)-2-thiazolyl]benzenesulfonamide in water at 37°C is 0.1 mg/100 cm^3 solution (2 x 10^{-6} mol dm^{-3}, compiler).

AUXILIARY INFORMATION

METHOD/APPARATUS/PROCEDURE:	SOURCE AND PURITY OF MATERIALS:
Excess sulfonamide in water was heated and stirred on a steam bath for 30 min. The suspension was then agitated for 24 h in a thermostat at 37°C. A sample of the satd soln was withdrawn through a glass filter, dild, and analyzed by the Marshall method (1) using a General Electric recording spectrophotometer for comparing the colors developed with those of the standards.	The sulfonamide was prepd by the authors. Mp 304-5°C (cor, dec). Anal: %C 51.1 (calcd 51.1); %H 3.9 (3.6) ; %N 13.6 (13.8). Purity of the water was not specified. ESTIMATED ERROR: Nothing specified REFERENCES: 1. Bratton, A. C.; Marshall, E. K., Jr. *J. Pharmacol.* 1939, *66*, 4.

COMPONENTS:

(1) Benzenesulfonamide, 4-amino-N-(4-methoxy-1,2,5-thiadiazol-3-yl)-; (sulfametrole); $C_9H_{10}N_4O_3S_2$; [32909-92-5]

(2) Phosphoric acid, disodium salt; Na_2HPO_4; [7558-94-4]

(3) Phosphoric acid, monopotassium salt; KH_2PO_4; [7778-77-0]

(4) Water; H_2O; [7732-18-5]

ORIGINAL MEASUREMENTS:

Hekster, Y. A.; Vree, T. B.; Damsma, J. E.; Friese, W. T.

J. Antimicrob. Chemother. 1981, *8*, 133-44.

VARIABLES:

pH

PREPARED BY:

R. Piekos

EXPERIMENTAL VALUES:

pH	Solubility at 25°C mg/1	10^3 mol dm^{-3} [a]
5.5	460	1.61
7.5	1700	5.94

[a] Calculated by compiler

AUXILIARY INFORMATION

METHOD/APPARATUS/PROCEDURE:

Satd solns of sulfametrole were prepd in phosphate buffers of pH 5.5 and 7.5 at room temp (25°C). The concn of the solute was measured by means of a Spectra Physics 3500B high-performance liquid chromatograph equipped with a column oven (Model 748) and a Pye-Unicam LC-UV spectrophotometric detector. The detector was connected to a 1-mV recorder. A stainless steel column (10 cm x 4.6 mm i.d.) was packed with Lichrosorb RPS, 5 μm, obtained from Chrompack. An injection loop of 100 μl was used. The oven temp was 40°C. Detection of sulfametrole was performed at 260 nm.

SOURCE AND PURITY OF MATERIALS:

Sulfametrole was obtained from Warrick Nederland. The compd was 100% pure according to the HPLC chromatogram. The source and purity of the remaining materials were not specified.

ESTIMATED ERROR:

The detection limit of the solute by HPLC was 0.5 mg/1 (authors).

The error in temperature and pH were not specified.

REFERENCES:

COMPONENTS:

(1) Acetamide, N-[(4-aminophenyl)sulfonyl]-N-(4-methoxy-1,2,5-thiadiazol-3-yl)- (N^1-acetylsulfametrole) $C_{11}H_{12}N_4O_4S_2$; [84930-17-6]

(2) Phosphoric acid, disodium salt; Na_2HPO_4; [7558-94-4]

(3) Phosphoric acid, monopotassium salt; KH_2PO_4; [7778-77-0]

(4) Water; H_2O; [7732-18-5]

ORIGINAL MEASUREMENTS:

Hekster, Ch. A.; Vree, T. B.
Antibiotics, Chemother. 1982, *31*, 22-118.

VARIABLES:

pH

PREPARED BY:

R. Piekos

EXPERIMENTAL VALUES:

pH	Solubility at 25°C mg/l	10^4 mol dm^{-3} [a]
5.5	34	1.00
7.5[b]	24	0.73

[a] Calculated by compiler.

[b] Erroneous pH value of 7.0 is given in the article.

AUXILIARY INFORMATION

METHOD/APPARATUS/PROCEDURE:

The earlier developed method (1) was used (personal communication). Satd solns of N^1-acetylsulfametrole were prepd in phosphate buffers of pH 5.5 and 7.5 at 25°C. The concn of the solute was measured by means of a Spectra Physics 3500B high-performance liquid chromatograph equipped with a Model 748 column oven and a Pye-Unicam LC-UV spectrophotometric detector.

SOURCE AND PURITY OF MATERIALS:

Neither source nor the purity of the materials was specified.

ESTIMATED ERROR:

Soly: the detection limit of the solute by HPLC was 0.5 mg/l (authors).
Errors in temp. and pH were not specified.

REFERENCES:

1. Hekster, Y. A.; Vree, T. B.; Damsma, J. E.; Friesen, W. T. *J. Antimicrob. Chemother.* 1981, *8*, 133.

COMPONENTS:

(1) Acetamide, N-[4-[[(4-methoxy-1,2,5-thiadiazol-3-yl)amino]sulfonyl]phenyl]-(N^4-acetylsulfametrole); $C_{11}H_{12}N_4O_4S_2$; [79962-97-3]

(2) Phosphoric acid, disodium salt; Na_2HPO_4; [7558-94-4]

(3) Phosphoric acid, monopotassium salt; KH_2PO_4; [7778-77-0]

(4) Water; H_2O; [7732-18-5]

ORIGINAL MEASUREMENTS:

Hekster, Y. A.; Vree, T. B.; Damsma, J. E.; Friesen, W. T.

J. Antimicrob. Chemother. 1981, *8*, 133-44.

VARIABLES:

pH

PREPARED BY:

R. Piekos

EXPERIMENTAL VALUES:

pH	Solubility at 25°C mg/l	10^3 mol dm^{-3} [a]
5.5	1100	3.350
7.5	6000	18.273

[a]Calculated by compiler

AUXILIARY INFORMATION

METHOD/APPARATUS/PROCEDURE:

Satd solns of N^4-acetylsulfametrole were prepd in phosphate buffers of pH 5.5 and 7.5 at room temp (25°C). The concn of the solute was measured by means of a Spectra Physics 3500B high-performance liquid chromatograph equipped with a column oven (Model 748) and a Pye-Unicam LC-UV spectrophotometric detector. The detector was connected to a 1-mV recorder. A stainless steel column (10 cm x 4.6 mm i.d.) was packed with Lichrosorb RPS, 5 μm, obtained from Chrompack. An injection loop of 100 μl was used. The oven temp was 40°C. Detection of the solute was performed at 260 nm.

SOURCE AND PURITY OF MATERIALS:

N^4-acetylsulfametrole was obtained from Chemie, Linz. The compd was 100% pure according to the HPLC chromatogram. The source and purity of the remaining materials were not specified.

ESTIMATED ERROR:

The detection limit of the solute by HPLC was 0.5 mg/l (authors). The error in temperature and pH were not specified.

REFERENCES:

COMPONENTS:

(1) Acetamide, N-[[4-(acetylamino)phenyl]-sulfonyl]-N-(4-methoxy-1,2,5-thiadiazol-3-yl)- (N^1, N^4-diacetylsulfametrole); $C_{13}H_{14}N_4O_5S_2$; [84930-18-7]

(2) Phosphoric acid, disodium salt; Na_2HPO_4; [7558-94-4]

(3) Phosphoric acid, monopotassium salt; KH_2PO_4; [7778-77-0]

(4) Water; H_2O; [7732-18-5]

ORIGINAL MEASUREMENTS:

Hekster, Ch. A.; Vree, T. B.

Antibiotics Chemother. 1982, *31*, 22-118.

VARIABLES:

pH

PREPARED BY:

R. Piekos

EXPERIMENTAL VALUES:

pH	Solubility at 25°C	
	mg/l	10^5 mol dm^{-3} [a]
5.5	22.7	6.12
7.5[b]	20.5	5.53

[a] Calculated by compiler.

[b] Erroneous pH value of 7.0 is given in the article.

AUXILIARY INFORMATION

METHOD/APPARATUS/PROCEDURE:

The earlier developed method (1) was used (personal communication). Satd solns of N^1, N^4-diacetylsulfametrole were prepd in phosphate buffers of pH 5.5 and 7.5 at 25°C. The concn of the solute was measured by means of a Spectra Physics 3500B high-performance liquid chromatograph equipped with a Model 748 column oven and a Pye-Unicam LC-UV spectrophotometric detector.

SOURCE AND PURITY OF MATERIALS:

Neither source nor the purity of the materials was specified.

ESTIMATED ERROR:

Soly: the detection limit of the solute by HPLC was 0.5 mg/l (authors).
The errors in temp and pH were not specified.

REFERENCES:

1. Hekster, Y.A.; Vree, T. B.; Damsma, J.E.; Friesen, W. T. *J. Antimicrob. Chemother.* 1981, *8*, 133.

COMPONENTS:	ORIGINAL MEASUREMENTS:
(1) Benzensulfonamide, 4-amino-N-1,3,4-thiadiazol-2-yl-; $C_8H_8N_4O_2S_2$; [16806-29-4] (2) Water; H_2O; [7732-18-5]	Roblin, R.O., Jr.; Williams, J. H.; Winnek, P. S.; English, J. P. *J. Am. Chem. Soc.* 1940, *62*, 2002-5.
VARIABLES: One temperature: 37°C	PREPARED BY: R. Piekos

EXPERIMENTAL VALUES:

Solubility of 4-amino-N-1,3,4-thiadiazol-2-ylbenzenesulfonamide in water at 37°C is 73 mg/100 cm^3 solution (2.85×10^{-3} mol dm^{-3}, compiler).

AUXILIARY INFORMATION

METHOD/APPARATUS/PROCEDURE:	SOURCE AND PURITY OF MATERIALS:
Excess sulfonamide in water was heated and stirred on a steam bath for 30 min. The suspension was then agitated for 24 h in a thermostat at 37°C. A sample of the satd soln was withdrawn through a glass filter, dild, and analyzed by the Marshall method (1) using a General Electric recording spectrophotometer for comparing the colors developed with those of the standards.	The sulfonamide, mp 216-18°C (dec, cor), was prepd by the authors. Anal: %C 38.1 (calcd 37.5); %H 2.9 (3.1); %N 21.0 (21.8). Purity of the water was not specified.
	ESTIMATED ERROR: Nothing specified.
	REFERENCES: 1. Bratton, A. C.; Marshall, E.K., Jr. *J. Pharmacol* 1939, *66*, 4.

COMPONENTS:	EVALUATOR:
(1) Benzenesulfonamide, 4-amino-N-(5-methyl-1,3,4-thiadiazol-2-yl)-(sulfamethylthiadiazole); $C_9H_{10}N_4O_2S_2$; [144-82-1] (2) Water	Anthony N. Paruta Department of Pharmaceutics University of Rhode Island Kingston, Rhode Island, USA and Ryszard Piekos Faculty of Pharmacy, University of Gdansk Gdansk, Poland 1986

CRITICAL EVALUATION:

At 310K, three values were available (1-3) for the aqueous solubility of this compound: Durel and Allinne (1) reported a value of 3.25 x 10^{-3} mol dm^{-3}. Kaneniwa and Watari (2) gave a value of 3.27 x 10^{-3} mol dm^{-3} , a value of 3.27 x mol^{-3} dm^{-3} in 1978 (3) and in 1980 (4) with Hanano a value of 3.27 x 10^{-3} mol dm^{-3} was given. Since all these values were produced by the same workers (2-4) using identical methodologies, the value given was only considered once. The simple average of the two values (1) and (2-4) were taken and a recommended value of 3.26 x 10^{-3} mol dm^{-3} is given in water at 310K.

REFERENCES:

(1) Durel, M.P.; Allinne, M. *Bull. Soc. Med. Hop. Paris III* 1941, 251-9.
(2) Watari, N.; Kaneniwa, N. *Chem. Pharm. Bull.* 1976, *24(11)*, 2577-84.
(3) Kaneniwa, N.; Watari, N. *Cehm. Pharm. Bull.* 1978, *26(3)*, 813-26.
(4) Watari N.; Kaneniwa, N.; Hanano, M. *Int. J. Pharm.* 1980, *6(2)*, 155-66.

COMPONENTS:	ORIGINAL MEASUREMENTS:
(1) Benzenesulfonamide, 4-amino-N-(5-methyl-1,3,4-thiadiazol-2-yl)- (sulfamethylthiadiazole); $C_9H_{10}N_4O_2S_2$; [144-82-1] (2) Water; H_2O; [7732-18-5]	Durel, M. P.; Allinne, M. *Bull. Soc. Med. Hop. Paris III* 1941, *251-9.*
VARIABLES:	**PREPARED BY:**
One temperature: 37°C	R. Piekos

EXPERIMENTAL VALUES:

Solubility of sulfamethylthiadiazole in water at 37°C is 0.88 g/liter (3.25×10^{-3} mol dm^{-3}, compiler).

AUXILIARY INFORMATION

METHOD/APPARATUS/PROCEDURE:

A mixt of sulfamethylthiadiazole and water was agitated for 24 hours at 37°C .

SOURCE AND PURITY OF MATERIALS:

Source and purity of sulfamethylthiadiazole were not specified.

Distilled water was used.

ESTIMATED ERROR:

Nothing specified.

REFERENCES:

COMPONENTS:	ORIGINAL MEASUREMENTS:
(1) Benzenesulfonamide, 4-amino-N-(5-methyl-1,3,4-thiadiazol -2-yl)- (sulfamethizole); $C_9H_{10}N_4O_2S_2$; [144-82-1] (2) Water; H_2O; [7732-18-5]	Watari, N.; Kaneniwa, N. *Chem. Pharm. Bull.* 1976, *24(11)*, 2577-84.
VARIABLES:	**PREPARED BY:**
One temperature: 37°C	R. Piekos

EXPERIMENTAL VALUES:

Total solubility of sulfamethizole in water at 37°C is 0.884 mg/ml solution (3.27 x 10^{-3} mol dm^{-3}, compiler).

AUXILIARY INFORMATION

METHOD/APPARATUS/PROCEDURE:

An excess of sulfamethizole, required to saturate water, was placed in a flask contg 25 ml of water. The flask was shaken (2 strokes/s) at the amplitude of 3 cm in a thermostatically controlled water bath at 37°C. One-ml sample was removed every 6 h (total equilibration period 3-5 days) using a warmed Millipore filter syringe with a filter pore size of 0.45 μ (Millipore HAWP 01300) and the filtrate was dild with water and assayed spectrophotometrically (1).

SOURCE AND PURITY OF MATERIALS:

Commercial sulfamethizole of the Japanese Pharmacopeia grade and distd water were used.

ESTIMATED ERROR:

Soly: not specified

Temp: ±0.05°C (authors).

REFERENCES:

1. Kaneniwa, N.; Watari, N. *Chem. Pharm. Bull.* 1974, *22*, 1699.

COMPONENTS:

(1) Benzenesulfonamide, 4-amino-N-(5-methyl-1,3,4-thiadiazol-2-yl)-(sulfamethizole); $C_9H_{10}N_4O_2S_2$; [144-82-1]

(2) Water; H_2O; [7732-18-5]

ORIGINAL MEASUREMENTS:

Kaneniwa, N.; Watari, N.
Chem. Pharm. Bull. 1978, *26(3)*, 813-26.

VARIABLES:

One temperature: 37°C

PREPARED BY:

R. Piekos

EXPERIMENTAL VALUES:

Solubility of sulfamethizole in water at 37°C is 0.884 mg/ml solution (3.27×10^{-3} mol dm^{-3}, compiler).

AUXILIARY INFORMATION

METHOD/APPARATUS/PROCEDURE:

An excess of sulfamethizole was placed in a flask contg 25 ml of water. The flask was shaken (2 strokes/s at the amplitude of 3 cm) in a thermostatically controlled water bath at 37°C. One-ml sample was withdrawn every 6 h (total equilibration period was 3-5 days) using a warmed Millipore filter syringe with a filter pore size of 0.45 μ (Millipore HAWP 01300) and the filtrate was dild with water and assayed spectrophotometrically (1).

SOURCE AND PURITY OF MATERIALS:

Commercial sulfamethizole of the Japanese Pharmacopeia grade and distd water were used.

ESTIMATED ERROR:

Soly: not specified.
Temp: ±0.05°C (authors).

REFERENCES:

1. Kaneniwa, N.; Watari, N.
Chem. Pharm. Bull. 1974, *22*, 1699.

COMPONENTS:

(1) Benzenesulfonamide, 4-amino-N-(5-methyl-1,3,4-thiadiazol-2-yl)-(sulfamethizole); $C_9H_{10}N_4O_2S_2$; [144-82-1]

(2) Water; H_2O; [7732-18-5]

ORIGINAL MEASUREMENTS:

Watari, N.; Kaneniwa, N.; Hanano, M.
Int. J. Pharm. 1980, *6(2)*, 155-66.

VARIABLES:

One temperature: 37°C

PREPARED BY:

R. Piekos

EXPERIMENTAL VALUES:

Solubility of sulfamethizole in water at 37°C is 88.4 mg/100 ml (3.27×10^{-3} mol dm^{-3}, compiler).

AUXILIARY INFORMATION

METHOD/APPARATUS/PROCEDURE:

The earlier developed method was employed (1), whereby an excess of sulfamethizole, required to saturate medium, was placed in a flask contg 25 ml of water. The flask was shaken (2 strokes/s) at an amplitude of 3 cm, in a thermostatically controlled bath. One-ml sample was removed every 6 h (total equilibration time was 3-5 days) using a warmed Millipore filter syringe with a filter pore soze pf 0.45 μ (Millipore HAWP 01300) and the filtrate was dild with water and assayed spectrophotometrically.

SOURCE AND PURITY OF MATERIALS:

Sulfamethizole was of the Japanese Pharmacopeia grade.

Distilled water was used.

ESTIMATED ERROR:

Soly: not specified.

Temp: ±0.05°C (authors).

REFERENCES:

1. Kaneniwa, N. ; Watari, N.
Chem. Pharm. Bull. 1974, *22*, 1699.

COMPONENTS:

(1) Benzenesulfonamide, 4-amino-N-(5-methyl-1,3,4-thiadiazol-2-yl)- (sulfamethizole); $C_9H_{10}N_4O_2S_2$; [144-82-1]

(2) Water; H_2O; [7732-18-5]

ORIGINAL MEASUREMENTS:

Goto, S.; Komatsu, M.; Tagawa, K.; Kawata, M.
Chem. Pharm. Bull. 1983, *31(1)*, 256-61.

VARIABLES:

Temperature

PREPARED BY:

R. Piekos

EXPERIMENTAL VALUES:

$t/^{\circ}C$	Solubility	
	g/l	10^3 mol dm^{-3} [a]
37	0.87	3.22
55	2.10	7.77

[a] Calculated by compiler

AUXILIARY INFORMATION

METHOD/APPARATUS/PROCEDURE:

A 3 g sample of sulfamethizole powder was accurately weighed into a 20-ml ampul and 10 ml of water was added. The ampul was sealed, placed in a const temp (37° or 55°C) bath and allowed to stand for several days. The equilibrium concn of the solute was measured spectrophotometrically at 542 nm after diazotization with the 0.1% Tsuda reagent (1).

SOURCE AND PURITY OF MATERIALS:

Sulfamethizole had mp 207-11°C.
The purity of water was not specified.

ESTIMATED ERROR:

Nothing specified.

REFERENCES:

1. Tsuda, K.; Matsunaga, S. *Yakugaku Zasshi* 1942, *62*, 362.

COMPONENTS:	ORIGINAL MEASUREMENTS:
(1) Benzenesulfonamide, 4-amino-N-(5-methyl-1,3,4-thiadiazol-2-yl)- (sulfamethizole); $C_9H_{10}N_4O_2S_2$; [144-82-1] (2) Hydrochloric acid; HCl; [7647-01-0] (3) Water; H_2O; [7732-18-5]	Ogata, H.; Shibazaki, T.; Inoue, T.; Ejima, A. *Chem. Pharm. Bull.* 1979, *27(6)*, 1281-6.
VARIABLES:	PREPARED BY:
One temperature: 37°C	R. Piekos

EXPERIMENTAL VALUES:

Solubility of sulfamethizole in 0.1N HCl at 37°C is 9.172 mg/ml (3.393×10^{-2} mol dm^{-3}, compiler).

AUXILIARY INFORMATION

METHOD/APPARATUS/PROCEDURE:	SOURCE AND PURITY OF MATERIALS:
A centrifuge tube contg 30 ml of 0.1N HCl and 0.5-3.0 g of the sulfamethizole powder was tigthly sealed and shaken at 37°C. The concn of the dissolved drug was detd spectrophotometrically following filtration through a Millipore filter (type EH, pore size 0.5 μm), and the procedure was repeated every 24 h until a const concn was obtained.	Comm available 250-mg uncoated tablets of sulfamethizole were used. Hydrochloric acid was of reagent grade. ESTIMATED ERROR: Nothing specified REFERENCES:

COMPONENTS:

(1) Benzenesulfonamide, 4-amino-N-(5-methyl-1,3,4-thiadiazol-2-yl)-(sulfamethizole); $C_9H_{10}N_4O_2S_2$; [144-82-1]

(2) Hydrochloric acid; HCl; [7647-01-0]

(3) Sodium chloride; NaCl; [7647-14-5]

(4) Water; H_2O; [7732-18-5]

ORIGINAL MEASUREMENTS:

Nicklasson, M.; Brodin, A.; Nyqvist, H. *Acta Pharm. Suec.* 1981, *18*, 119-28.

VARIABLES:

One temperature: 37°C; one pH: 1.20

PREPARED BY:

R. Piekos

EXPERIMENTAL VALUES:

Solubility of sulfamethizole in a HCl - NaCl buffer solution of pH 1.20 (ionic strength 0.2) at 37°C is 5.62 mg/ml[a] (2.08×10^{-2} mol dm^{-3}, compiler).

[a]Numerical value given by one of the authors (M. N.)

AUXILIARY INFORMATION

METHOD/APPARATUS/PROCEDURE:

Sulfamethizole, taken in excess of a quantity required for satn, was added to the HCl-NaCl buffer soln of pH 1.20. The suspension was equilibrated at 37°C for 18-24 h using a magnetic stirrer. No degradation of the drug was observed at the pH indicated. Five-ml samples were filtered through 0.1-μm polycarbonate filters (Nuclepore® Co.), and dild to suitable concn for spectrophotometry. Concns were determined using a Pye Unicam SP8-100 spectrophotometer. Samples were assayed at wavelengths of max absorption, taking into consideration changes in spectra due to ionization.

SOURCE AND PURITY OF MATERIALS:

Sulfamethizole of commercial grade was used (source not specified).

HCl and NaCl were anal grade reagents (source not specified).

Purity of the water was not specified.

ESTIMATED ERROR:

Soly: mean of 2 detns is given (authors).
pH : precision ±0.01 pH unit (authors).
Temp: ±0.5°C (authors).

REFERENCES:

COMPONENTS:

(1) Benzenesulfonamide, 4-amino-N-(5-methyl-1,3,4-thiadiazol-2-yl)- (sulfamethylthiadiazole); $C_9H_{10}N_4O_2S_2$; [144-82-1]

(2) Phosphoric acid, disodium salt; Na_2HPO_4; [7558-94-4]

(3) Water; H_2O; [7732-18-5]

ORIGINAL MEASUREMENTS:

Krüger-Thiemer, E.

Arch. Dermatol. Syphilis 1942, *183*, 90-116.

VARIABLES:

One temperature: ca 20°C; one pH: 8.74

PREPARED BY:

R. Piekos

EXPERIMENTAL VALUES:

Solubility of sulfamethylthiadiazole in a 0.705 M (10%) Na_2HPO_4 solution of pH 8.74, at room temperature (about 20°C), is 1.625 g% (6.011×10^{-2} mol dm^{-3} solution, compiler).

AUXILIARY INFORMATION

METHOD/APPARATUS/PROCEDURE:

Sulfamethylthiadiazole (0.5 g) was dissolved in 10 cm^3 of the 0.705 M (10%) Na_2HPO_4 solution of pH 8.74, shaken for 2 h at room temp (about 20°C), and filtered. A 1-cm^3 aliquot of the filtrate was withdrawn, cooled, acidified with 2N HCl, and the sulfonamide content was detd colorimetrically by the method of Marshall modified by Kimmig (1) using an Authenrieth colorimeter. The pH was detd on an ultraionograph using a glass electrode.

SOURCE AND PURITY OF MATERIALS:

Sulfamethylthiadiazole was the product manufd by Schering under the name Tetracid. The source and purity of the remaining materials was not specified.

ESTIMATED ERROR:

Soly: precision ±5% (author).

Temp: not specified.

pH : ±0.05 pH unit (author).

REFERENCES:

1. Kimmig, J. *Arch. Dermatol.* 1938, *176*, 722; *Erg. Hyg.* 1941, *24*, 398.

COMPONENTS:	ORIGINAL MEASUREMENTS:
(1) Benzenesulfonamide, 4-amino-N-(5-methyl-1,3,4-thiadiazole-2-yl)- (sulfamethylthiadiazol); $C_9H_{10}N_4O_2S_2$; [144-82-1] (2) Phosphoric acid, monopotassium salt; KH_2PO_4; [7778-77-0] (3) Water; H_2O; [7732-18-5]	Krüger-Thiemer, E. *Arch. Dermatol. Syphilis* 1942, *183*, 90-116.
VARIABLES:	**PREPARED BY:**
One temperature: ca 20°C; one pH: 4.37	R. Piekos

EXPERIMENTAL VALUES:

Solubility of sulfamethylthiadiazole in a 0.735 M (10%) KH_2PO_4 solution of pH 4.37, at room temperature (about 20°C), is 0.027 g% (9.99×10^{-2} mol dm^{-3} solution, compiler).

AUXILIARY INFORMATION

METHOD/APPARATUS/PROCEDURE:

Sulfamethylthiadiazole (0.5 g) was dissolved in 10 cm^3 of the 0.735 M (10%) KH_2PO_4 soln, shaken for 2 h at room temp (about 20°C), and filtered. A 1-cm^3 aliquot of the filtrate was withdrawn, cooled, acidified with 2N HCl, and the sulfonamide content was detd colorimetrically by the method of Marshall modified by Kimmig (1) using an Authenrieth colorimeter. The pH was detd on an ultraionograph using a glass electrode.

SOURCE AND PURITY OF MATERIALS:

Sulfamethylthiadiazole was the product manufd by Schering under the name Tetracid. The source and purity of the remaining reagents were not specified.

ESTIMATED ERROR:

Soly: precision ±5% (author)

Temp: not specified

pH : ±0.05 pH unit (author)

REFERENCES:

1. Kimmig, J. *Arch. Dermatol.* 1938, *176*, 722; *Erg. Hyg.* 1941, *24*, 398.

COMPONENTS:

(1) Benzenesulfonamide, 4-amino-N-(5-methyl-1,3,4-thiadiazole-2-yl)- (sulfamethylthiadiazole); $C_9H_{10}N_4O_2S_2$; [144-82-1]

(2) Phosphoric acid, disodium salt; Na_2HPO_4; [7558-94-4]

(3) Phosphoric acid, monopotassium salt; KH_2PO_4; [7778-77-0]

(4) Water; H_2O; [7732-18-5]

ORIGINAL MEASUREMENTS:

Krüger-Thiemer, E.

Arch. Dermatol. Syphilis 1942, *183*, 90-116.

VARIABLES:

Temperature, pH

PREPARED BY:

R. Piekos

EXPERIMENTAL VALUES:

Composition of 1/15 M phosphate buffer solutions			pH	Solubility			
				Room temp (ca 20°C)		37°C	
Na_2HPO_4	KH_2PO_4	%Content		g%	10^2 mol dm^{-3} solution	g%	10^2 mol dm^{-3} solution[a]
1.0	99.0	0.91	4.944	0.058	0.214	-	-
10.0	90.0	0.91	5.906	0.155	0.573	0.212	0.784
61.1	38.9	0.93	7.005	0.823	3.044	0.913	3.377
9.5	0.5	0.733[b]	7.51	1.235	4.569	-	-
94.7	5.3	0.95	8.018	1.232	4.557	-	-

[a] Calculated by compiler

[b] Molar content; 10% buffer solution

AUXILIARY INFORMATION

METHOD/APPARATUS/PROCEDURE:

Sulfamethylthiadiazole (0.5 g) was dissolved in 10 cm^3 of a buffer soln, shaken for 2 h at 20°C (or left for 48 h at 37°C), and filtered at respective temp. A 1-cm^3 aliquot of the filtrate was then withdrawn, cooled (dild for expts at 37°C), acidified with 1 cm^3 of 2N HCl, and the sulfonamide content was detd colorimetrically by the method of Marshall modified by Kimmig (1) using an Authenrieth colorimeter. The pH was detd on an ultraionograph using a glass electrode.

SOURCE AND PURITY OF MATERIALS:

Sulfamethylthiadiazole was the product manufd by Schering under the name Tetracid. The source and purity of the remaining materials were not specified.

ESTIMATED ERROR:

Soly: precision ±5% (author)

Temp: not specified

pH : ±0.05 pH unit (author)

REFERENCES:

1. Kimmig, J. *Arch. Dermatol.* 1938, *176*, 722; *Erg. Hyg.* 1941, *24*, 398.

COMPONENTS:

(1) Benzenesulfonamide, 4-amino-N-(5-methyl-1,3,4-thiadiazol-2-yl)- (sulfamethylthiadiazole); $C_9H_{10}N_4O_2S_2$; [144-82-1]

(2) Phosphoric acid, disodium salt; Na_2HPO_4; [7558-94-4]

(3) Phosphoric acid, monopotassium salt; KH_2PO_4; [7778-77-0]

(4) Water; H_2O; [7732-18-5]

ORIGINAL MEASUREMENTS:

Bandelin, F. J.; Malesh, W.

J. Am. Pharm. Assoc. Sci. Ed. <u>1959</u>, *48*, 177-81.

VARIABLES:

pH

PREPARED BY:

R. Piekos

EXPERIMENTAL VALUES:

Solubility of sulfamethylthiadiazole in buffers of varying mixtures of $Na_2HPO_4 \cdot 7H_2O$ (71.6 g/l distilled water; 0.27 mol dm^{-3}, compiler) and KH_2PO_4 (36.3 g/l distilled water; 0.27 mol dm^{-3}, compiler) at 37°C.

Initial pH	Solubility	
	mg/100 ml	10^2 mol dm^{-3} [a]
4.5	105	0.388
5.0	125	0.462
5.5	200	0.739
6.0	470	1.738
6.5	1000	3.699
7.0	1990	7.361
8.0	9250	34.218

[a] Calculated by compiler

AUXILIARY INFORMATION

METHOD/APPARATUS/PROCEDURE:

Solns were prepd by adding an excess of sulfamethylthiadiazole to 10 ml of buffer soln at each pH level in 18 x 150-mm test tubes, stoppering the tubes and placing them in a water bath at 37°C with gentle agitation for 24 h. The mixt was then filtered and a 1-ml aliquot was accurately pipetted into a volumetric flask for diln and analysis. The balance was retained for pH detn to ascertain any change in pH value. The sulfonamide was assayed colorimetrically by the method of Bratton and Marshall as described in detail by Biamonte and Schneller (1). A standard curve was prepd using accurately prepd standard solutions.

SOURCE AND PURITY OF MATERIALS:

Neither source nor purity of the reagents were specified. Distilled water was used.

ESTIMATED ERROR:

Soly: av values of duplicate runs are reported (authors).

Temp and pH: not specified.

REFERENCES:

1. Biamonte, A. R.; Schneller, G. E. *J. Am. Pharm. Assoc. Sci. Ed.;* <u>1952</u>, *41*, 341.

COMPONENTS:

(1) Benzenesulfonamide, 4-amino-N-(5-methyl-1,3,4-thiadiazol-2-yl)- (sulfamethizole); $C_9H_{10}N_4O_2S_2$; [144-82-1]

(2) Phosphoric acid, disodium salt; Na_2HPO_4; [7558-94-4]

(3) Phosphoric acid, monopotassium salt; KH_2PO_4; [7778-77-0]

(4) Water; H_2O; [7732-18-5]

ORIGINAL MEASUREMENTS:

Hekster; Y. A.; Vree, T. B.; Damsma, J.E.; Friesen, W. T.; *J. Antimicrob. Chemother.* 1981, *8*, 133-44.

VARIABLES:

pH

PREPARED BY:

R. Piekos

EXPERIMENTAL VALUES:

pH	Solubility at 25°C mg/l	10^3 mol dm^{-3} [a]
5.5	1555	5.752
7.5	5022	18.578

[a] Calculated by compiler

AUXILIARY INFORMATION

METHOD/APPARATUS/PROCEDURE:

Satd solns of sulfamethizole were prepd in phosphate buffers of pH 5.5 and 7.5 at room temp (25°C). The concn of the solute was measured by means of a Spectra Physics 3500B high-performance liquid chromatograph equipped with a column oven (Model 748) and a Pye-Unicam LC-UV spectrophotometric detector. The detector was connected to a 1-mV recorder. A stainless steel column (10 cm x 4.6 mm i.d.) was packed with Lichrosorb RPS, 5 µm, obtained from Chrompack. An injection loop of 100 µl was used. The oven temp was 40°C. Detection of sulfamethizole was performed at 260 nm.

SOURCE AND PURITY OF MATERIALS:

The source and purity of the materials were not specified.

ESTIMATED ERROR:

The detection limit of the solute by HPLC was 0.5 mg/l (authors). The error in temperature and pH was not specified.

REFERENCES:

COMPONENTS:

(1) Benzenesulfonamide, 4-amino-N-(5-methyl-1,3,4-thiadiazol -2-yl)- (sulfamethizole); $C_9H_{10}N_4O_2S_2$; [144-82-1]

(2) Phosphoric acid; H_3PO_4; [7664-38-2]

(3) Phosphoric acid; monosodium salt; NaH_2PO_4; [7558-80-7]

(4) Sodium chloride; NaCl; [7647-14-5]

(5) Water; H_2O; [7732-18-5]

ORIGINAL MEASUREMENTS:

Nicklasson, M.; Brodin, A.; Nyqvist, H. *Acta Pharm. Suec.* 1981, *18*, 119-28.

VARIABLES:

pH

PREPARED BY:

R. Piekos

EXPERIMENTAL VALUES:

pH	Solubility in a H_3PO_4 - NaH_2PO_4 - NaCl buffer solution (ionic strength 0.2) at 37°C[a] mg/ml	10^3 mol dm^{-3} [b]
1.93	1.73	6.40
3.15	0.76	2.81

[a] Numerical values given by one of the authors (M.N.)

[b] Calculated by compiler

AUXILIARY INFORMATION

METHOD/APPARATUS/PROCEDURE:

Sulfamethizole, taken in excess of a quantity required for satn, was added to the H_3PO_4-NaH_2PO_4-NaCl buffer soln and the suspension was equilibrated at 37°C for 18-24 h using a magnetic stirrer. No degradation of the drug was observed at the pH values indicated. Five-ml samples were filtered through 0.1-μm polycarbonate filters (Nuclepore® Co.), and dild to suitable concn for spectrophotometry. Concns were determined using a Pye Unicam SP8-100 spectrophotometer. Samples were assayed at wavelengths of max absorption, taking into consideration changes in spectra due to ionization.

SOURCE AND PURITY OF MATERIALS:

Sulfamethizole of commercial grade was used (source not specified).
The remaining materials were anal grade regents (source not specified).
Purity of the water was not specified.

ESTIMATED ERROR:

Soly: mean of 2 detns is given (authors).
pH : precision ±0.01 pH unit (authors).
Temp: ±0.5°C (authors).

REFERENCES:

COMPONENTS:

(1) Benzenesulfonamide, 4-amino-N-(5-methyl-1,3,4-thiadiazol-2-yl)- (sulfamethizole); $C_9H_{10}N_4O_2S_2$; [144-82-1]

(2) Phosphoric acid, disodium salt; Na_2HPO_4; [7558-94-4]

(3) 1,2,3-Propanecarboxylic acid, 2-hydroxy- (citric acid); $C_6H_8O_7$; [77-92-9]

(4) Sodium chloride; NaCl; [7647-14-5]

(5) Water; H_2O; [7732-18-5]

ORIGINAL MEASUREMENTS:

Nicklasson, M.; Brodin, A.; Nyqvist, H.

Acta Pharm. Suec. 1981, *18*, 119-28.

VARIABLES:

pH

PREPARED BY:

R. Piekos

EXPERIMENTAL VALUES:

pH	Solubility in a citric acid - Na_2HPO_4- NaCl buffer solution (ionic strength 0.2) at 37°C[a]	
	mg/ml	mol dm^{-3} [b]
4.50	0.77	2.85×10^{-3}
5.32	1.66	6.14×10^{-3}
6.10	9.55	3.53×10^{-2}
7.38	134.9	0.4990

[a]Numerical values given by one of the authors (M.N.).

[b]Calculated by compiler.

AUXILIARY INFORMATION

METHOD/APPARATUS/PROCEDURE:

Sulfamethizole, taken in excess of a quantity required for satn, was added to the citirc acid-Na_2HPO_4-NaCl buffer soln and the suspension was equilibrated at 37°C for 18-24 h using magnetic stirrer. No degradation of the drug was observed at the pH values indicated. Five-ml samples were filtered through 0.1-μm polycarbonate filters (Nuclepore® Co.), and dild to suitable concn for spectrophotometry. Concns were determined using a Pye Unicam SP8-100 spectrophotometer. Samples were assayed at wavelengths of max absorption, taking into consideration changes in spectra due to ionization.

SOURCE AND PURITY OF MATERIALS:

Sulfamethizole of commercial grade was used (source not specified).

The remaining materials were anal grade reagents (source not specified).

Purity of the water was not specified.

ESTIMATED ERROR:

Soly: mean of 2 detns is given (authors).

pH : precision ±0.01 pH unit (authors).

Temp: ±0.5°C (authors).

REFERENCES:

COMPONENTS:

(1) Benzenesulfonamide, 4-amino-N-(5-methyl-1,3,4-thiadiazol-2-yl)- (sulfamethylthiadiazole); $C_9H_{10}N_4O_2S_2$; [144-82-1]

(2) Phosphoric acid, disodium salt; Na_2HPO_4; [7558-94-4]

(3) 1,2,3-Propanetricarboxylic acid, 2-hydroxy- (citric acid); $C_6H_8O_7$; [77-92-9]

(4) Water; H_2O; [7732-18-5]

ORIGINAL MEASUREMENTS:

Alric, R.; Puech, R.

J. Pharmacol. (Paris) 1971, *2(2)*, 141-54.

VARIABLES:

One temperature: 37°C; one pH: 4

PREPARED BY:

R. Piekos

EXPERIMENTAL VALUES:

Intrinsic solubility[a] of sulfamethylthiadiazole in a solution 0.025 M in Na_2HPO_4 and 0.05 M in citric acid, of pH 4, at 37°C is (33.2 ± 0.8 x 10^{-4} mol liter^{-1} ,compiler)

[a]Under "intrinsic solubility" a minimum on the solubility - pH curve is meant which corresponds to the limiting concentration of the undissociated form of sulfamethylthiadiazole.

AUXILIARY INFORMATION

METHOD/APPARATUS/PROCEDURE:

The soln was equilibrated for 48 h in a thermostat under occasional stirring. Samples were withdrawn through a 1-μ membrane filter, dild with 0.155M NaOH soln to ensure total dissocn of sulfamethylthiadiazole, and the sulfonamide was assayed by UV spectrophotometry.

SOURCE AND PURITY OF MATERIALS:

Nothing specified.

ESTIMATED ERROR:

Soly: std error of 8 measurements was ±0.8 x 10^{-4} mol liter^{-1} (authors). pH: accuracy ±0.5 pH unit (authors).
Temp: ±0.1°C (authors).

REFERENCES:

COMPONENTS:

(1) Benzenesulfonamide, 4-amino-N-(5-methyl-1,3,4-thiadiazol-2-yl)- (sulfamethylthiadiazole); $C_9H_{10}N_4O_2S_2$; [144-82-1]

(2) Calcium chloride; $CaCl_2$; [10043-52-4]

(3) Magnesium chloride; $MgCl_2$; [7786-30-3]

(4) Phosphoric acid, monoammonium salt; $NH_4H_2PO_4$; [7722-76-1]

(5) Potassium chloride; KCl; [7447-40-7]

(6) Sodium chloride; NaCl; [7647-14-5]

(7) Urea; CH_4N_2O; [57-13-6]

(8) Water; H_2O; [7732-18-5]

ORIGINAL MEASUREMENTS:

Bandelin, F. J.; Malesh, W.

J. Am. Pharm. Assoc. Sci. Ed. 1959, *48*, 177-81.

PREPARED BY:

R. Piekos

VARIABLES:

pH at 37°C

EXPERIMENTAL VALUES:

Solubility of sulfamethylthiadiazole in a solution containing $CaCl_2$ 0.143, $MgCl_2$ 0.121, $NH_4H_2PO_4$ 0.300, KCl 1.660, NaCl 2.950 and urea 20 g/dm^3 (synthetic urea, Mosher Vehicle) at 37°C.

Equilibrium pH	Solubility mg/100 ml	Solubility 10^2 mol/dm^3 [a]
4.5	120	0.444
5.0	150	0.555
5.5	260	0.962
6.0	620	2.293
6.5	1980	7.324
6.9	8400	31.074

[a] Calculated by compiler

AUXILIARY INFORMATION

METHOD/APPARATUS/PROCEDURE:

Excess sulfamethylthiadiazole was added to aliquots of synthetic urine solns and 1% H_3PO_4 or 1% NaOH solns were used to adjust the pH to the required value. The solns were agitated for 24 h with addn of acid or base to keep them at the desired pH level until equilibrium was attained. Then the solns were filtered and in aliquots the sulfonamide was assayed spectrophotometrically by the method described by Biamonte and Schneller (1).

SOURCE AND PURITY OF MATERIALS:

Nothing specified

ESTIMATED ERROR:

Soly: average values of 2 detns were given.

Temp: not specified.

pH : not specified.

REFERENCES:

1. Biamonte, A. R.; Schneller, G. E. *J. Am. Pharm. Assoc. Sci. Ed.* 1952, *41*, 341.

COMPONENTS:

(1) Benzenesulfonamide, 4-amino-N-(5-methyl-1,3,4-thiadiazol-2-yl)- (sulfamethizole); [144-82-1]

(2) Sulfuric acid monododecyl ester, sodium salt (Na lauryl sulfate); $C_{12}H_{25}NaO_4S$; [151-21-3]

(3) Water; H_2O; [7732-18-5]

ORIGINAL MEASUREMENTS:

Watari, N.; Kaneniwa, N.

Chem. Pharm. Bull. 1976, *24(11)*, 2577-84.

VARIABLES:

Concentration of Na lauryl sulfate

PREPARED BY:

R. Piekos

EXPERIMENTAL VALUES:

Concentration of Na lauryl sulfate	Total solubility of sulfamethizole at 37°C	
% w/v	mg/ml solution	10^3 mol dm^{-3} [a]
0.01	0.864	3.196
0.05	0.877	3.504
0.10	0.899	3.325
0.25	1.00	3.70
0.50	1.36	5.03
1.00	1.86	6.88
2.00	2.64	9.77
3.00	3.42	12.65
4.00	4.05	14.98
6.00	5.45	20.16

[a] Calculated by compiler

AUXILIARY INFORMATION

METHOD/APPARATUS/PROCEDURE:

An excess of sulfamethizole was added to 15 ml of the Na lauryl sulfate soln contained in a 50-ml flask and the flask was shaken (2 strokes/s at the amplitude of 3 cm) in a thermostatically controlled water bath at 37°C. One-ml sample was removed every 6 h (total equilibration period was 3-5 days) using a warmed Millipore filter syringe with a filter pore size of 0.45 μ (Millipore HAWP 01300) and the filtrate was dild with water and assayed spectrophotometrically (1).

SOURCE AND PURITY OF MATERIALS:

Commercial sulfamethizole of the Japanese Pharmacopeia grade and distd water were used.

Na lauryl sulfate was of the reagent grade (Wako Pure Chemical Industries, Ltd. lot No. PA10233) and used without further purification.

ESTIMATED ERROR:

Soly: not specified.

Temp: ±0.05°C (authors).

REFERENCES:

1. Kaneniwa, N. ; Watari, N. *Chem. Pharm. Bull.* 1974, *22*, 1699.

COMPONENTS:	ORIGINAL MEASUREMENTS:
(1) Benzenesulfonamide, 4-amino-N-(5-methyl-1,3,4-thiadiazol-2-yl)-(sulfamethizole); $C_9H_{10}N_4O_2S_2$; [144-82-1] (2) Ethanol; C_2H_6O; [64-17-5]	Sekikawa, H.; Nakano, M.; Arita, T. *Chem. Pharm. Bull.* 1978, *26(1)*, 118-26.
VARIABLES: Temperature	PREPARED BY: R. Piekos

EXPERIMENTAL VALUES:

t/°C	Solubility[a] 10^2 mol dm^{-3} solution
10	2.44
20	3.24
30	4.28
40	5.84
50	7.90

[a] Original data are presented graphically.
The numerical values are given by the authors.

AUXILIARY INFORMATION

METHOD/APPARATUS/PROCEDURE:

After attaining equilibrium, sample solns were removed by a syringe and filtered quickly through a membrane filter (pore size 0.2 μ) and sulfamethizole was assayed spectrophotometrically at 284 nm using a Hitachi Type 200-20 spectrophotometer.

SOURCE AND PURITY OF MATERIALS:

Sulfamethizole (Eisai Co.) was of the Japanese Pharmacopeia IX grade. Abs EtOH was obtained by drying and distn of EtOH following the conventional procedures.

ESTIMATED ERROR:

Nothing specified.

REFERENCES:

COMPONENTS:

(1) Benzenesulfonamide, 4-amino-N-(5-methyl-1,3,4-thiadiazol-2-yl)- (sulfamethizole); $C_9H_{10}N_4O_2S_2$; [144-82-1]

(2) 2-Pyrrolidinone-, 1-ethenyl-, polymers (poly-vinyl pyrrolidone)); $(C_6H_9NO)_x$; [9003-39-8] K-15

(3) Ethanol; C_2H_6O; [64-17-5]

ORIGINAL MEASUREMENTS:

Sekikawa, H.; Nakano, M.; Arita, T. *Chem. Pharm. Bull.* 1978, *26(1)*, 118-26.

VARIABLES:

Temperature

PREPARED BY:

R. Piekos

EXPERIMENTAL VALUES:

t/°C	M x 10^2 sulfamethizole solubilized by 1M vinyl pyrrolidone equivalent
10.0	5.51
20.0	6.58
30.0	7.85
40.0	9.75
50.0	12.1

AUXILIARY INFORMATION

METHOD/APPARATUS/PROCEDURE:

After attaining equilibrium, sample solns were removed by a syringe and filtered quickly through a membrane filter, (pore size 0.2 μ) and sulfamethizole was assayed spectrophotometrically at 284 nm using a Hitachi Type 200-20 spectrophotometer. No significant absorbance was found for poly-vinyl pyrrolidone.

SOURCE AND PURITY OF MATERIALS:

Poly(vinyl pyrrolidone) K-15 was from Daiichi Pure Chemicals Co., Tokyo. Sulfamethizole (Esai Co.) was of the Japanese Pharmacopeia IX grade. Abs EtOH was obtained by drying and distn of EtOH following the conventional procedures.

ESTIMATED ERROR:

Nothing specified.

REFERENCES:

COMPONENTS:

(1) Benzenesulfonamide, 4-amino-N-(5-methyl-1,3,4-thiadiazol-2-yl)-, monosilver salt (Ag sulfamethizole); $C_9H_9AgN_4O_2S_2$; [24342-31-2]

(2) 4-Morpholinepropanesulfonic acid; $C_7H_{15}NO_4S$; [1132-61-2]

(3) 4-Morpholinepropanesulfonic acid, sodium salt; $C_7H_{14}NNaO_4S$; [71119-22-7]

(4) Potassium nitrate; KNO_3; [7757-79-1]

(5) Water; H_2O; [7732-18-5]

ORIGINAL MEASUREMENTS:

Nesbitt, R. U., Jr.; Sandmann, B. J.

J. Pharm. Sci. 1978, *67(7)*, 1012-17.

VARIABLES:

Hydronium-ion concentration

PREPARED BY:

R. Piekos

EXPERIMENTAL VALUES:

Equilibrium values of S^2 (S = total molar solubility) versus $[H_3O^+]$ for for Ag sulfamethizole in 0.05M 4-morpholinepropanesulfonic acid buffer at 0.1M ionic strength (KNO_3) and 25±0.1°C.

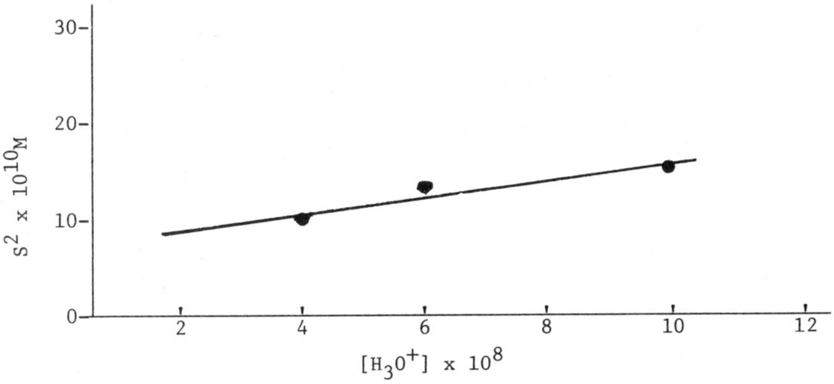

AUXILIARY INFORMATION

METHOD/APPARATUS/PROCEDURE:

Mixt of 100 mg Ag sulfamethizole and 25 or 27 ml of the 4-morpholinepropanesulfonic acid buffer were placed in paraffin-coated vials, adjusted to an ionic strength 0.1M with KNO_3, and rotated end over end in a thermostated bath until equilibrium soly was obtained (3-7 days). After filtration through 20M glass filtering crucibles, the solns were analyzed at 25±0.1°C in paraffin-coated beakers for Ag^+ ions with a silver-ion selective electrode No. 94-16, Orion Res., Cambridge, Mass) standardized at the temp indicated and 0.1M ionic strength. The pH was measured with a triple-purpose pH electrode (Corning Sci. Instruments, Medfield, Mass) standardized using buffers meeting NBS requirements. The buffers were prepd with a total molar concn of 0.05M and adjusted to an ionic strength of 0.1M with KNO_3.

SOURCE AND PURITY OF MATERIALS:

All reagents used were anal or USP grade. Ag sulfamethizole was prepd by the method of Rosenzweig and Fuchs (1) and recrystd from ammonia (2). Water had a sp cond of (1-10) x 10^{-7} $ohm^{-1}cm^{-1}$.

The buffer soln was from US Biochem. Corp., Cleveland, Ohio (purity not specified).

ESTIMATED ERROR:

Soly: not specified.

Temp: ±0.1°C (authors).

pH : accuracy ±0.001 pH unit (authors).

REFERENCES:

1. Rosenzweig, S.; Fuchs, W. *U.S. pat.* 2,536,095 (1951).
2. Sandmann, B.J.; Nesbitt, R. U., Jr.; Sandmann, R. A. *J. Pharm. Sci.* 1974, *63*, 948.

COMPONENTS:

(1) Benzenesulfonamide, 4-amino-N-(5-methyl-1,3,4-thiadiazol-2-yl)-, monosilver salt (Ag sulfamethizole); $C_9H_9AgN_4O_2S_2$; [24342-31-2]

(2) Nitric acid; HNO_3; [53081-02-0]

(3) Potassium nitrate; KNO_3; [7757-79-1]

(4) Water; H_2O; [7732-18-5]

ORIGINAL MEASUREMENTS:

Nesbitt, R. U., Jr.; Sandman, B. J.

J. Pharm. Sci. 1978, *67(7)*, 1012-17.

VARIABLES:

pH

PREPARED BY:

R. Piekos

EXPERIMENTAL VALUES:

Comparison of Total Silver Sulfamethizole Molar Solubility, S, Determined by the Method of Known Subtraction with the Molar Concentration of the Silver Ion Determined by Direct Potentiometry of Identical Samples at 25±0.1°C, 0.1M Ionic Strength, in Nitric Acid Buffer

	pH 1.931		pH 2.565	
	$S \times 10^4$	$[Ag^+] \times 10^9$	$S \times 10^4$	$[Ag^+] \times 10^4$
	4.073	4.052	1.477	1.486
	4.077	4.068	1.491	1.486
	4.088	4.068	1.477	1.492
	4.062	4.036	1.459	1.475
	4.120	4.099	1.482	1.486
	4.080	4.021	1.476	1.475
Mean	4.084	4.057	1.477	1.483

AUXILIARY INFORMATION

METHOD/APPARATUS/PROCEDURE:

Mixts of 100 mg of Ag sulfamethizole and 25 or 27 ml of the nitric acid buffer were placed in paraffin-coated vials, adjusted to an ionic strength 0.1M with KNO_3, and rotated end over end in a thermostated bath until equilibrium soly was obtained (3-7 days). After filtration through 20M glass filtering crucibles, the solns were analyzed at 25±0.1°C in paraffin-coated beakers for Ag^+ ions with a silver-ion selective electrode (No.94-16, Orion Res., Cambridge, Mass) standardized at the temp indicated and 0.1M ionic strength. The pH was measured with a triple-purpose pH electrode (Corning Sci. Instruments, Medfield, Mass) standardized using buffers meeting NBS requirements. The nitric acid buffers were prepd by diln to 0.1M HNO_3 and were adjusted to an ionic strength of 0.1M with KNO_3

SOURCE AND PURITY OF MATERIALS:

All reagents were anal or USP grade. Ag sulfamethizole was prepd by the method of Rosenzweig and Fuchs (1) and recrytd from ammonia (2). Water had a sp cond of $(1 - 10) \times 10^{-7}$ $ohm^{-1}cm^{-1}$.

The source of the reagents was not specified.

ESTIMATED ERROR: Soly: when tested by one way analysis of variance, the means displayed in the Table were found not to be statistically different at the 1% confidence level (authors).

REFERENCES:

1. Rosenzweig, S.; Fuchs, W. *U. S. pat.* 2,536,095 (1951).
2. Sandmann, R.A.; Nesbitt, R.U., Jr.; Sandmann, R. A. *J. Pharm. Sci.* 1974, *63*, 948.

COMPONENTS:

ORIGINAL MEASUREMENTS:

Nesbitt, R. U., Jr.; Sandman, B. J.

J. Pharm. Sci. 1978, *67(7)*, 1012-17.

VARIABLES:

pH

PREPARED BY:

R. Piekos

EXPERIMENTAL VALUES: Continued from previous page

Calculation of the solubility product of Ag sulfamethizole[a], K_s, at 25±0.1°C and 0.1M ionic strength

pH	f_o	S^2	K_s[a]
1.965	1.583×10^{-4}	1.733×10^{-7}	2.74×10^{-11}
2.102	2.526×10^{-4}	1.051×10^{-7}	2.65×10^{-11}
2.345	5.448×10^{-4}	4.693×10^{-8}	2.56×10^{-11}
2.598	1.132×10^{-3}	2.508×10^{-8}	2.85×10^{-11}
		Mean	$(2.70\pm0.12)10^{-11}$

[a]K_s reported as mean ±SD

[a]from eq. $K_s = f_o S^2$, where $f_o = \left(1 + \frac{[H_3O^+]}{K_2} + \frac{[H_3O^+]^2}{K_1\ K_2}\right)^{-1}$

S is the total molar solubility, and K_1 and K_2 are the apparent dissociation constants of the N^4- (amino) and N^1- (amido) hydrogens of sulfamethizole,respectively.

AUXILIARY INFORMATION

METHOD/APPARATUS/PROCEDURE:

SOURCE AND PURITY OF MATERIALS:

ESTIMATED ERROR:

REFERENCES:

COMPONENTS:

ORIGINAL MEASUREMENTS:

Nesbitt, R. U., Jr.; Sandmann, B. J.

J. Pharm. Sci. 1978, *67(7)*, 1012-17.

VARIABLES:

pH

PREPARED BY:

R. Piekos

EXPERIMENTAL VALUES: Continued from previous page

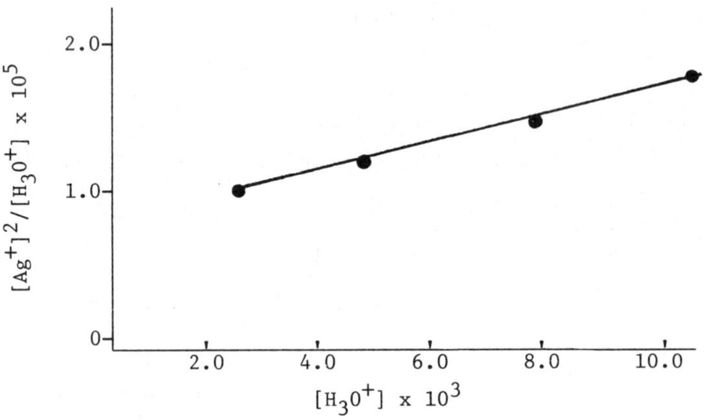

Equilibrium values of $[Ag^+]^2/[H_3O^+]$ versus $[H_3O^+]$ for silver sulfamethizole in nitric acid buffer at 0.1M ionic strength and 25±0.1°C

AUXILIARY INFORMATION

METHOD/APPARATUS/PROCEDURE:

SOURCE AND PURITY OF MATERIALS:

ESTIMATED ERROR:

REFERENCES:

COMPONENTS:

ORIGINAL MEASUREMENTS:

Nesbitt, R. U., Jr.; Sandmann, B. J.

J. Pharm. Sci. 1977, *66(4)*, 519-22.

VARIABLES:

pH

PREPARED BY:

R. Piekos

EXPERIMENTAL VALUES: Continued from previous page

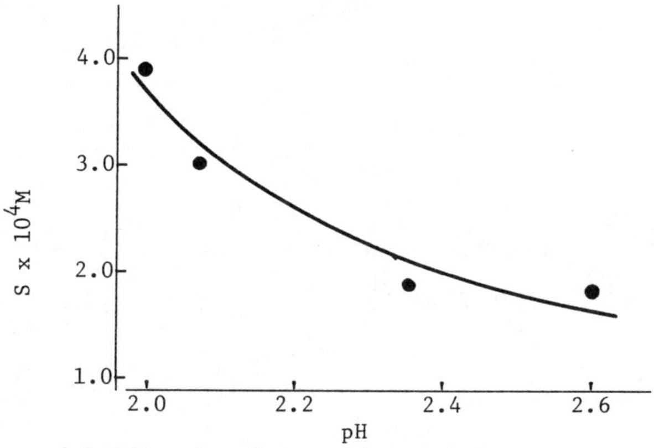

Molar solubility, S, of Ag sulfamethizole versus pH at 0.1M ionic strength and 25±0.1°C. Key: 0 calculated from equation

$$S^2 = [Ag^+]^2 = \frac{[H_3O]^{+\ 2}\ K_s}{K_1 \quad K_2} + \frac{[H_3O^+]\ K_s}{K_2}$$

AUXILIARY INFORMATION

METHOD/APPARATUS/PROCEDURE:

SOURCE AND PURITY OF MATERIALS:

ESTIMATED ERROR:

REFERENCES:

COMPONENTS:	ORIGINAL MEASUREMENTS:
(1) Acetamide, N-[4-[[(5-methyl -1,3,4-thiadiazol-2-yl)amino]sulfonyl]phenyl]-(acetyl sulfamethythiadiazole); $C_{11}H_{12}N_4O_3S_2$; [39719-87-4] (2) Water; H_2O; [7732-18-5]	Durel, M. P.; Allinne, M. *Bull. Soc. Med. Hop. Paris III* <u>1941</u>, 251-9.
VARIABLES:	**PREPARED BY:**
One temperature: 37°C	R. Piekos

EXPERIMENTAL VALUES:

Solubility of acetyl sulfamethythiadiazole in water at 37°C is 0.10 g/liter (3.2×10^{-4} mol dm^{-3}, compiler).

AUXILIARY INFORMATION

METHOD/APPARATUS/PROCEDURE:

The mixt of acetyl sulfamethylthiadiazole and water was agitated for 24 hours at 37°C.

SOURCE AND PURITY OF MATERIALS:

Source and purity of acetyl sulfamethylthiadiazole were not specified.

Distilled water was used.

ESTIMATED ERROR:

Nothing specified.

REFERENCES:

COMPONENTS:

(1) Acetamide, N-[4-[[(5-methyl-1,3,4-thiadiazol-2-yl)amino]sulfonyl]phenyl]-(acetyl sulfamethylthiadiazole); $C_{11}H_{12}N_4O_3S_2$; [39719-87-4]

(2) Phosphoric acid, disodium salt; Na_2HPO_4; [7558-94-4]

(3) Water; H_2O; [7732-18-5]

ORIGINAL MEASUREMENTS:

Krüger-Thiemer, E.
Arch. Dermatol. Syphilis 1942, *183*, 90-116.

VARIABLES:

One temperature: ca 20°C; one pH: 8.74

PREPARED BY:

R. Piekos

EXPERIMENTAL VALUES:

Solubility of acetyl sulfamethylthiadiazole in a 0.705M (10%) Na_2HPO_4 solution of pH 8.74 at room temperature (about 20°C) is 1.250 g% (4.002×10^{-2} mol dm^{-3} solution, compiler).

AUXILIARY INFORMATION

METHOD/APPARATUS/PROCEDURE:

Acetyl sulfamethylthiadiazole (0.5 g) was dissolved in 10 cm^3 of the 0.705M (10%) Na_2HPO_4 soln, shaken for 2 h at room temp (about 20°C), and filtered. The filtrate was treated with equal vol of 2N HCl and refluxed for 15 min. After proper diln, a 1-cm^3 aliquot was withdrawn, acidified, cooled, and the sulfonamide content was detd colorimetrically (as sulfamethylthiadiazole) by the Marshall method modified by Kimmig (1) using an Authenrieth colorimeter. The pH was detd on an ultraionograph using a glass electrode.

SOURCE AND PURITY OF MATERIALS:

Acetyl sulfamethylthiadiazole (source not specified) gave no coloration upon diazotization of its satd soln, thus showing absence of sulfamethylthiadiazole. The source and purity of the remaining materials were not specified.

ESTIMATED ERROR:

Soly: precision ±5% (author)
Temp: not specified.
pH : ±0.05 pH unit (author)

REFERENCES:

1. Kimmig, J. *Arch. Dermatol.* 1938, *176*, 722; *Erg. Hyg.* 1941, *24*, 398.

COMPONENTS:	ORIGINAL MEASUREMENTS:
(1) Acetamide, N-[4-[[(5-methyl-1,3,4-thiadiazol-2-yl)amino]sulfonyl]phenyl]- (acetyl sulfamethylthiadiazole); $C_{11}H_{12}N_4O_3S_2$; [39719-87-4] (2) Phosphoric acid, monopotassium salt; KH_2PO_4; [7778-77-0] (3) Water; H_2O; [7732-18-5]	Krüger-Thiemer, E. *Arch. Dermatol. Syphilis* 1942, *183*, 90-116.
VARIABLES:	PREPARED BY:
One temperature: ca 20°C; one pH: 4.37	R. Piekos

EXPERIMENTAL VALUES:

Solubility of acetyl sulfamethylthiadiazole in a 0.735M (10%) KH_2PO_4 solution of pH 4.37 at room temperature (about 20°C) is 0.0066 g% (2.11×10^{-4} mol dm^{-3} solution, compiler).

AUXILIARY INFORMATION

METHOD/APPARATUS/PROCEDURE:

Acetyl sulfamethylthiadiazole (0.5 g) was dissolved in the 0.735M (10%) KH_2PO_4 soln, shaken for 2 h at room temp (about 20°C), and filtered. The filtrate was treated with equal vol of 2N HCl, and refluxed for 15 min. After proper diln, a 1-cm^3 aliquot was withdrawn, acidified, cooled, and the sulfonamide content was detd colorimetrically by the Marshall method modified by Kimmig (1) using an Authenrieth colorimeter. The pH was detd on an ultraionograph using a glass electrode.

SOURCE AND PURITY OF MATERIALS:

Acetyl sulfamethylthiadiazole (source not specified) gave no coloration upon diazotization of its satd soln, thus showing absence of sulfamethylthiadiazole. The source and purity of the remaining materials were not specified.

ESTIMATED ERROR:

Soly: precision ±5% (author)

Temp: not specified

pH : ±0.05 pH unit (author)

REFERENCES:

1. Kimmig, J. *Arch. Dermatol.* 1938, *176*, 722; *Erg. Hyg.* 1941, *24*, 398.

COMPONENTS:

(1) Acetamide, N-[4-[[(5-methyl-1,3,4-thiadiazol-2-yl)amino]sulfonyl]phenyl]- (acetyl sulfamethylthiadiazole); $C_{11}H_{12}N_4O_3S_2$; [39719-87-4]

(2) Phosphoric acid, disodium salt; Na_2HPO_4; [7558-94-4]

(3) Phosphoric acid, monopotassium salt; KH_2PO_4; [7778-77-0]

(4) Water; H_2O; [7732-18-5]

ORIGINAL MEASUREMENTS:

Krüger-Thiemer, E.

Arch. Dermatol. Syphilis 1942, *183*, 90-116.

VARIABLES:

Temperature; pH

PREPARED BY:

R. Piekos

EXPERIMENTAL VALUES:

Composition of 1/15M phosphate buffer solution			pH	Solubility			
				Room temp (ca 20°C)		37°C	
Na_2HPO_4	KH_2PO_4	%Content		g%	10^3 mol dm^{-3} solution	g%	10^3 mol dm^{-3} solution
1.0	99.0	0.91	4.944	0.0073	0.23	-	-
10.0	90.0	0.91	5.906	0.022	0.70	0.022	0.70
61.1	38.9	0.93	7.005	0.197	6.31	0.274	8.77
9.5	0.5	0.733[b]	7.51	0.726	23.24	-	-
94.7	5.3	0.95	8.018	0.455	14.57	-	-

[a] Calculated by compiler

[b] Molar content; 10% buffer solution

AUXILIARY INFORMATION

METHOD/APPARATUS/PROCEDURE:

Acetyl sulfamethylthiadiazole (0.5 g) was dissolved in 10 cm^3 of a buffer soln, shaken for 2 h at 20°C (or left for 48 h at 37°C), and filtered at respective temp. The filtrate was treated with equal vol of 2N HCl and refluxed for 15 min. After proper diln, a 1-cm^3 aliquot was withdrawn, acidified, cooled, and the sulfonamide content was detd colorimetrically (as sulfamethylthiadiazole) by the Marshall method modified by Kimmig (1) using an Authenrieth colorimeter. The pH was detd on an ultraionograph using a glass electrode.

SOURCE AND PURITY OF MATERIALS:

Acetyl sulfamethylthiadiazole (source not specified) gave no coloration upon diazotization of its satd soln, thus showing absence of sulfamethylthiadiazole. The source and purity of the remaining materials were not specified.

ESTIMATED ERROR:

Soly: precision ±5% (author).

Temp: not specified.

pH : ±0.05 pH unit (author).

REFERENCES:

1. Kimmig, J. *Arch. Dermatol.* 1938, *176*, 722; *Erg. Hyg.* 1941, *24*, 398.

COMPONENTS:

(1) Acetamide, N-[4-[[(5-methyl-1,3,4-thiadiazol-2-yl)amino]sulfonyl]phenyl]-(acetyl sulfamethylthiadiazole); $C_{11}H_{12}N_4O_3S_2$; [39719-87-4]

(2) Phosphoric acid, disodium salt; Na_2HPO_4; [7558-94-4]

(3) Phosphoric acid, monopotassium salt; KH_2PO_4; [7778-77-0]

(4) Water; H_2O; [7732-18-5]

VARIABLES:

pH

ORIGINAL MEASUREMENTS:

Bandelin, F. J.; Malesh, W.

J. Am. Pharm. Assoc., Sci. Ed. 1959, *48*, 177-81.

PREPARED BY:

R. Piekos

EXPERIMENTAL VALUES:

Solubility of acetyl sulfamethylthiadiazole in buffers of varying mixtures of $Na_2HPO_4 \cdot 7H_2O$ (71.6 g/l distilled water; 0.27 mol dm^{-3}, compiler) and KH_2PO_4 (36.3 g/l distilled water; 0.27 mol dm^{-3}, compiler) at 37°C.

Equilibrium pH	Solubility (based on sulfamethylthiadiazole)	
	mg/100 ml	10^2 mol dm^{-3} [a]
4.5	41	0.151
5.0	50	0.185
5.5	71	0.262
6.0	102	0.377
6.3	260	0.962
6.6	630	2.33
7.3	2400	8.878

[a] Calculated by compiler

AUXILIARY INFORMATION

METHOD/APPARATUS/PROCEDURE:

Solns were prepd by adding an excess of acetyl sulfamethylthiadiazole to 10 ml of buffer soln at each pH level in 18 x 150-mm test tubes, stoppering the tubes, placing them in water bath at 37°C with gentle agitation for 24 h. The solute was then hydrolyzed with 5% H_2SO_4 for 1 h to liberate the free sulfonamide. One-ml aliquot of the hydrolyzate was accurately pipetted into a volumetric flask for diln and analysis. The sulfonamide was assayed colorimetrically by the method of Bratton and Marshall as described in detail by Biamonte and Schneller (1). A standard curve was prepd using accurately prepd standard solutions.

SOURCE AND PURITY OF MATERIALS:

Neither source nor purity of the reagents were specified. Distilled water was used.

ESTIMATED ERROR:

Soly: ave values of duplicate runs are reported (authors).

Temp and pH: not specified.

REFERENCES:

1. Biamonte, A. R.; Schneller, G. E. *J. Am. Pharm. Assoc., Sci. Ed.* 1952, *41*, 341.

COMPONENTS:

(1) Acetamide, N-[4-[[(5-methyl-1,3,4-thiadiazole-2-yl)amino]sulfonyl]phenyl]-(N^4-acetylsulfamethizole); $C_{11}H_{12}N_4O_3S_2$; [39719-87-4]

(2) Phosphoric acid, disodium salt; Na_2HPO_4; [7558-94-4]

(3) Phosphoric acid, monopotassium salt; KH_2PO_4; [7778-77-0]

(4) Water; H_2O; [7732-18-5]

ORIGINAL MEASUREMENTS:

Hekster, Y.A.; Vree, T. B.; Damsma, J. E. Friesen, W. T. *J. Antimicrob. Chemother.* 1981, *8*, 133-44.

VARIABLES:

pH

PREPARED BY:

R. Piekos

EXPERIMENTAL VALUES:

pH	Solubility at 25°C	
	mg/l	10^3 mol dm^{-3} [a]
5.5	200	0.640
7.5	3000	9.604

[a] Calculated by compiler

AUXILIARY INFORMATION

METHOD/APPARATUS/PROCEDURE:

Satd solns of N^4-acetylsulfamethizole were prepd in phosphate buffers of pH 5.5 and 7.5 at room temp (25°C). The concn of the solute was measured by means of a Spectra Physics 3500B high-performance liquid chromatograph equipped with a column oven (Model 748) and a Pye-Unicam LC-UV spectrophotometric detector. The detector was connected to a 1-mV recorder. A stainless steel column (10 cm x 4.6 mm i.d.) was packed with Lichrosorb RPS, 5 μm, obtained from Chrompack. An injection loop of 100 μl was used. The oven temp was 40°C. Detection of the solute was performed at 260 nm.

SOURCE AND PURITY OF MATERIALS:

The source and purity of the materials were not specified.

ESTIMATED ERROR:

The detection limit of the solute by HPLC was 0.5 mg/l (authors). The error in temperature and pH were not specified.

REFERENCES:

COMPONENTS:

(1) Acetamide, N-[4-[[(5-methyl-1,3,4-thiadiazol-2-yl)amino]sulfonyl]phenyl]- (acetyl sulfamethylthiadiazole); $C_{11}H_{12}N_4O_3S_2$; [39719-87-4]

(2) Calcium chloride; $CaCl_2$; [10043-52-4]

(3) Magnesium chloride; $MgCl_2$; [7786-30-3]

(4) Phosphoric acid, monoammonium salt; $NH_4H_2PO_4$; [7722-76-1]

(5) Potassium chloride; KCl; [7447-40-7]

(6) Sodium chloride; NaCl; [7647-14-5]

(7) Urea; CH_4N_2O; [57-13-6]

(8) Water; H_2O; [7732-18-5]

ORIGINAL MEASUREMENTS:

Bandelin, F. J.; Malesh, W.

J. Am. Pharm. Assoc., Sci. Ed. 1959, *48*, 177-81.

PREPARED BY:

R. Piekos

VARIABLES:

pH at 37°C

EXPERIMENTAL VALUES:

Solubility of acetyl sulfamethylthiadiazole in a solution containing $CaCl_2$ 0.143, $MgCl_2$ 0.121, $NH_4H_2PO_4$ 0.300, KCl 1.660, NaCl 2.950 and urea 20 g/dm^3 (synthetic urine, Mosher Vehicle) at 37°C.

Equilbiruum pH	Solubility	
	mg/100 ml	mol/dm^3 [a]
	(as sulfamethylthiadiazole)	
4.5	10	3.7×10^{-4}
5.0	21	7.8×10^{-4}
5.5	45	1.7×10^{-3}
6.0	145	5.4×10^{-3}
6.5	380	1.4×10^{-2}
7.0	995	3.7×10^{-2}

[a]Calculated by compiler

AUXILIARY INFORMATION

METHOD/APPARATUS/PROCEDURE:

Excess acetyl sulfamethylthiadiazole was added to aliquots of synthetic urine solns and 1% H_3PO_4 or 1% NaOH solns were used to adjust the pH to the required value. The solns were agitated for 24 h with addn of acid or base to keep them at the desired pH level until equilibrium was attained. Then the solns were filtered and in aliquots the sulfonamide was assayed spectrophotometrically by the method described by Biamonte and Schneller (1). Before detn the soln was refluxed with 5% H_2SO_4 for 1 h to liberate the fee amino compd.

SOURCE AND PURITY OF MATERIALS:

Nothing specified.

ESTIMATED ERROR:

Soly: average values of 2 detns were given.

Temp: not specified.

pH : not specified.

REFERENCES:

1. Biamonte, A. R.; Schneller, G. E., *J. Am. Pharm. Assoc., Sci. Ed.* 1952, *41*, 341.

COMPONENTS:	EVALUATOR:
(1) Benzenesulfonamide, 4-amino-N-(5-ethyl-1,3,4-thiadiazol-2-yl)-(sulfaethylthiadiazole); $C_{10}H_{12}N_4O_2S_2$; [94-19-9]	Anthony N. Paruta Department of Pharmaceutics University of Rhode Island Kingston, Rhode Island, USA and Ryszard Piekos Faculty of Pharmacy, University of Gdansk Gdansk, Poland 1986
(2) Aqueous phosphate buffers	

CRITICAL EVALUATION:

For the above compound, there were three reports (1-3) which determined the solubility in water at 293K and 310K at five pH levels as shown in Table I.

Table I: Solubility of Sulfaethylthiadiazole in water at various pH's and temperatures

		10^3 mol dm^{-3}	
Reference	pH	293K	310K
1	4.9[a]	1.48	-
3	5.0[b]	-	11.4
1	5.91[a]	2.95	4.64
2	5.9[a]	-	5.13
3	6.0[b]	-	26.7
1	7.0[a]	17.80	22.93
2	7.1[a]	-	21.45
3	7.0[b]	-	207.5
1	7.51[a]	44.91	-
3	7.5[b]	-	256.7
1	8.02[a]	32.64	-
3	8.0[b]	-	597.8

a = buffer concentration at 0.066 mol dm^{-3}
b = buffer concentration at 0.27 mol dm^{-3}

The data of Bandelin and Malesh (3) reported solubility over a pH range of 5-8 in phosphate buffers of 0.27 mol dm^{-3} concentration substantially greater than in the other data (1,2). The data, while showing the expected large increases in solubility with pH, refer only to initial pH values. At concentrations reported here, especially those about 0.1 mol dm^{-3} (~ pH 6.5), the dissolved amount should affect the final pH of the equilibrated solution. This would occur at pH values greater than about 5.5 (pK_a) by the production of highly soluble anionic species affecting the pH value through the ionic strength effect. The values given by Krüger-Thiemer (1) and Langecker (2) are for 0.066 mol dm^{-3} phosphate buffer. There are two sets of values that merit consideration, those at pH 5.9 and pH 7.0 (1,2). If it can be assumed that the solubility at 310K and a pH 5.5 (≅ pK_a) is about 2 x 10^{-3} mol dm^{-3} then at pH 5.9, about 2.5 times as many highly water solubble anions are formed leading to a value of about 5 x 10^{-3} mol dm^{-3}. The average of the two values (1,2) lead to a tentative solubility value at a pH = 5.9 in phosphate buffer of 4.88 x 10^{-3} mol dm^{-3}. At a pH of 7, there would be about 31 fold increase in anions, however, the values only indicate about a 10-11 fold increase. Although the values at a pH 7 (1,2) are reasonable in magnitude they could not be reconciled with each other and were not considered further. None of the data at 293K was duplicated by any two authors and are shown for completeness and data enhancement trend (except for pH 7.5) as a function of pH.

REFERENCES:

(1) Krüger-Thiemer, E. *Arch. Dermatol. Syphillis* 1942, *183*, 90-116.
(2) Langecker, H. *Arch. Exptl. Path. Pharmakol.* 1948, *205*, 291-301.
(3) Bandelin, F.J.; Malesh, W. *J. Am. Pharm. Assoc., Sci. Ed.* 1959, *48*, 177-81.

COMPONENTS:	ORIGINAL MEASUREMENTS:
(1) Benzenesulfonamide, 4-amino-N-(5-ethyl-1,3,4-thiadiazol-2-yl)- (sulfaethylthiadiazole); $C_{10}H_{12}N_4O_2S_2$; [94-19-9] (2) Water; H_2O; [7732-18-5]	Durel, M. P.; Allinne, M. *Bull. Soc. Med. Hop. Paris III* 1941, 251-9.
VARIABLES:	**PREPARED BY:**
One temperature: 37°C	R. Piekos

EXPERIMENTAL VALUES:

Solubility of sulfaethylthiadiazole in water at 37°C is 0.40 g/liter (1.41×10^{-3} mol dm^{-3}, compiler).

AUXILIARY INFORMATION

METHOD/APPARATUS/PROCEDURE:	SOURCE AND PURITY OF MATERIALS:
A mixture of sulfaethylthiadiazole and water was agitated for 24 hours at 37°C.	Source and purity of sulfaethylthiadiazole were not specified. Distilled water was used.
	ESTIMATED ERROR:
	Nothing specified.
	REFERENCES:

COMPONENTS:

(1) Benzenesulfonamide, 4-amino-N-(5-ethyl-1,3,4-thiadiazol-2-yl)- (sulfaethylthiadiazole); $C_{10}H_{12}N_4O_2S_2$; [94-19-9]

(2) Water; H_2O; [7732-18-5]

ORIGINAL MEASUREMENTS:

Langecker, H.

Arch. Exptl. Path. Pharmakol. 1948, *205*, 291-301.

VARIABLES:

One temperature: 37°C

PREPARED BY:

R. Piekos

EXPERIMENTAL VALUES:

Solubility of sulfaethylthiadiazole in water at 37°C is 60 mg%

(2.11×10^{-3} mol dm^{-3}, compiler).

AUXILIARY INFORMATION

METHOD/APPARATUS/PROCEDURE:

An excess of sulfaethylthiadiazole in water was boiled and left for 24 h in a vessel protected from access of CO_2. The concn of the sulfonamide was detd colorimetrically by the method of Bratton and Marshall (1) using a Havemann colorimeter (2), as well as by microanal detn of the solid residue.

SOURCE AND PURITY OF MATERIALS:

Source and purity of the materials were not specified.

ESTIMATED ERROR:

Nothing specified.

REFERENCES:

1. Bratton, A. G.; Marshall, E. K.,Jr. *J. Biol. Chem.* 1939, *128*, 537.
2. Havemann, R. *Klin. Wochenschr.* 1940, p. 503.

COMPONENTS:	ORIGINAL MEASUREMENTS:
(1) Benzenesulfonamide, 4-amino-N-(5-ethyl-1,3,4-thiadiazol-2-yl)- (sulfaethylthiadiazole); $C_{10}H_{12}N_4O_2S_2$; [94-19-9] (2) Sodium chloride; NaCl; [7647-14-5] (3) Water; H_2O; [7732-18-5]	Langecker, H. *Arch. Exptl. Path. Pharmakol.* 1948, *205*, 291-301.
VARIABLES:	**PREPARED BY:**
One temperature: 37°C	R. Piekos

EXPERIMENTAL VALUES:

Solubility of sulfaethylthiadiazole in a 0.9% w/w NaCl solution at 37°C is 62 mg% (2.2 x 10^{-3} mol dm^{-3}, compiler).

AUXILIARY INFORMATION

METHOD/APPARATUS/PROCEDURE:

An excess of sulfaethylthiadiazole in the 0.9% w/w NaCl soln was boiled for 1 h in a sealed ampul followed by keeping the ampul at 37°C. The concn of the sulfonamide was assayed colorimetrically by the method of Bratton and Marshall (1) using a Havemann colorimeter (2), as well as by microanal detn of the solid residue.

SOURCE AND PURITY OF MATERIALS:

Source and purity of the materials were not specified.

ESTIMATED ERROR:

Nothing specified.

REFERENCES:

1. Bratton, A. G.; Marshall, E. K., Jr. *J. Biol. Chem.* 1939, *128*, 537.
2. Havemann, R. *Klin. Wochenschr.* 1940, p. 503.

COMPONENTS:

(1) Benzenesulfonamide, 4-amino-N-(5-ethyl-1,3,4-thiadiazol-2-yl)- (sulfaethylthiadiazole); $C_{10}H_{12}N_4O_2S_2$; [94-19-9]

(2) Phosphoric acid, disodium salt; Na_2HPO_4; [7558-94-4]

(3) Water; H_2O; [7732-18-5]

ORIGINAL MEASUREMENTS:

Krüger-Thiemer, E.

Arch. Dermatol. Syphilis 1942, *183*, 90-116.

VARIABLES:

One temperature: ca 20°C; one pH: 8.74

PREPARED BY:

R. Piekos

EXPERIMENTAL VALUES:

Solubility of sulfaethylthiadiazole in a 0.705M (10%) Na_2HPO_4 solution of pH 8.74 at room temperature (about 20°C) is 1.820 g% (6.400 x 10^{-2} mol dm^{-3} solution, compiler).

AUXILIARY INFORMATION

METHOD/APPARATUS/PROCEDURE:

Sulfaethylthiadiazole (0.5 g) was dissolved in 10 cm^3 of the 0.705M (10%) Na_2HPO_4 solution of pH 8.74, shaken for 2 h at room temp (about 20°C), and filtered. A 1-cm^3 aliquot of the filtrate was withdrawn, cooled, acidified with 1 cm^3 of 2N HCl, and the sulfonamide content was detd colorimetrically by the method of Marshall modified by Kimmig (1) using an Authenrieth colorimeter. The pH was detd on an ultraionograph using a glass electrode.

SOURCE AND PURITY OF MATERIALS:

Sulfaethylthiadiazole was the product manufd by Schering under the name Globucid. The source and purity of the remaining materials were not specified.

ESTIMATED ERROR:

Soly: precision ±5% (author).

Temp: not specified.

pH : ±0.05 pH unit (author).

REFERENCES:

1. Kimmig, J. *Arch. Dermatol.* 1938, *176*, 722; *Erg. Hyg.* 1941, *24*, 398.

COMPONENTS:

(1) Benzenesulfonamide, 4-amino-N-(5-ethyl-1,3,4-thiadiazol-2-yl)- (sulfaethylthiadiazole); $C_{10}H_{12}N_4O_2S_2$; [94-19-9]

(2) Phosphoric acid, monopotassium salt; KH_2PO_4; [7778-77-0]

(3) Water; H_2O; [7732-18-5]

ORIGINAL MEASUREMENTS:

Krüger-Thiemer, E.

Arch. Dermatol. Syphilis 1942, *183*, 90-116.

VARIABLES:

One temperature: ca 20°C; one pH: 4.37

PREPARED BY:

R. Piekos

EXPERIMENTAL VALUES:

Solubility of sulfaethylthiadiazole in a 0.735M (10%) KH_2PO_4 solution of pH 4.37 at room temperature (about 20°C) is 0.0167 g% (5.87 x 10^{-4} mol dm^{-3} solution, compiler).

AUXILIARY INFORMATION

METHOD/APPARATUS/PROCEDURE:

Sulfaethylthiadiazole (0.5 g) was dissolved in 10 cm^3 of the 0.735M (10%) KH_2PO_4 soln of pH 4.37, shaken for 2 h at room temp (about 20°C), and filtered. A 1-cm^3 aliquot of the filtrate was withdrawn, cooled, acidified with 1 cm^3 of 2N HCl, and the sulfonamide content was detd colorimetrically by the method of Marshall modified by Kimmig (1) using an Authenrieth colorimeter. The pH was detd on an utraiongraph using a glass electrode.

SOURCE AND PURITY OF MATERIALS:

Sulfaethylthiadiazole was the product manufd by Schering under the name Globucid. The source and purity of the remaining materials were not specified.

ESTIMATED ERROR:

Soly: precision ±5% (author).

Temp: not specified.

pH : ±0.05 pH unit (author)

REFERENCES:

1. Kimmig, J. *Arch. Dermatol.* 1938, *176*, 722; *Erg. Hyg.* 1941, *24*, 398.

COMPONENTS:

(1) Benzenesulfonamide, 4-amino-N-(5-ethyl-1,3,4-thiadiazol-2-yl)- (sulfaethylthiadiazole); $C_{10}H_{12}N_4O_2S_2$; [94-19-9]

(2) Phosphoric acid, disodium salt; Na_2HPO_4; [7558-94-4]

(3) Phosphoric acid, monopotassium salt; KH_2PO_4; [7778-77-0]

(4) Water; H_2O; [7732-18-5]

ORIGINAL MEASUREMENTS:

Krüger-Thiemer, E.

Arch. Dermatol. Syphilis 1942, *183*, 90-116.

VARIABLES:

Temperature; pH

PREPARED BY:

R. Piekos

EXPERIMENTAL VALUES:

Composition of 1/15M phosphate buffer solution			pH	Solubility			
				Room temp (ca 20°C)		37°C	
Na_2HPO_4	KH_2PO_4	%Content		g%	10^2 mol dm^{-3} solution[a]	g%	10^2 mol dm^{-3} solution[a]
1.0	99.0	0.91	4.944	0.042	0.148	-	-
10.0	90.0	0.91	5.906	0.084	0.295	0.132	0.464
61.1	38.9	0.93	7.005	0.506	1.780	0.652	2.293
9.5	0.5	0.733[b]	7.51	1.277	4.491	-	-
94.7	5.3	0.95	8.018	0.928	3.264	-	-

[a]Calculated by compiler.

[b]Molar content; 10% buffer solution.

AUXILIARY INFORMATION

METHOD/APPARATUS/PROCEDURE:

Sulfaethylthiadiazole (0.5 g) was dissolved in 10 cm^3 of a buffer soln, shaken for 2 h at 20°C (or left for 48 h at 37°C), and filtered at respective temp. A 1-cm^3 aliquot of the filtrate was then withdrawn, cooled, (dild for expts at 37°C), acidified with 1 cm^3 of 2N HCl, and the sulfonamide content was detd colorimetrically by the method of Marshall modified by Kimmig (1) using an Authenrieth colorimeter. The pH was detd on an ultraionograph using a glass electrode.

SOURCE AND PURITY OF MATERIALS:

Sulfaethylthiadiazole was the product manufd by Schering under the name Globucid.

The source and purity of the remaining materials were not specified.

ESTIMATED ERROR:

Soly: precision ±5 (author)

Temp: not specified

pH : ±0.05 pH unit (author)

REFERENCES:

1. Kimmig, J. *Arch. Dermatol.* 1938, *176*, 722; *Erg. Hyg.* 1941, *24*, 398.

COMPONENTS:

(1) Benzenesulfonamide, 4-amino-N-(5-ethyl-1,3,4-thiadiazol-2-yl)- (sulfaethylthiadiazole); $C_{10}H_{12}N_4O_2S_2$; [94-19-9]

(2) Phosphoric acid, disodium salt; Na_2HPO_4; [7558-94-4]

(3) Phosphoric acid, monopotassium salt; KH_2PO_4; [7778-77-0]

(4) Water; H_2O; [7732-18-5]

ORIGINAL MEASUREMENTS:

Langecker, H.

Arch. Exptl. Path. Pharmakol. 1948, *205*, 291-301.

VARIABLES:

pH

PREPARED BY:

R. Piekos

EXPERIMENTAL VALUES:

pH of the 1/15M phosphate buffer	Solubility at 37°C	
	mg%	10^3 mol dm^{-3} [a]
5.7	146	5.13
5.9	146[b]	5.13
6.6	500	17.58
7.1	610	21.45

[a] Calculated by compiler.

[b] Measured at 20°C.

AUXILIARY INFORMATION

METHOD/APPARATUS/PROCEDURE:

An excess of sulfaethylthiadiazole was added to a buffer soln and boiled for 1 h in a sealed ampul followed by keeping the ampul at 37°C. The concn of the sulfonamide was detd colorimetrically by the method of Bratton and Marshall (1) using a Havemann colorimeter (2), as well as by microanal detn of the solid residue.

SOURCE AND PURITY OF MATERIALS:

Source and purity of the materials were not specified.

ESTIMATED ERROR:

Nothing specified.

REFERENCES:

1. Bratton, A. G.; Marshall, E. K., Jr. *J. Biol. Chem.* 1939, *128*, 537.
2. Havemann, R. *Klin. Wochenschr.* 1940, p. 503.

COMPONENTS:	ORIGINAL MEASUREMENTS:
(1) Benzenesulfonamide, 4-amino-N-(5-ethyl-1,3,4-thiadiazole-2-yl)- (sulfaethylthiadiazole); $C_{10}H_{12}N_4O_2S_2$; [94-19-9] (2) Phosphoric acid, disodium salt; Na_2HPO_4; [7558-94-4] (3) Phosphoric acid, monopotassium salt; KH_2PO_4; [7778-77-0] (4) Water; H_2O; [7732-18-5]	Bandelin, F. J.; Malesh, W. *J. Am. Pharm. Assoc., Sci. Ed.* 1959, *48*, 177-81.
VARIABLES: pH	PREPARED BY: R. Piekos

EXPERIMENTAL VALUES:

Solubility of sulfaethylthiadiazole in buffers of varying mixtures of $Na_2HPO_4 \cdot 7H_2O$ (71.6 g/l distilled water; 0.27 mol dm^{-3}, compiler) and KH_2PO_4 (36.3 g/l distilled water; 0.27 mol dm^{-3}, compiler) at 37°C.

Initial pH	Solubility mg/100 ml	Solubility mol dm^{-3} [a]
5.0	325	0.0114
5.5	465	0.0163
6.0	760	0.0267
6.5	2250	0.0791
7.0	5900	0.2075
7.5	7300	0.2567
8.0	17,000	0.5978

[a]Calculated by compiler.

AUXILIARY INFORMATION

METHOD/APPARATUS/PROCEDURE:

Solns were prepd by adding an excess of sulfaethylthiadiazole to 10 ml of buffer soln at each pH level in 18 x 150-mm test tubes, stoppering the tubes and placing them in a water bath at 37°C with gentle agitation for 24 h. The mixt was then filtered and a 1-ml aliquot was accurately pipetted into a volumetric flask for diln and analysis. The balance was retained for pH detn to ascertain any change in pH value. The sulfonamide was assayed colorimetrically by the method of Bratton and Marshall as described in detail by Biamonte and Schneller (1). A standard curve was prepd using accurately prepd standard solutions.

SOURCE AND PURITY OF MATERIALS:

Neither source nor purity of the reagents were specified. Distilled water was used.

ESTIMATED ERROR:

Soly: av values of duplicate runs are reported (authors).

Temp and pH: not specified.

REFERENCES:

1. Biamonte, A. R.; Schneller, G. E. *J. Am. Pharm. Assoc., Sci. Ed.* 1952, *41*, 341.

COMPONENTS:

(1) Benzenesulfonamide, 4-amino-N-(5-ethyl-1,3,4-thiadiazol-2-yl)- (sulfaethylthiadiazole); $C_{10}H_{12}N_4O_2S_2$; [94-19-9]

(2) Phosphoric acid, disodium salt; Na_2HPO_4; [7558-94-4]

(3) Phosphoric acid, monopotassium salt; KH_2PO_4; [7778-77-0]

(4) Water; H_2O; [7732-18-5]

ORIGINAL MEASUREMENTS:

Riess, W.

Intern. Congr. Chemotherapy, Proc. 3rd. Stuttgart 1963, *1*, 627-32.

VARIABLES:

One temperature: 20°C; one pH: 7.4

PREPARED BY:

R. Piekos

EXPERIMENTAL VALUES:

Solubility of sulfaethylthiadiazole in M/15 phosphate buffer (pH 7.4) at 20°C is 1500 mg% (5.275×10^{-2} mol dm^{-3}, compiler).

AUXILIARY INFORMATION

METHOD/APPARATUS/PROCEDURE:

Sörensen buffer solns of pH varying between 7 and 8 were prepd, satd with sulfaethylthiadiazole at 20°C, their pH was measured at equilibrium , and the sulfaethylthiadiazole was assayed colorimetrically. The measured pH values were plotted against concn, and the soly at pH 7.4 was detd by interpolation (personal communication).

SOURCE AND PURITY OF MATERIALS:

Nothing specified.

ESTIMATED ERROR:

Nothing specified.

REFERENCES:

COMPONENTS:

(1) Benzenesulfonamide, 4-amino-N-(5-ethyl-1,3,4-thiadiazol-2-yl)- (sulfaethylthiadiazole); $C_{10}H_{12}N_4O_2S_2$; [94-19-9]

(2) Phosphoric acid, disodium salt; Na_2HPO_4; [7558-94-4]

(3) Phosphoric acid, monopotassium salt; KH_2PO_4; [7778-77-0]

(4) Water; H_2O; [7732-18-8]

ORIGINAL MEASUREMENTS:

Hekster, Ch. A.; Vree, T. B.

Antibiotics Chemother. 1982, *31*, 22-118.

VARIABLES:

pH

PREPARED BY:

R. Piekos

EXPERIMENTAL VALUES:

pH	Solubility at 25°C	
	mg/l	10^3 mol dm^{-3} [a]
5.5	489	1.72
7.5[b]	7,110	25.00

AUXILIARY INFORMATION

METHOD/APPARATUS/PROCEDURE:

The earlier developed method (1) was used (personal communication). Satd solns of sulfaethylthiadiazole were prepd in phosphate buffers of pH 5.5 and 7.5 at 25°C. The concn of the solute was measured by means of a Spectra Physics 3500B high-performance liquid chromatograph equipped with a Model 748 column oven and a Pye-Unicam LC-UV spectrophotometric detector.

SOURCE AND PURITY OF MATERIALS:

Neither source nor the purity of the materials was specified.

ESTIMATED ERROR:

Soly: the detection limit of the solute by HPLC was 0.5 mg/l (authors).

The errors in temp and pH were not specified.

REFERENCES:

1. Hekster, Y. A.; Vree, T. B.; Damsma, J. E.; Friesen, W. T. *J. Antimicrob. Chemother.* 1981, *8*, 133.

COMPONENTS:

(1) Benzenesulfonamide, 4-amino-N-(5-ethyl-1,3,4-thiadiazol-2-yl)- (sulfaethylthiadiazole); $C_{10}H_{12}N_4O_2S_2$; [94-19-9]
(2) Calcium chloride; $CaCl_2$; [10043-52-4]
(3) Magnesium chloride; $MgCl_2$; [7786-30-3]
(4) Phosphoric acid, monoammonium salt; $NH_4H_2PO_4$; [7722-76-1]
(5) Potassium chloride; KCl; [7447-40-7]
(6) Sodium chloride; NaCl; [7647-14-5]
(7) Urea; CH_4N_2O; [57-13-6]
(8) Water; H_2O; [7732-18-5]

ORIGINAL MEASUREMENTS:

Bandelin, F. J.; Malesh, W.
J. Am. Pharm. Assoc., Sci. Ed. 1959, *48*, 177-81.

VARIABLES:

pH at 37°C

PREPARED BY:

R. Piekos

EXPERIMENTAL VALUES:

Solublity of sulfaethylthiadiazole in a solution containing $CaCl_2$ 0.143, $MgCl_2$ 0.121, $NH_4H_2PO_4$ 0.300, KCl 1.660, NaCl 2.950 and urea 20 g/dm^3 (synthetic urine, Mosher Vehicle) at 37°C.

Equilibrium pH	Solubility	
	mg/100 ml	10^2 mol/dm^3 [a]
4.4	360	1.27
4.7	380	1.34
5.2	440	1.55
5.6	480	1.69
6.35	600	2.11
6.7	1875	6.59

[a]Calculated by compiler.

AUXILIARY INFORMATION

METHOD/APPARATUS/PROCEDURE:

Excess sulfaethylthiadiazole was added to aliquots of synthetic urine solns and 1% H_3PO_4 or 1% NaOH solns were used to adjust the pH to the required value. The solns were agitated for 24 h with addn of acid or base to keep them at the desired pH level until equilibrium was attained. Then the solns were filtered and in aliquots the sulfonamide was assayed spectrophotometrically by the method described by Biamonte and Schneller (1).

SOURCE AND PURITY OF MATERIALS:

Nothing specified

ESTIMATED ERROR:

Soly: average values of 2 detns were given.
Temp: not specified.
pH : not specified.

REFERENCES:

1. Biamonte, A. R.; Schneller, G. E. *J. Am. Pharm. Assoc., Sci. Ed.* 1952, *41*, 341.

COMPONENTS:	ORIGINAL MEASUREMENTS:
(1) Benzenesulfonamide, 4-amino-N-(5-ethyl-1,3,4-thiadiazol-2-yl)- (sulfaethylthiadiazole); $C_{10}H_{12}N_4O_2S_2$; [94-19-9] (2) Methane, trichloro- (chloroform); $CHCl_3$; [67-66-3]	Riess, W. *Intern. Congr. Chemotherapy, Proc., 3rd. Stuttgart* 1963, *1*, 627-32.
VARIABLES:	**PREPARED BY:**
One temperature: 20°C	R. Piekos

EXPERIMENTAL VALUES:

Solubility of sulfaethylthiadiazole in chloroform at 20°C is 109 mg%
(3.83×10^{-3} mol dm^{-3} solution, compiler).

AUXILIARY INFORMATION

METHOD/APPARATUS/PROCEDURE:

Nothing specified.

SOURCE AND PURITY OF MATERIALS:

Nothing specified.

ESTIMATED ERROR:

Nothing specified.

REFERENCES:

COMPONENTS:	ORIGINAL MEASUREMENTS:
(1) Acetamide. N-[4[[(5-ethyl-1,3,4-thiadiazol-2-yl)amino]sulfonyl]phenyl]-(acetyl sulfaethylthiadiazole); $C_{12}H_{14}N_4O_3S_2$; [1037-51-0] (2) Water; H_2O; [7732-18-5]	Durel, M. P.; Allinne, M. *Bull. Soc. Med. Hop. Paris III* <u>1941</u>, 251-9.
VARIABLES:	**PREPARED BY:**
One temperature: 37°C	R. Piekos

EXPERIMENTAL VALUES:

Solubility of acetyl sulfaethylthiadiazole in water at 37°C is 0.20 g/liter (6.1×10^{-4} mol dm^{-3}, compiler).

AUXILIARY INFORMATION

METHOD/APPARATUS/PROCEDURE:	SOURCE AND PURITY OF MATERIALS:
A mixt of acetyl sulfaethylthiadiazole and water was agitated for 24 hours at 37°C.	Source and purity of acetyl sulfaethylthiadiazole was not specified. Distilled water was used.
	ESTIMATED ERROR:
	Nothing specified.
	REFERENCES:

COMPONENTS:

(1) Acetamide, N-[4-[[(5-ethyl-1,3,4- thiadiazol-2-yl)amino]sulfonyl]phenyl]- (acetyl sulfaethylthiadiazole); $C_{12}H_{14}N_4O_3S_2$; [1037-51-0]

(2) Water; H_2O; [7732-18-5]

ORIGINAL MEASUREMENTS:

Langecker, H.
Arch. Exptl. Path. Pharmakol. 1948, *205*, 291-301.

VARIABLES:

pH

PREPARED BY:

R. Piekos

EXPERIMENTAL VALUES:

pH	Solubility at 37°C mg%	10^4 mol dm^{-3} [a]
5.2	12	3.7
6.0	16	4.9

[a] Calculated by compiler

AUXILIARY INFORMATION

METHOD/APPARATUS/PROCEDURE:

An excess of acetyl sulfaethylthiadiazole in water was boiled for 1 h in a sealed ampul followed by keeping the ampul at 37°C . Before the assaying, the solute was treated with 2.6N NaOH soln (1) to cleave the acetyl group and the sulfaethylthiadiazole was detd colorimetrically by the method of Bratton and Marshall (2) using a Havemann colorimeter (3), as well as by microanal detd of the solid residue.

SOURCE AND PURITY OF MATERIALS:

Source and purity of the materials were not specified.

ESTIMATED ERROR:

Nothing specified.

REFERENCES:

1. Scudi, J.V. *J. Lab. Clin. Med.* 1940, *25*, 404.
2. Bratton, A. G.; Marshall, E.K., Jr. *J. Biol. Chem.* 1939, *128*, 537.
3. Havemann, R. *Klin. Wochenschr.* 1940, p. 503.

COMPONENTS:

(1) Acetamide, N-[4-[[(5-ethyl-1,3,4-thiadiazol-2-yl)amino]sulfonyl]phenyl]- (acetyl sulfaethylthiadiazole); $C_{12}H_{14}N_4O_3S_2$; [1037-51-0]

(2) Phosphoric acid, disodium salt; Na_2HPO_4; [7558-94-4]

(3) Water; H_2O; [7732-18-5]

ORIGINAL MEASUREMENTS:

Krüger-Thiemer, E.

Arch. Dermatol. Syphilis 1942, *183*, 90-116.

VARIABLES:

One temperature: ca 20°C; one pH: 8.74

PREPARED BY:

R. Piekos

EXPERIMENTAL VALUES:

Solubility of acetyl sulfaethylthiadiazole in a 0.705M (10%) Na_2HPO_4 solution of pH 8.74 at room temperature (about 20°C) is 1.840 g% (5.637×10^{-2} mol dm^{-3} solution, compiler).

AUXILIARY INFORMATION

METHOD/APPARATUS/PROCEDURE:

Acetyl sulfaethylthiadiazole (0.5 g) was dissolved in 10 cm^3 of the 0.705M (10%) Na_2HPO_4 soln, shaken for 2 h at room temp (about 20° C), and filtered. The filtrate was treated with equal vol of 2N HCl, and refluxed for 15 min. After proper diln, a 1-cm^3 aliquot was withdrawn, acidified, cooled, and the sulfonamide content was detd colorimetrically (as sulfaethylthiadiazole) by the Marshall method modified by Kimmig (1) using an Authenrieth colorimeter. The pH was detd on an ultraionograph using a glass electrode.

SOURCE AND PURITY OF MATERIALS:

Acetyl sulfaethylthiadiazole (source not specified) gave no coloration upon diazotization of its satd soln, thus showing absence of sulfaethylthiadiazole. The source and purity of the remaining materials were not specified.

ESTIMATED ERROR:

Soly: precision ±5% (author).

Temp: not specified.

pH : ±0.05 pH unit (author).

REFERENCES:

1. Kimmig, J. *Arch. Dermatol.* 1938, *176*, 722; *Erg. Hyg.* 1941, *24*, 398.

COMPONENTS:

(1) Acetamide, N-[4-[[(5-ethyl-1,3,4-thiadiazol-2-yl)amino]sulfonyl]phenyl]- (acetyl sulfaethylthiadiazole); $C_{12}H_{14}N_4O_3S_2$; [1037-51-0]

(2) Phosphoric acid, monopotassium salt; KH_2PO_4; [7778-77-0]

(3) Water; H_2O; [7732-18-5]

ORIGINAL MEASUREMENTS:

Krüger-Thiemer, E.

Arch. Dermatol. Syphilis 1942, *183*, 90-116.

VARIABLES:

One temperature: ca 20°C; one pH: 4.37

PREPARED BY:

R. Piekos

EXPERIMENTAL VALUES:

Solubility of acetyl sulfaethylthiadiazole in a 0.735M (10%) KH_2PO_4 solution of pH 4.37 at room temperature (about 20°C) is 0.0063 g% (1.9×10^{-4} mol dm^{-3} solution, compiler).

AUXILIARY INFORMATION

METHOD/APPARATUS/PROCEDURE:

Acetyl sulfaethylthiadiazole (0.5 g) was dissolved in 10 cm^3 of the 0.735M (10%) KH_2PO_4 soln, shaken for 2 h at room temp (about 20° C), and filtered. The filtrate was treated with equal vol of 2N HCl and refluxed for 15 min. After proper diln, a 1-cm^3 aliquot was withdrawn, acidified, cooled, and the sulfonamide content was detd colorimetrically (as sulfaethylthiadiazole) by the Marshall method modified by Kimmig (1) using an Authenrieth colorimeter. The pH was detd on an ultraionograph using a glass electrode.

SOURCE AND PURITY OF MATERIALS:

Acetyl sulfaethylthiadiazole (source not specified) gave no coloration upon diazotization of its satd soln, thus showing absence of sulfaethylthiadiazole. The source and purity of the remaining materials was not specified.

ESTIMATED ERROR:

Soly: precision ±5% (author).

Temp: not specified.

pH : ±0.05 pH unit (author).

REFERENCES:

1. Kimmig, J. *Arch. Dermatol.* 1938, *176*, 722; *Erg. Hyg.* 1941, *24*, 398.

COMPONENTS:

(1) Acetamide, N-[4-[[(5-ethyl-1,3,4-thiadiazole-2-yl)amino]sulfonyl]phenyl]-(acetyl sulfaethylthiadiazole); $C_{12}H_{14}N_4O_3S_2$; [1037-51-0]

(2) Phosphoric acid, disodium salt; Na_2HPO_4; [7558-94-4]

(3) Phosphoric acid, monopotassium salt; KH_2PO_4; [7778-77-0]

(4) Water; H_2O; [7732-18-5]

ORIGINAL MEASUREMENTS:

Krüger-Thiemer, E.

Arch. Dermatol. Syphilis 1942, *183*, 90-116.

VARIABLES:

Temperature; pH

PREPARED BY:

R. Piekos

EXPERIMENTAL VALUES:

Composition of 1/15M phosphate buffer solutions			pH	Solubility			
				Room temp (ca 20°C)		37°C	
Na_2HPO_4	KH_2PO_4	%Content		g%	10^3 mol dm^{-3} solution[a]	g%	10^3 mol dm^{-3} solution[a]
1.0	99.0	0.91	4.944	0.0128	0.392	-	-
10.0	90.0	0.91	5.906	0.0530	1.600	0.112	3.43
61.1	38.9	0.93	7.005	0.3910	12.0	0.750	22.98
9.5	0.5	0.733[b]	7.51	1.1100	34.01	-	-
94.7	5.3	0.95	8.018	0.8790	26.9	-	-

[a] Calculated by compiler

[b] Molar content; 10% buffer solution

AUXILIARY INFORMATION

METHOD/APPARATUS/PROCEDURE:

Acetyl sulfaethylthiadiazole (0.5 g) was dissolved in 10 cm^3 of a buffer soln, shaken for 2 h at 20°C (or left for 48 h at 37°C), and filtered at respective temp. The filtrate was treated with equal vol of 2N HCl and refluxed for 15 min. After proper diln, a 1-cm^3 aliquot was withdrawn, acidified, cooled, and the sulfonamide content was detd colorimetrically (as sulfaethylthiadiazole) by the Marshall method modified by Kimmig (1) using an Authenrieth colorimeter. The pH was detd on an ultraionograph using a glass electrode.

SOURCE AND PURITY OF MATERIALS:

Acetyl sulfaethylthiadiazole (source not specified) gave no coloration upon diazotization of its satd soln, thus showing absence of sulfaethylthiadiazole. The source and purity of the remaining materials were not specified.

ESTIMATED ERROR:

Soly: precision ±5% (author).

Temp: not specified.

pH : ±0.05 pH unit (author).

REFERENCES:

1. Kimmig, J. *Arch. Dermatol.* 1938, *176*, 722; *Erg. Hyg.* 1941, *24*, 398.

COMPONENTS:

(1) Acetamide, N-[4-[[(5-ethyl-1,3,4-thiadiazol-2-yl)amino]sulfonyl]phenyl]- (acetyl sulfaethylthiadiazole); $C_{12}H_{14}N_4O_3S_2$; [1037-51-0]

(2) Phosphoric acid, disodium salt; Na_2HPO_4; [7558-94-4]

(3) Phosphoric acid, monopotassium salt; KH_2PO_4; [7778-77-0]

(4) Water; H_2O; [7732-18-5]

ORIGINAL MEASUREMENTS:

Bandelin, F. J.; Malesh, W.

J. Am. Pharm. Assoc., Sci. Ed. 1959, *48*, 177-81.

VARIABLES:

pH

PREPARED BY:

R. Piekos

EXPERIMENTAL VALUES:

Solubility of acetyl sulfaethylthiadiazole in buffers of varying mixtures of $Na_2HPO_4 \cdot 7H_2O$ (71.6 g/l distilled water; 0.27 mol dm^{-3}, compiler) and KH_2PO_4 (36.3 g/l distilled water; 0.27 mol dm^{-3}, compiler) at 37°C.

Equilibrium pH	Solubility (based on sulfaethylthiadiazole)	
	mg/100 ml	10^2 mol dm^{-3} [a]
4.5	140	0.492
4.6	162	0.570
5.2	212	0.745
5.6	300	1.055
6.2	510	1.794
6.6	740	2.602
6.8	1175	4.132

[a] Calculated by compiler

AUXILIARY INFORMATION

METHOD/APPARATUS/PROCEDURE:

Solns were prepd by adding an excess of acetyl sulfaethylthiadiazole to 10 ml of buffer soln at each pH level in 18 x 150-mm test tubes, stoppering the tubes, and placing them in water bath at 37°C with gentle agitation for 24 h. The solute was then hydrolyzed with 5% H_2SO_4 for 1 h to liberate the free sulfonamide. One-ml aliquot of the hydrolyzate was accurately pipetted into a volumetric flask for diln and analysis. The sulfonamide was assayed colorimetrically by the method of Bratton and Marshall as described in detail by Biamonte and Schneller (1). A standard curve was prepd using accurately prepd standard solutions.

SOURCE AND PURITY OF MATERIALS:

Neither source nor purity of the reagents were specified. Distilled water was used.

ESTIMATED ERROR:

Soly: av values of ducplicate runs are reported (authors).

Temp and pH: not specified.

REFERENCES:

1. Biamonte, A. R.; Schneller, G. E. *J. Am. Pharm. Assoc., Sci. Ed.* 1952, *41*, 341.

COMPONENTS:

(1) Acetamide, N-[4-[[(5-ethyl-1,3,4-thiadiazol-2-yl)amino]sulfonyl]phenyl]-(acetyl sulfaethylthiadiazole); $C_{12}H_{14}N_4O_3S_2$; [1037-51-0]

(2) Phosphoric acid, disodium salt; Na_2HPO_4; [7558-94-4]

(3) Phosphoric acid, monopotassium salt; KH_2PO_4; [7778-77-0]

(4) Water; H_2O; [7732-18-5]

ORIGINAL MEASUREMENTS:

Hekster, Ch. A.; Vree, T. B.

Antibiotics Chemother. 1982, *31*, 22-118.

VARIABLES:

pH

PREPARED BY:

R. Piekos

EXPERIMENTAL VALUES:

pH	Solubility at 25°C	
	mg/l	10^3 mol dm^{-3} [a]
5.5	392	1.20
7.5[b]	7,850	24.05

[a]Calculated by compiler

[b]Erroneous pH value of 7.0 is given in the article

AUXILIARY INFORMATION

METHOD/APPARATUS/PROCEDURE:

The earlier developed method (1) was used (personal communication). Satd solns of acetyl sulfaethylthiadiazole were prepd in phosphate buffers of pH 5.5 and 7.5 at 25°C. The concn of the solute was measured by means of a Spectra Physics 3500B high-performance liquid chromatograph equipped with a Model 748 column oven and a Pye-Unicam LC-UV spectrophotometric detector.

SOURCE AND PURITY OF MATERIALS:

Neither source nor the purity of the materials was not specified.

ESTIMATED ERROR:

Soly: the detection limit of the solute by HPLC was 0.5 mg/l (authors).

The errors in temp and pH were not specified.

REFERENCES:

1. Hekster, Y. A.; Vree, T. B.; Damsma, J. E.; Friesen, W. T. *J. Antimicrob. Chemother.* 1981, *8*, 133.

COMPONENTS:

(1) Acetamide, N-[4-[[(5-ethyl-1,3,4-thiadiazol-2-yl)amino]sulfonyl]phenyl]-(acetyl sulfaethylthiadiazole); $C_{12}H_{14}N_4O_3S_2$; [1037-51-0]
(2) Calcium chloride; $CaCl_2$; [10043-52-4]
(3) Magnesium chloride; $MgCl_2$; [7786-30-3]
(4) Phosphoric acid, monoammonium salt; $NH_4H_2PO_4$; [7722-76-1]
(5) Potassium chloride; KCl; [7447-40-7]
(6) Sodium chloride; NaCl; [7647-14-5]
(7) Urea; CH_4N_2O; [57-13-6]
(8) Water; H_2O; [7732-18-5]

ORIGINAL MEASUREMENTS:

Bandelin, F. J.; Malesh, W.
J. Am. Pharm. Assoc., Sci. Ed. 1959, *48*, 177-81.

PREPARED BY:

R. Piekos

VARIABLES:

pH at 37°C

EXPERIMENTAL VALUES:

Solubility of acetyl sulfaethylthiadiazole in a solution containing $CaCl_2$ 0.143, $MgCl_2$ 0.121, $NH_4H_2PO_4$ 0.300, KCl 1.660, NaCl 2.950 and urea 20 g/dm^3 (synthetic urine, Mosher Vehicle) at 37°C

Equilbrium pH	Solubility mg/100 ml as sulfaethylthiadiazole	Solubility 10^2 mol dm^{3} [a]
4.5	225	0.69
5.0	230	0.70
5.5	250	0.77
6.0	350	1.07
6.5	650	1.99
7.0	1140	3.49

[a]Calculated by compiler

AUXILIARY INFORMATION

METHOD/APPARATUS/PROCEDURE:

Excess acetyl sulfaethylthiadiazole was added to aliquots of synthetic urine solns and 1% H_3PO_4 or 1% NaOH solns were used to adjust the pH to the required value. The solns were agitated for 24 h with addn of acid or base to keep them at the desired pH level until equilibrium was attained. Then the solns were filtered and in aliquots the acetyl sulfonamide was assayed spectrophotometrically by the method described by Biamonte and Schneller (1). Before detn the soln was refluxed with 5% H_2SO_4 for 1 h to liberate the free amino compound.

SOURCE AND PURITY OF MATERIALS:

Nothing specified.

ESTIMATED ERROR:

Soly: average values of 2 detns were given.
Temp: not specified.
pH : not specified.

REFERENCES:

1. Biamonte, A. R.; Schneller, G. E. *J. Am. Pharm. Assoc., Sci. Ed.* 1952, *41*, 341.

COMPONENTS:

(1) Benzenesulfonamide, 4-amino-N-(5-propyl-1,3,4-thiadiazol-2-yl); $C_{11}H_{14}N_4O_2S_2$; [71119-32-9]

(2) Phosphoric acid, disodium salt; Na_2HPO_4; [7558-94-4]

(3) 1,2,3-Propanetricarboxylic acid, 2-hydroxy- (citric acid); $C_6H_8O_7$; [77-92-9]

(4) Water; H_2O; [7732-18-5]

ORIGINAL MEASUREMENTS:

Alric, R.; Puech, R.

J. Pharmacol. (Paris) 1971, *2(2)*, 141-54.

VARIABLES:

One temperature: 37°C; one pH: 3.5

PREPARED BY:

R. Piekos

EXPERIMENTAL VALUES:

Intrinsic solubility[a] of 4-amino-N-(5-propyl-1,3,4-thiadiazol-2-yl)benzenesulfonamide in a solution 0.025M in Na_2HPO_4 and 0.05M in citric acid, of pH 3.5, at 37°C is $(8.98 \pm 0.23) \times 10^{-4}$ mol liter^{-1}.

[a]Under "intrinsic solubility" a minimum on the solubility - pH curve is meant which corresponds to the limiting concentration of the undissociated form of the sulfonamide.

AUXILIARY INFORMATION

METHOD/APPARATUS/PROCEDURE:

The soln was equilibrated for 48 h in a thermostat under occasional stirring. Samples were withdrawn through a 1-μ membrane filter, dild with 0.155M NaOH soln to ensure total dissocn of the sulfonamide, and its content was detd by UV spectrophotometry.

SOURCE AND PURITY OF MATERIALS:

Nothing specified.

ESTIMATED ERROR:

Soly: std error of 8 measurements was ±0.23 x 10^{-4} mol liter^{-1} (authors).

pH : accuracy ±0.5 pH unit (authors).

Temp: ±0.1°C (authors).

REFERENCES:

COMPONENTS:

(1) Benzenesulfonamide, 4-amino-N-[5-(2-propyl)-1,3,4-thiadiazol-2-yl]-; $C_{11}H_{14}N_4O_2S_2$; [80-34-2]

(2) Phosphoric acid, disodium salt; Na_2HPO_4; [7558-94-4]

(3) 1,2,3-Propanetricarboxylic acid, 2-hydroxy- (citric acid); $C_6H_8O_7$; [77-92-9]

(4) Water; H_2O; [7732-18-5]

ORIGINAL MEASUREMENTS:

Alric, R.; Puech, R.

J. Pharmacol. (Paris) 1971, *2(2)*, 141-54.

VARIABLES:

One temperature: 37°C; one pH: 4

PREPARED BY:

R. Piekos

EXPERIMENTAL VALUES:

Intrinsic solubility[a] of 4-amino-N-[5-(2-propyl)-1,3,4-thiadiazol-2-yl]benzenesulfonamide in a solution 0.025M in Na_2HPO_4 and 0.05M in citric acid, of pH 4, at 37°C is (7.33 ± 0.20) x 10^{-4} mol $liter^{-1}$.

[a]Under "intrinsic solubility" a minimum on the solubility - pH curve is meant which corresponds to the limiting concentration of the undissociated form of the sulfonamide.

AUXILIARY INFORMATION

METHOD/APPARATUS/PROCEDURE:

The soln was equilibrated for 48 h in a thermostat under occassional stirring. Samples were withdrawn through a 1-μ membrane filter, dild with 0.155M NaOH soln to ensure total dissocn of the sulfonamide, and its content was detd by UV spectrophotometry.

SOURCE AND PURITY OF MATERIALS:

Nothing specified.

ESTIMATED ERROR:

Soly: std error of 8 measurements was ±0.20 x 10^{-4} mol $liter^{-1}$ (authors).
pH : accuracy ±0.5 pH unit (authors).
Temp: ±0.1°C (authors).

REFERENCES:

COMPONENTS:

(1) Benzenesulfonamide, 4-amino-N-[5-butyl-1,3,4-thiadiazol-2-yl]-; $C_{12}H_{16}N_4O_2S_2$; [71119-31-8]

(2) Phosphoric acid, disodium salt; Na_2HPO_4; [7558-94-4]

(3) 1,2,3-Propanetricarboxylic acid, 2-hydroxy- (Citric acid); $C_6H_8O_7$; [77-92-9]

(4) Water; H_2O; [7732-18-5]

ORIGINAL MEASUREMENTS:

Alric, R.; Puech, R.

J. Pharmacol. (Paris) 1971, *2(2)*, 141-54.

VARIABLES:

One temperature: 37°C; one pH: 3.5

PREPARED BY:

R. Piekos

EXPERIMENTAL VALUES:

Intrinsic solubility[a] of 4-amino-N-[5-butyl-1,3,4-thiadiazol-2-yl]benzenesulfonamide in a solution 0.025M in Na_2HPO_4 and 0.05M in citric acid, of pH 3.5, at 37°C is (2.71 ± 0.06) x 10^{-4} mol $liter^{-1}$.

[a]Under "intrinsic solubility" a minimum on the solubility - pH curve is meant which corresponds to the limiting concentration of the undissociated form of the sulfonamide.

AUXILIARY INFORMATION

METHOD/APPARATUS/PROCEDURE:

The soln was equilibrated for 48 h in a thermostat under occassional stirring. Samples were withdrawn through a 1-μ membrane filter, dild with 0.155M NaOH soln to ensure total dissocn of the sulfonamide, and its content was detd by UV spectrophotometry.

SOURCE AND PURITY OF MATERIALS:

Nothing specified.

ESTIMATED ERROR:

Soly: std error of 8 measurements was ±0.06 x 10^{-4} mol $liter^{-1}$ (authors).

pH : accuracy of ±0.5 pH unit (authors).

Temp: ±0.1°C (authors).

REFERENCES:

COMPONENTS:

(1) Benzenesulfonamide, 4-amino-N-[5-(2-methyl-2-propyl)-1,3,4-thiadiazol-2-yl]-; $C_{12}H_{16}N_4O_2S_2$; [535-65-9]

(2) Phosphoric acid, disodium salt; Na_2HPO_4; [7558-94-4]

(3) 1,2,3-Propanetricarboxylic acid, 2-hydroxy- (citric acid); $C_6H_8O_7$; [77-92-9]

(4) Water; H_2O; [7732-18-5]

ORIGINAL MEASUREMENTS:

Alric, R.; Puech, R.

J. Pharmacol. (Paris) 1971, *2(2)*, 141-54.

VARIABLES:

One temperature: 37°C; one pH: 3.5

PREPARED BY:

R. Piekos

EXPERIMENTAL VALUES:

Intrinsic solubility[a] of 4-amino-N-[5-(2-methyl-2-propyl)-1,3,4-thiadiazol-2-yl]benzenesulfonamide in a solution 0.025M in Na_2HPO_4 and 0.05M in citric acid, of pH 3.5, at 37°C is (1.82 ± 0.05) x 10^{-4} mol $liter^{-1}$.

[a]Under "intrinsic solubility" a minimum on the solubility - pH curve is meant which corresponds to the limiting concentration of the undissociated form of the sulfonamide.

AUXILIARY INFORMATION

METHOD/APPARATUS/PROCEDURE:

The soln was equilibrated for 48 h in a thermostat under occassional stirring. Samples were withdrawn through a 1-μ membrane filter, dild with 0.155M NaOH soln to ensure total dissocn of the sulfonamide, and its content was detd by UV spectrophotometry.

SOURCE AND PURITY OF MATERIALS:

Nothing specified.

ESTIMATED ERROR:

Soly: std error of 8 measurements was ±0.05 x 10^{-4} mol $liter^{-1}$ (authors).

pH : accuracy ±0.5 pH unit (authors).

Temp: ±0.1°C (authors).

REFERENCES:

COMPONENTS:	ORIGINAL MEASUREMENTS:
(1) Benzenesulfonamide, 4-amino-N-(5-pentyl-1,3,4-thiadiazol-2-yl)-; $C_{13}H_{18}N_4O_2S_2$; [71119-30-7] (2) Phosphoric acid, disodium salt; Na_2HPO_4; [7558-94-4] (3) 1,2,3-Propanetricarboxylic acid, 2-hydroxy- (citric acid); $C_6H_8O_7$; [77-92-9] (4) Water; H_2O; [7732-18-5]	Alric, R.; Puech, R. *J. Pharmacol. (Paris)* 1971, *2(2)*, 141-54.
VARIABLES:	PREPARED BY:
One temperature: 37°C; one pH: 3.5	R. Piekos

EXPERIMENTAL VALUES:

Intrinsic solubility[a] of 4-amino-N-(5-pentyl-1,3,4-thiadiazol-2-yl)benzenesulfonamide in a solution of 0.025M in Na_2HPO_4 and 0.05M in citric acid, of pH 3.5, at 37°C is (1.12 ± 0.04) x 10^{-4} mol $liter^{-1}$.

[a]Under "intrinsic solubility" a minimum on the solubility - pH curve is meant which corresponds to the limiting concentration of the undissociated form of the sulfonamide.

AUXILIARY INFORMATION

METHOD/APPARATUS/PROCEDURE:	SOURCE AND PURITY OF MATERIALS:
The soln was equilibrated for 48 h in a thermostat under occasional stirring. Samples were withdrawn through a 1-μ membrane filter, dild with 0.155M NaOH soln to ensure total dissocn of the sulfonamide, and its content was detd by UV spectrophotometry.	Nothing specified.
	ESTIMATED ERROR:
	Soly: std error of 8 measurements was ±0.04 x 10^{-4} mol $liter^{-1}$ (authors). Temp: ±0.1°C (authors). pH : accuracy ±0.5 pH unit (authors).
	REFERENCES:

COMPONENTS:

(1) Benzenesulfonamide, 4-amino-N-[5-(3-methylbutyl)-1,3,4-thiadiazol-2-yl]-; $C_{13}H_{18}N_4O_2S_2$; [71119-29-4]

(2) Phosphoric acid, disodium salt; Na_2HPO_4; [7558-94-4]

(3) 1,2,3-Propanetricarboxylic acid, 2-hydroxy- (citric acid); $C_6H_8O_7$; [77-92-9]

(4) Water; H_2O; [7732-18-5]

ORIGINAL MEASUREMENTS:

Alric, R.; Puech, R.

J. Pharmacol. (Paris) 1971, *2(2)*, 141-54.

VARIABLES:

One temperature: 37°C; one pH: 3.5

PREPARED BY:

R. Piekos

EXPERIMENTAL VALUES:

Intrinsic solubility[a] of 4-amino-N-[5-(3-methylbutyl)-1,3,4-thiadiazol-2-yl]benzenesulfonamide in a solution 0.025M in Na_2HPO_4 and 0.05M in citric acid, of pH 3.5, at 37°C is (0.90 ± 0.06) x 10^{-4} mol $liter^{-1}$.

[a]Under "intrinsic solubility" a minimum on the solubility - pH curve is meant which corresponds to the limiting concentration of the undissociated form of the sulfonamide.

AUXILIARY INFORMATION

METHOD/APPARATUS/PROCEDURE:

The soln was equilibrated for 48 h in a thermostat under occasional stirring. Samples were withdrawn through a 1-μ membrane filter, dild with 0.155M NaOH soln to ensure total dissocn of the sulfonamide, and its content was detd by UV spectrophotometry.

SOURCE AND PURITY OF MATERIALS:

Nothing specified.

ESTIMATED ERROR:

Soly: std error of 8 measurements was ±0.06 x 10^{-4} mol $liter^{-1}$ (authors).

pH : accuracy ±0.5 pH unit (authors).

Temp: ±0.1°C (authors).

REFERENCES:

COMPONENTS:	ORIGINAL MEASUREMENTS:
(1) Benzenesulfonamide, 4-amino-N-(5-amino-1,3,4-thiadiazol-2-yl)-; $C_8H_9N_5O_2S_2$; [71119-25-0] (2) Water; H_2O; [7732-18-5]	Anderson, G. W.; Faith, H. E.; Marson, H.W. Winnek, P. S.; Roblin, R. O. Jr. *J. Am. Chem. Soc.* 1942, *64*, 2902-5.
VARIABLES:	PREPARED BY:
One temperature: 37°C	R. Piekos

EXPERIMENTAL VALUES:

Solubility of 4-amino-N-(5- amino-1,3,4-thiadiazol-2-yl)benzenesulfonamide in water at 37°C is 36.3 mg/100 cm^3 solution (1.34×10^{-3} mol dm^{-3}, compiler).

AUXILIARY INFORMATION

METHOD/APPARATUS/PROCEDURE:

Excess sulfonamide in water was heated and stirred on a steam bath for 30 min. The suspension was then agitated for 24 h in a thermostat. A sample of the satd soln was withdrawn through a glass filter, dild, and analyzed by the Marshall method (1) using a General Electric recording spectrophotometer for comparing the colors developed with those of the standards.

SOURCE AND PURITY OF MATERIALS:

The sulfonamide, mp 259°C (cor), was prepd by the authors. Anal. %C 35.3 (calcd 35.4); %H 3.5 (3.7); %N 25.5 (25.8). Purity of the water was not specified.

ESTIMATED ERROR:

Nothing specified.

REFERENCES:

1. Bratton, A. C.; Marshall, E. K. Jr. *J. Pharmacol.* 1939, *66*, 4.

COMPONENTS:

(1) Benzenesulfonamide, 4-(4,5-dihydro-3-methyl-5-oxo-1H-pyrazol-1-yl)-; $C_{10}H_{11}N_3O_3S$; [13269-73-3]

(2) Water; H_2O; [7732-18-5]

ORIGINAL MEASUREMENTS:

Roblin, R. O., Jr.; Williams, J. H.; Winnek, P. S.; English, J. P.

J. Am. Chem. Soc. 1940, *62*, 2002-5.

VARIABLES:

One temperature: 37°C

PREPARED BY:

R. Piekos

EXPERIMENTAL VALUES:

Solubility of 4-(4,5-dihydro-3-methyl-5-oxo-1H-pyrazol-1-yl)benzenesulfonamide in water at 37°C is 45.9 mg/100 cm^3 solution (1.81×10^{-3} mol dm^{-3}, compiler).

AUXILIARY INFORMATION

METHOD/APPARATUS/PROCEDURE:

Excess sulfonamide in water was heated and stirred on a steam bath for 30 min. The suspension was then agitated for 24 h in a thermostat at 37°C. A sample of the satd soln was withdrawn through a glass filter, dild, and analyzed by the Marshall method (1) using a General Electric recording spectrophotometer for comparing the colors developed with those of the standards.

SOURCE AND PURITY OF MATERIALS:

The sulfonamide, mp 166-7°C (cor), was prepd by the authors. Anal: %C 47.5 (calcd 47.4); %H 4.4 (4.4); %N 16.6 (16.6).

Purity of the water was not specified.

ESTIMATED ERROR:

Nothing specified.

REFERENCES:

1. Bratton, A. C.; Marshall, E. K., Jr. *J. Pharmacol.* 1939, *66*, 4.

SYSTEM INDEX

Page numbers preceded by E refer to evaluation texts whereas page numbers not preceded by E refer to compiled tables.

REGISTRY NUMBER INDEX

Page numbers preceded by E refer to evaluation texts whereas page numbers not preceded by E refer to compiled tables.

AUTHOR INDEX

Page numbers preceded by E refer to evaluation texts whereas page numbers not preceded by E refer to compiled tables.

SOLUBILITY DATA SERIES

Volume 1	H. L. Clever, *Helium and Neon*
Volume 2	H. L. Clever, *Krpton, Xenon and Radon*
Volume 3	M. Salomon, *Silver Azide, Cyanide, Cyanamides, Cyanate, Selenocyanate and Thiocyanate*
Volume 4	H. L. Clever, *Argon*
Volume 5/6	C. L. Young, *Hydrogen and Deuterium*
Volume 7	R. Battino, *Oxygen and Ozone*
Volume 8	C. L. Young, *Oxides of Nitrogen*
Volume 9	W. Hayduk, *Ethane*
Volume 10	R. Battino, *Nitrogen and Air*
Volume 11	B. Scrosati and C. A. Vincent, *Alkali Metal, Alkaline Earth Metal and Ammonium Halides. Amide Solvents*
Volume 12	C. L. Young, *Sulfur Dioxide, Chlorine, Fluorine and Chlorine Oxides*
Volume 13	S. Siekierski, T. Mioduski and M. Salomon, *Scandium, Yttrium, Lanthanum and Lanthanide Nitrates*
Volume 14	H. Miyamoto, M. Salomon and H. L. Clever, *Alkaline Earth Metal Halates*
Volume 15	A. F. M. Barton, *Alcohols with Water*
Volume 16/17	E. Tomlinson and A. Regosz, *Antibiotics: I. ß-Lactam Antibiotics*
Volume 18	O. Popovych, *Tetraphenylborates*
Volume 19	C. L. Young, *Cumulative Index: Volumes 1–18*
Volume 20	A. L. Horvath and F. W. Getzen, *Halogenated Benzenes, Toluenes and Phenols with Water*
Volume 21	C. L. and Young P. G. T. Fogg, *Ammonia, Amines, Phosphine, Arsine, Stibine, Silane, Germane and Stannane in Organic Solvents*
Volume 22	T. Mioduski and M. Salomon, *Scandium, Yttrium, Lanthanum and Lanthanide Halides in Nonaqueous Solvents*
Volume 23	T. P. Dirkse, *Copper, Silver, Gold, and Zinc, Cadmium, Mercury Oxides and Hydroxides*
Volume 24	W. Hayduk, *Propane, Butane and 2-Methylpropane*
Volume 25	C. Hiraymana, Z. Galus and C. Guminski, *Metals in Mercury*
Volume 26	M. R. Masson, H. D. Lutz and B. Engelen, *Sulfites, Selenites and Tellurites*
Volume 27/28	H. L. Clever and C. L. Young, *Methane*
Volume 29	H. L. Clever, *Mercury in Liquids, Compressed Gases, Molten Salts and Other Elements*
Volume 30	H. Miyamoto and M. Salomon, *Alkali Metal Halates, Ammonium Iodate and Iodic Acid*
Volume 31	J. Eysseltová and T. P. Dirkse, *Alkali Metal Orthophosphates*
Volume 32	P. G. T. Fogg and C. L. Young, *Hydrogen Sulfide, Deuterium Sulfide and Hydrogen Selenide*
Volume 33	P. Franzosini, *Molten Alkali Metal Alkanoates*
Volume 34	A. N. Paruta and R. Piekos, *4-Aminobenzenesulfonamides. Part I: Non-cyclic Substituents*
Volume 35	A. N. Paruta and R. Piekos, *4-Aminobenzenesulfonamides. Part II: 5-Membered Heterocyclic Substituents*